DATE

VOLUME FIVE HUNDRED AND SEVENTY TWO

METHODS IN
ENZYMOLOGY

Visualizing RNA Dynamics
in the Cell

METHODS IN ENZYMOLOGY

Editors-in-Chief

ANNA MARIE PYLE
*Departments of Molecular, Cellular and Developmental
Biology and Department of Chemistry
Investigator, Howard Hughes Medical Institute
Yale University*

DAVID W. CHRISTIANSON
*Roy and Diana Vagelos Laboratories
Department of Chemistry
University of Pennsylvania
Philadelphia, PA*

Founding Editors

SIDNEY P. COLOWICK and NATHAN O. KAPLAN

VOLUME FIVE HUNDRED AND SEVENTY TWO

METHODS IN ENZYMOLOGY

Visualizing RNA Dynamics in the Cell

Edited by

GRIGORY S. FILONOV
Department of Pharmacology
Weill Medical College, Cornell University
1300 York Avenue, Room LC-523, Box 70
New York, NY 10065
USA

SAMIE R. JAFFREY
Tri-Institutional Chemical Biology Program at Weill-Cornell
Medical College, Rockefeller University, Memorial
Sloan-Kettering Cancer Center; Weill Medical College,
Cornell University, New York, NY, United States

AMSTERDAM • BOSTON • HEIDELBERG • LONDON
NEW YORK • OXFORD • PARIS • SAN DIEGO
SAN FRANCISCO • SINGAPORE • SYDNEY • TOKYO
Academic Press is an imprint of Elsevier

Academic Press is an imprint of Elsevier
50 Hampshire Street, 5th Floor, Cambridge, MA 02139, USA
525 B Street, Suite 1800, San Diego, CA 92101-4495, USA
The Boulevard, Langford Lane, Kidlington, Oxford OX5 1GB, UK
125 London Wall, London, EC2Y 5AS, UK

First edition 2016

Copyright © 2016 Elsevier Inc. All rights reserved.

No part of this publication may be reproduced or transmitted in any form or by any means, electronic or mechanical, including photocopying, recording, or any information storage and retrieval system, without permission in writing from the publisher. Details on how to seek permission, further information about the Publisher's permissions policies and our arrangements with organizations such as the Copyright Clearance Center and the Copyright Licensing Agency, can be found at our website: www.elsevier.com/permissions.

This book and the individual contributions contained in it are protected under copyright by the Publisher (other than as may be noted herein).

Notices

Knowledge and best practice in this field are constantly changing. As new research and experience broaden our understanding, changes in research methods, professional practices, or medical treatment may become necessary.

Practitioners and researchers must always rely on their own experience and knowledge in evaluating and using any information, methods, compounds, or experiments described herein. In using such information or methods they should be mindful of their own safety and the safety of others, including parties for whom they have a professional responsibility.

To the fullest extent of the law, neither the Publisher nor the authors, contributors, or editors, assume any liability for any injury and/or damage to persons or property as a matter of products liability, negligence or otherwise, or from any use or operation of any methods, products, instructions, or ideas contained in the material herein.

ISBN: 978-0-12-802292-4
ISSN: 0076-6879

For information on all Academic Press publications
visit our website at https://www.elsevier.com

Publisher: Zoe Kruze
Acquisition Editor: Zoe Kruze
Editorial Project Manager: Sarah Lay
Production Project Manager: Magesh Kumar Mahalingam
Cover Designer: Maria Ines Cruz

Typeset by SPi Global, India

CONTENTS

Contributors   xi
Preface   xv

1. RNA Imaging with Multiplexed Error-Robust Fluorescence In Situ Hybridization (MERFISH)   1
J.R. Moffitt and X. Zhuang

  1. Introduction   2
  2. MERFISH Overview   4
  3. The Design of Oligonucleotide Probes   13
  4. Probe Construction   22
  5. Sample Preparation and Staining   27
  6. MERFISH Imaging   31
  7. MERFISH Data Analysis   39
  8. Summary   48
  Acknowledgments   48
  References   48

2. Imaging Single mRNA Dynamics in Live Neurons and Brains   51
H.C. Moon and H.Y. Park

  1. Introduction   52
  2. Neuron Culture Imaging   55
  3. Brain Slice Imaging   61
  Acknowledgments   63
  References   63

3. Monitoring of RNA Dynamics in Living Cells Using PUM-HD and Fluorescent Protein Reconstitution Technique   65
H. Yoshimura and T. Ozawa

  1. Introduction   66
  2. Principle of PUM-HD-Based RNA Probes   70
  3. Development of PUM-HD-Based Probes   73
  4. Microscopy Setup and Visualization of Single-Molecule RNA in Living Cells   76
  5. Example of RNA Visualization by Using PUM-HD-Based RNA Probes   80
  6. Conclusion   82
  Acknowledgment   84
  References   84

4. Applications of Hairpin DNA-Functionalized Gold Nanoparticles for Imaging mRNA in Living Cells   87
S.R. Jackson, A.C. Wong, A.R. Travis, I.E. Catrina, D.P. Bratu, D.W. Wright, and A. Jayagopal

1. Introduction	88
2. Protocol for hAuNP-Guided Imaging of mRNA in Living Cells	92
3. Key Results	96
4. Conclusions	99
Acknowledgments	101
References	101

5. In Vivo RNA Visualization in Plants Using MS2 Tagging   105
E.J. Peña and M. Heinlein

1. Introduction	106
2. Plasmid Constructs: Cloning of SL-Tagged RNA and NLS:MCP:FP into Plant Expression Vectors	111
3. Transient Expression in *N. benthamiana* Leaves	114
4. Imaging and Time-Lapse Image Acquisition	116
5. Summary and Perspectives	119
Acknowledgments	119
References	119

6. TRICK: A Single-Molecule Method for Imaging the First Round of Translation in Living Cells and Animals   123
J.M. Halstead, J.H. Wilbertz, F. Wippich, T. Lionnet, A. Ephrussi, and J.A. Chao

1. Introduction	124
2. Design of TRICK Reporter mRNAs	125
3. TRICK Experiment in Mammalian Cells	128
4. Microscopy	132
5. Data Collection	138
6. Analysis	140
7. TRICK Experiment in HeLa Cells to Determine Fraction of Untranslated mRNAs	145
8. TRICK Experiment in *Drosophila*	148
9. Outlook	152
Acknowledgments	153
References	153

7. Fluctuation Analysis: Dissecting Transcriptional Kinetics with Signal Theory — 159
A. Coulon and D.R. Larson

1. Introduction — 160
2. Computing and Averaging Correlation Functions — 165
3. Interpretation of Correlation Functions — 176
4. Common Issues and Pitfalls — 182
5. Conclusion — 189
References — 190

8. IMAGEtags: Quantifying mRNA Transcription in Real Time with Multiaptamer Reporters — 193
J. Ray, I. Shin, M. Ilgu, L. Bendickson, V. Gupta, G.A. Kraus, and M. Nilsen-Hamilton

1. Introduction — 194
2. IMAGEtags — 196
3. Visualizing Gene Expression with IMAGEtags — 199
4. Measurement of IMAGEtag RNA Level by RT-qPCR — 205
5. Synthesis of Ligands — 207
6. Cloning Repetitive Sequences — 209
7. Conclusions — 211
Acknowledgments — 212
References — 212

9. A Method for Expressing and Imaging Abundant, Stable, Circular RNAs In Vivo Using tRNA Splicing — 215
C.A. Schmidt, J.J. Noto, G.S. Filonov, and A.G. Matera

1. Introduction — 216
2. Design and Generation of tricRNA Vectors — 218
3. In Vivo Expression of tricRNAs — 226
4. In-Gel Imaging of tricRNAs — 229
5. Cellular Imaging of tricRNAs — 233
6. Concluding Remarks — 234
Acknowledgments — 235
References — 235

10. RNA-ID, a Powerful Tool for Identifying and Characterizing Regulatory Sequences 237
C.E. Brule, K.M. Dean, and E.J. Grayhack

1. Introduction 238
2. Features of the RNA-ID System 239
3. Applications of the RNA-ID System 241
4. Analysis of cis-Regulatory Sequences in the RNA-ID Reporter 244
5. Verification of Regulatory Sequences 250
6. Conclusions and Extensions of the RNA-ID Reporter System 251
Acknowledgments 252
References 252

11. Fluorescent Protein-Based Quantification of Alternative Splicing of a Target Cassette Exon in Mammalian Cells 255
N.G. Gurskaya, D.B. Staroverov, and K.A. Lukyanov

1. Introduction 256
2. Description of the Method 257
3. Materials 260
4. Procedure 261
5. Concluding Remarks 266
Acknowledgments 267
References 267

12. IRAS: High-Throughput Identification of Novel Alternative Splicing Regulators 269
S. Zheng

1. Introduction 270
2. Method Design 272
3. Construction of Dual-Fluorescence Minigene Reporters 276
4. Generation of Stable Cell Clones 278
5. Library and Array Construction 280
6. Cell-Based High-Throughput Screens 281
7. Data Acquisition 282
8. Background Fluorescence Correction and Normalization 285
9. Data Analysis 285
10. Validation 287
11. Limitations and Perspectives 287
References 288

13. Analysis of Nonsense-Mediated mRNA Decay at the Single-Cell Level Using Two Fluorescent Proteins   291

N.G. Gurskaya, A.P. Pereverzev, D.B. Staroverov, N.M. Markina, and K.A. Lukyanov

1. Introduction	292
2. Description of the Method	294
3. Materials	296
4. Procedure	297
5. Further Possible Modifications of Genetic Constructs	309
6. Concluding Remarks	311
Acknowledgments	312
References	312

14. Developing Fluorogenic Riboswitches for Imaging Metabolite Concentration Dynamics in Bacterial Cells   315

J.L. Litke, M. You, and S.R. Jaffrey

1. Introduction	316
2. Structure-Guided Design of Spinach Riboswitch Sensors	319
3. Identifying Spinach Riboswitch Sensors In Vitro	323
4. Live-Cell Imaging of Metabolites with Spinach Riboswitch Sensors	327
5. Summary and Concluding Remarks	331
Acknowledgments	332
References	332

Author Index	*335*
Subject Index	*351*

CONTRIBUTORS

L. Bendickson
Ames Laboratory, US DOE; Iowa State University, Ames, IA, United States

D.P. Bratu
Hunter College; Program in Molecular, Cellular, and Developmental Biology, and Program in Biochemistry, The Graduate Center, City University of New York, New York, NY, United States

C.E. Brule
Center for RNA Biology, University of Rochester School of Medicine and Dentistry, Rochester, NY, United States

I.E. Catrina
Hunter College, City University of New York, New York, NY, United States

J.A. Chao
Friedrich Miescher Institute for Biomedical Research, Basel, Switzerland

A. Coulon
Institut Curie, PSL Research University, Laboratoire Physico-Chimie, CNRS UMR 168; Sorbonne Universités, UPMC Univ Paris 06, Paris, France

K.M. Dean
Center for RNA Biology, University of Rochester School of Medicine and Dentistry, Rochester, NY, United States

A. Ephrussi
European Molecular Biology Laboratory, Heidelberg, Germany

G.S. Filonov
Weill Cornell Medical College, New York, NY, United States

E.J. Grayhack
Center for RNA Biology, University of Rochester School of Medicine and Dentistry, Rochester, NY, United States

V. Gupta
The Scripps Research Institute, Jupiter, FL, United States

N.G. Gurskaya
Shemyakin-Ovchinnikov Institute of Bioorganic Chemistry, Moscow; Nizhny Novgorod State Medical Academy, Nizhny Novgorod, Russia

J.M. Halstead
Friedrich Miescher Institute for Biomedical Research, Basel, Switzerland

M. Heinlein
Institut de Biologie Moléculaire des Plantes du CNRS, Université de Strasbourg, Strasbourg, France

M. Ilgu
Aptalogic Inc., Ames, IA, United States

S.R. Jackson
Vanderbilt University, Nashville, TN, United States

S.R. Jaffrey
Tri-Institutional Chemical Biology Program at Weill-Cornell Medical College, Rockefeller University, Memorial Sloan-Kettering Cancer Center; Weill Medical College, Cornell University, New York, NY, United States

A. Jayagopal
Pharma Research and Early Development (pRED), F. Hoffman-La Roche Ltd., Basel, Switzerland

G.A. Kraus
Ames Laboratory, US DOE; Iowa State University, Ames, IA, United States

D.R. Larson
Laboratory of Receptor Biology and Gene Expression, National Cancer Institute, NIH, Bethesda, MD, United States

T. Lionnet
Transcription Imaging Consortium, HHMI Janelia Research Campus, Ashburn, VA, United States

J.L. Litke
Tri-Institutional Chemical Biology Program at Weill-Cornell Medical College, Rockefeller University, Memorial Sloan-Kettering Cancer Center; Weill Medical College, Cornell University, New York, NY, United States

K.A. Lukyanov
Shemyakin-Ovchinnikov Institute of Bioorganic Chemistry, Moscow; Nizhny Novgorod State Medical Academy, Nizhny Novgorod, Russia

N.M. Markina
Shemyakin-Ovchinnikov Institute of Bioorganic Chemistry, Moscow, Russia

A.G. Matera
Integrative Program for Biological and Genome Sciences, Curriculum in Genetics and Molecular Biology, University of North Carolina, Chapel Hill, NC, United States

J.R. Moffitt
Howard Hughes Medical Institute, Harvard University, Cambridge, MA, United States

H.C. Moon
Seoul National University, Seoul, South Korea

M. Nilsen-Hamilton
Ames Laboratory, US DOE; Iowa State University; Aptalogic Inc., Ames, IA, United States

J.J. Noto
Integrative Program for Biological and Genome Sciences, Curriculum in Genetics and Molecular Biology, University of North Carolina, Chapel Hill, NC, United States

T. Ozawa
School of Science, The University of Tokyo, Tokyo, Japan

H.Y. Park
Seoul National University, Seoul, South Korea

E.J. Peña
Instituto de Biotecnología y Biología Molecular, CCT—La Plata CONICET, Fac. Cs. Exactas, U.N.L.P., La Plata, Argentina

A.P. Pereverzev
Shemyakin-Ovchinnikov Institute of Bioorganic Chemistry, Moscow, Russia

J. Ray
Cornell University, Ithaca, NY, United States

C.A. Schmidt
Integrative Program for Biological and Genome Sciences, Curriculum in Genetics and Molecular Biology, University of North Carolina, Chapel Hill, NC, United States

I. Shin
National Forensic Service, Seoul, South Korea

D.B. Staroverov
Shemyakin-Ovchinnikov Institute of Bioorganic Chemistry, Moscow, Russia

A.R. Travis
Vanderbilt University, Nashville, TN, United States

J.H. Wilbertz
Friedrich Miescher Institute for Biomedical Research; University of Basel, Basel, Switzerland

F. Wippich
European Molecular Biology Laboratory, Heidelberg, Germany

A.C. Wong
Vanderbilt University, Nashville, TN, United States

D.W. Wright
Vanderbilt University, Nashville, TN, United States

H. Yoshimura
School of Science, The University of Tokyo, Tokyo, Japan

M. You
Weill Medical College, Cornell University, New York, NY, United States

S. Zheng
University of California, Riverside, CA, United States

X. Zhuang
Howard Hughes Medical Institute; Harvard University, Cambridge, MA, United States

PREFACE

The major function of the human genome is to encode a vast diversity of RNAs that control essentially every facet of cellular function. The RNAs include well-known RNAs, such as messenger RNAs, transfer RNAs, and ribosomal RNAs, as well as a dizzying set of other RNAs such as long noncoding RNAs, Y-RNAs, microRNAs, small nucleolar RNAs, vault RNAs, piwi-interacting RNAs, and numerous others. A major focus has been mRNAs and uncovering the relationship between their localization and function in normal and disease cells. However, for the large and mysterious class of long noncoding RNAs, studies that reveal their localization and trafficking can provide a first clue about their potential function in cells. For these reasons, methods in localizing RNAs in cells have been important for biomedical researchers. In recent years, these methods have undergone a renaissance with multiple innovations providing researchers with the ability to image RNAs in living cells in real time.

In addition to RNA trafficking, RNA undergoes diverse types of processing events in cells, such as splicing, degradation, and modification. These processes are typically measured by harvesting cells and performing assays such as Northern blotting. However, these approaches fail to reveal the spatial and temporal dynamics of these processes, which may be subjected to regulation. To overcome this, new methods are being developed to image these events in living cells. Methods to image RNA processing are still in their infancy, but are rapidly developing.

This volume is roughly divided into two parts. The first part highlights diverse methods for detecting and imaging different types of endogenous RNA in cells. The second part comprises techniques to study mRNA molecular biology and for using RNA itself as a reporter.

The volume starts with MERFISH (Chapter 1), an impressive technique described by Moffitt and Zhuang. This protocol can be used for visualization of hundreds to thousands of different RNAs in individual cells with single-molecule resolution. In the next chapter, Moon and Park describe an approach that utilizes the MS2-GFP fusion to image single β-actin mRNA dynamics in live cells of mouse brain slices. In Chapter 3, Yoshimura and Ozawa present a protocol that describes fluorescent-protein-based imaging of RNAs utilizing PUM-HD, a domain that can be modified to recognize different 8-base RNA sequences. Jackson et al. (Chapter 4) describe a way to utilize hairpin-DNA-functionalized gold nanoparticles as an efficient way to label and image mRNAs in live cells. Another application of a fluorescent protein fused to MS2 is described by Peña and Heinlein (Chapter 5). In this chapter, viral RNA is visualized in live plant cells. Chao and colleagues present a powerful technique to track untranslated RNAs and then to report on the first round of translation (Chapter 6). A separate chapter is presented by Coulon and Larson on the topic of fluctuation analysis, which is a way to analyze and interpret multicolor transcriptional time traces (Chapter 7). Ray et al. describe a protocol to use IMAGEtags to image bulk RNA in cells and to study transcription changes in real time (Chapter 8). Finally, Matera and colleagues present a protocol for expression of highly stable circular RNAs and their imaging using the Broccoli fluorescent RNA aptamer.

The second part of the volume describes the methods for studying RNA molecular biology. The first chapter in this series (Chapter 10) is compiled by Grayhack and colleagues and describes RNA-ID, an approach to identify and analyze sequences that regulate gene expression. The group of Lukyanov presents two protocols that use flow cytometry and fluorescence microscopy to report on alternative splicing or nonsense-mediated RNA decay (Chapters 11 and 13, respectively). As another approach to study splicing, Zheng describes IRAS, a fluorescence-based protocol to identify novel alternative splicing regulators (Chapter 12). The final chapter in this volume describes an innovative way to use RNA fluorescence to report on intracellular events. Litke et al. describe a protocol for development of fluorogenic riboswitches to serve as RNA-based biosensors (Chapter 14).

While all methods for imaging RNA and RNA processing are not presented here, we hope that these chapters will be of use to the scientific community and will help to stimulate further development of imaging tools that can provide new insights into the intricacies of RNA biology.

GRIGORY S. FILONOV
SAMIE R. JAFFREY

CHAPTER ONE

RNA Imaging with Multiplexed Error-Robust Fluorescence In Situ Hybridization (MERFISH)

J.R. Moffitt[*,1], X. Zhuang[*,†,1]
*Howard Hughes Medical Institute, Harvard University, Cambridge, MA, United States
†Harvard University, Cambridge, MA, United States
[1]Corresponding authors: e-mail address: lmoffitt@mcb.harvard.edu; zhuang@chemistry.harvard.edu

Contents

1. Introduction	2
2. MERFISH Overview	4
2.1 Combinatorial Barcoding and Sequential Readout	4
2.2 Error-Robust and -Correcting Codes	6
2.3 Two-Stage Hybridization	8
2.4 Performance of MERFISH	10
3. The Design of Oligonucleotide Probes	13
3.1 Design of Target Regions	14
3.2 Design of Readout Probes	16
3.3 Design of the Codebook	18
3.4 Assembly and Screening of Encoding Probes	20
3.5 Design of Priming Regions	20
4. Probe Construction	22
4.1 Required Reagents	23
4.2 Amplification of In Vitro Template	24
4.3 In Vitro Transcription	26
4.4 Reverse Transcription and Purification	26
5. Sample Preparation and Staining	27
5.1 Required Materials	28
5.2 Fixation and Permeabilization of Cells	29
5.3 Hybridization of Encoding Probes	30
6. MERFISH Imaging	31
6.1 Required Materials	32
6.2 Assembly and Operation of Flow System	33
6.3 Microscope Requirements	36
6.4 MERFISH Imaging Protocol	37
7. MERFISH Data Analysis	39
7.1 Identification of Fluorescent Spots	40
7.2 Correction of Image Offsets	41

7.3 Decoding Barcodes	42
7.4 Calculate MERFISH Performance	43
7.5 (Optional) Iterative Identification of Optimal Thresholds	47
8. Summary	48
Acknowledgments	48
References	48

Abstract

Quantitative measurements of both the copy number and spatial distribution of large fractions of the transcriptome in single cells could revolutionize our understanding of a variety of cellular and tissue behaviors in both healthy and diseased states. Single-molecule fluorescence in situ hybridization (smFISH)—an approach where individual RNAs are labeled with fluorescent probes and imaged in their native cellular and tissue context—provides both the copy number and spatial context of RNAs but has been limited in the number of RNA species that can be measured simultaneously. Here, we describe multiplexed error-robust fluorescence in situ hybridization (MERFISH), a massively parallelized form of smFISH that can image and identify hundreds to thousands of different RNA species simultaneously with high accuracy in individual cells in their native spatial context. We provide detailed protocols on all aspects of MERFISH, including probe design, data collection, and data analysis to allow interested laboratories to perform MERFISH measurements themselves.

1. INTRODUCTION

In situ hybridization of fluorescently labeled oligonucleotide probes to cellular RNAs has emerged as a powerful technique for the quantification of both the copy number of individual transcripts and the spatial distribution of these molecules within single cells (Femino, Fay, Fogarty, & Singer, 1998; Raj, van den Bogaard, Rifkin, van Oudenaarden, & Tyagi, 2008). In this approach, cells are fixed and permeabilized, and a small number of fluorescently labeled oligonucleotide probes, whose sequences are complementary to different regions of the RNA of interest, are introduced into the cell. Basepairing between these probes and their target RNA favors specific binding of these probes to their target, and the use of multiple probes per target concentrates many probes within the small volume of the cell occupied by each copy of that RNA, producing bright fluorescent spots that can be distinguished from cellular background and stray probes. These spots can then be counted to determine the copy number of the RNA of interest and their locations within individual cells and intact tissues.

With these technical abilities, single-molecule fluorescence in situ hybridization (smFISH) has emerged as a powerful tool in the study of a wide range of questions in cellular and tissue biology. For example, the ability to quantify the copy number of individual RNAs within single cells has driven a variety of exciting advances in our understanding of the natural cell-to-cell variation in gene expression within homogenous populations of cells as well as the expression differences which correlate with and likely drive the development of complex cell types within tissues (Larson, Singer, & Zenklusen, 2009; Munsky, Neuert, & van Oudenaarden, 2012; Padovan-Merhar & Raj, 2013; Raj & van Oudenaarden, 2008; Sanchez & Golding, 2013). Similarly, the ability to precisely image the spatial distribution of different RNAs within cells has revealed that the transcriptome of many cell types is spatially organized. Remarkably, this organization has emerged as an important regulator of a broad range of cellular phenotypes, from directional crawling of individual cells to the establishment of body plan in multicellular organisms (Balagopal & Parker, 2009; Besse & Ephrussi, 2008; Buxbaum, Haimovich, & Singer, 2014; Holt & Schuman, 2013; Lécuyer et al., 2009; Martin & Ephrussi, 2009; Rodriguez, Czaplinski, Condeelis, & Singer, 2008; St Johnston, 2005).

However, the number of RNA species that can be simultaneously imaged by smFISH has been limited. Most experiments stain and image only one RNA species at a time, and even the most advanced multiplexing efforts have only extended this approach to the simultaneous measurement of ~10–30 RNA species (Jakt, Moriwaki, & Nishikawa, 2013; Levesque & Raj, 2013; Levsky, Shenoy, Pezo, & Singer, 2002; Lubeck & Cai, 2012; Lubeck, Coskun, Zhiyentayev, Ahmad, & Cai, 2014). However, many interesting questions require much higher levels of multiplexing—levels that approach the transcriptome scale. For example, comprehensive or hypothesis-free studies of gene regulatory networks, transcriptional definitions of cell type, or systematic screens for cell-type markers, could all benefit from precise copy number measurement of hundreds to thousands of RNAs, if not the entire transcriptome, simultaneously within single cells. Similarly, efforts to systematically map the spatial organization of the transcriptome, to determine the molecular transport mechanisms that establish and dynamically regulate this spatial organization, and to elucidate the ultimate role that this organization plays in posttranscriptional regulation, would also benefit from the ability to map the spatial organization of large fractions of the transcriptome simultaneously within single cells.

To address this technical demand, we recently introduced a massively multiplexed smFISH imaging method that extends the benefits of this powerful technique toward the transcriptome scale. We term this approach multiplexed error-robust fluorescence in situ hybridization (MERFISH) (Chen, Boettiger, Moffitt, Wang, & Zhuang, 2015). MERFISH achieves large-scale multiplexing by assigning error-robust barcodes to different RNA species and then reading out these barcodes through successive rounds of hybridization and imaging on the same sample. In Section 2, we provide a more detailed discussion of the conceptual challenges to multiplexing smFISH, the way in which MERFISH addresses these challenges, and the published performance of MERFISH. In the subsequent sections, we provide detailed instructions, protocols, and tips on how to perform a MERFISH measurement, including details on how to design oligonucleotide probes (Section 3), how to construct these probes (Section 4), how to prepare samples (Section 5), how to execute a MERFISH measurement (Section 6), and, finally, how to analyze MERFISH data (Section 7). In many ways, MERFISH is only moderately more complicated than smFISH, and the goal of this chapter is to provide protocols and advice that will allow interested laboratories to perform MERFISH measurements themselves. As MERFISH is based on smFISH, we also recommend several reviews on smFISH for readers to learn about the basics of smFISH measurements (Batish, Raj, & Tyagi, 2011; Itzkovitz & van Oudenaarden, 2011; Youk, Raj, & Van Oudenaarden, 2010; Zenklusen & Singer, 2010).

2. MERFISH OVERVIEW
2.1 Combinatorial Barcoding and Sequential Readout

MERFISH significantly increases the multiplexing of smFISH measurements by utilizing a combinatorial labeling approach to associate unique barcodes with individual RNA species and then by reading out these barcodes through a series of sequential hybridization and imaging measurements. Combinatorial labeling approaches have been utilized previously to increase the multiplexing of smFISH measurements (Jakt et al., 2013; Levesque & Raj, 2013; Levsky et al., 2002; Lubeck & Cai, 2012). In these approaches, each RNA species is often identified not by a single color of fluorescently labeled smFISH probe but rather with a unique combination or "barcode" of colors. For example, one RNA might be identified by red-labeled probes alone while another might be identified by the combination of red-, green-, and blue-labeled probes. We can formalize this discussion by adopting a binary representation of this barcoding scheme with the presence or absence of a color being represented by a "1" or "0," respectively, within a given bit. For example, a red-labeled RNA might be represented as 100 while an RNA labeled with all three colors would be represented by 111. In this example, three colors imply three bits which, in turn, imply $2^3 = 8$ possible color combinations. This geometric scaling between the number of unique barcodes and the number of measured labels (ie, bits or colors) is what makes combinatorial labeling such a powerful approach to multiplexing.

However, to significantly increase the multiplexing of smFISH measurements, far more than three distinguishable colors would be needed. For example, the simultaneous identification of 1000 RNA species would require at least 10 colors! Unfortunately, distinguishing such a large number of colors via optical microscopy can be quite challenging. For this reason, MERFISH adopts a different approach: instead of measuring binary barcodes with different color combinations, each bit in the binary barcode is determined by the presence or absence of fluorescence in a single round of hybridization and imaging (Chen et al., 2015). Simply put, the barcode is built up one bit at a time via a series of smFISH measurements, all on the same sample. Fig. 1A illustrates the conceptual approach. In the first round, a subset of RNAs is stained. If fluorescent in this round, an RNA is assigned a "1" in the first bit. If it is not fluorescent, it is assigned a "0." A second set of probes is then hybridized to the same sample, and this sample is imaged again. Each RNA is then assigned a "1" or "0" in the *second* bit based on whether or not it is fluorescent in this second round of smFISH. This process is repeated across N rounds of smFISH imaging, allowing the construction of an N-bit barcode for each labeled RNA in the sample. These binary barcodes can then be used to identify the species of each labeled RNA. By adopting a sequential hybridization and imaging readout approach, it is possible to imagine identifying very large numbers of RNAs within a single sample (Chen et al., 2015; see Fig. 1B). For example, only 16 rounds of hybridization and imaging would be required to generate enough unique barcodes, ~65,000, to identify most if not all of the expressed RNA species within individual human cells! Combination of this sequential hybridization/imaging method with multicolor imaging will further increase the number of RNA species addressable or decrease the number of rounds of hybridization/imaging required for addressing the same number of RNA species. We note that a different approach to take advantage of color barcodes and sequential imaging to increase the number detectable RNA species has also been proposed, though in the reported experimental demonstration of this approach, the number of RNA species imaged in individual cells is still on the order of ~10–20 (Lubeck et al., 2014).

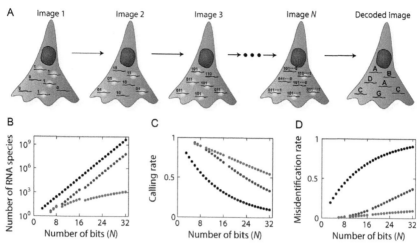

Fig. 1 Multiplexed, error-robust fluorescence in situ hybridization (MERFISH). (A) A schematic depiction of the process by which binary barcodes associated with each labeled RNA in a sample is readout. The presence or absence of fluorescence in each round of hybridization and imaging determines whether the barcode associated with each RNA in the sample has a "1" or "0" in the corresponding bit in the measured binary barcode. These barcodes are then used to identify each RNA, eg, species, A, B, C, etc. (B) The number of RNA species that can be encoded with a binary barcode vs the number of bits N in those barcodes. A binary encoding scheme that utilizes all possible binary barcodes of length N is depicted in black. A binary encoding scheme that utilizes a subset of all possible binary barcodes that are all separated by at least a Hamming Distance of 4—an encoding scheme known as the Extended Hamming Code—is depicted in *blue*. A binary encoding scheme that utilizes a modified Hamming Distance 4 encoding scheme, which consists of all barcodes from the Extended Hamming Code that have a Hamming Weight of 4, ie, only 4 "1" bits, is depicted in *magenta*. (C) The fraction of the binary barcodes that can be properly decoded into the correct RNA species—the calling rate—in the presence of modest readout errors as a function of the number of bits in the barcode for the same encoding schemes depicted in (B). (D) The fraction of binary barcodes that are misidentified as the wrong barcode and, thus, are decoded as the wrong RNA—the misidentification rate—as a function of the number of bits for the same encoding schemes depicted in (B). Panels (C) and (D) were calculated assuming an average "1" → "0" error rate of 10% and a "0" → "1" error rate of 4%, which correspond to the typical error rates observed in MERFISH measurements. *Reproduced with permission from Chen, K.H., Boettiger, A.N., Moffitt, J.R., Wang, S., & Zhuang, X. (2015). Spatially resolved, highly multiplexed RNA profiling in single cells. Science, 348, aaa6090.* (See the color plate.)

2.2 Error-Robust and -Correcting Codes

In practice, a substantial limitation to multiplexing arises from the small readout errors that are inherent to smFISH. There are two such errors. First, occasionally an RNA that should fluoresce in one imaging round does not accumulate enough fluorescently labeled probe to produce a bright enough signal to be called as an RNA. Second, stray probes or a bright autofluorescent spot in the cell can sometimes produce a spot bright enough to be called an RNA when it is not. In the context of reading out a binary barcode, these errors would correspond to misreading a "1" as a "0" or a "0" as a "1," respectively.

Such errors are infrequent—"1" to "0" errors typically occur at frequencies of ~10% while "0" to "1" errors occur a few-fold less frequently—and, thus, have little effect on the performance of traditional smFISH measurements (Raj et al., 2008). However, in combinatorial barcoding approaches, such as MERFISH, the effect of these errors compounds quickly as the length of the barcode increases, producing two forms of measurement errors (Chen et al., 2015; see Fig. 1C and D). First, the probability that no measurement errors are made during barcode readout drops with each additional bit in the barcode and, thus, the fraction of barcodes measured correctly—what we term the *calling rate*—decreases with increasing barcode length. More problematically, each readout error transforms one barcode into another, and such errors can lead to the misidentification of one RNA as another. As the length of the barcode increases, the rate at which RNAs are incorrectly identified—what we term the *misidentification rate*—increases quickly. Thus, while it is theoretically possible to measure enough barcodes to identify the entire human transcriptome in just 16 rounds of imaging, even with modest readout errors, the vast majority of these RNAs would be identified incorrectly!

To address this problem, we sought to use error-robust or error-correcting encoding schemes in MERFISH (Chen et al., 2015). The idea behind these encoding schemes is quite simple: instead of assigning each possible binary barcode to an RNA, many of the possible binary barcodes are left unassigned (Moon, 2005). If one of these unassigned barcodes is observed, then it is clear that an error has occurred. The Hamming Distance (HD) between two barcodes—the number of bits that must be switched to convert one barcode into another—is a useful concept in understanding the behaviors of these encoding schemes (Moon, 2005). If, out of all possible binary barcodes, only barcodes that are separated by at least a HD of 2 are used, then no single-bit readout error can transform one barcode into another, reducing the misidentification rate. Further increasing the minimum HD between used barcodes by leaving more possible barcodes unassigned will produce even larger reductions in the misidentification rate, see Fig. 1D. In addition to a reduction in the misidentification rate, some encoding schemes can also improve the calling rate through a process known as error correction. Consider an encoding scheme in which the minimum HD between all barcodes is 4. If a single error occurs in the readout of a barcode, it will produce a barcode with an HD of 1 from the actual correct barcode but an HD of at least 3 from all other possible barcodes. Thus, the measured barcode can not only be recognized as containing an error, it can also be associated with the correct barcode with reasonable certainty. The result, as illustrated in Fig. 1C, is a significant increase in the calling rate.

In the typical smFISH measurement, the rate at which a "1" is misread as a "0" is often larger than the rate of misreading a "0" as a "1." If left unaddressed, this asymmetry would introduce bias into measurements (Chen et al., 2015). Specifically, barcodes that contain more "1"s than other barcodes would tend to have more errors and would thus be prone to a lower calling rate and a higher misidentification rate than barcodes that contain fewer "1"s. We address this potential bias with encoding schemes known as constant Hamming Weight codes (Chen et al., 2015) (the Hamming Weight is defined as the number of "1"s in a binary barcode). In such codes, the only binary barcodes that are used are barcodes that have the same Hamming Weight, eg, only barcodes with four "1"s are used. Because of the asymmetries in error rates, the use of a low Hamming Weight code can remove barcodes subject to higher frequency of errors and, thus, raises calling rates and lowers misidentification rates, as seen in Fig. 1C and D.

2.3 Two-Stage Hybridization

Hybridization time is a significant challenge for an approach that adopts a sequential hybridization and imaging approach to readout barcodes. This challenge arises from the observation that hybridization of DNA oligonucleotides to cellular RNA is a slow process, typically requiring 6–24 h for efficient labeling (Shaffer, Wu, Levesque, & Raj, 2013) because sequences on the cellular RNAs could be occluded by base-pairing with other sequences or by binding of proteins. Thus, with typical hybridization approaches, measuring a 16-bit barcode could require nearly 2 weeks! To solve this problem, we perform two types of hybridizations in MERFISH (Chen et al., 2015). In the first hybridization, we stain the sample with single-stranded DNA oligonucleotides that we term *encoding probes*, see Fig. 2. These probes contain two types of hybridization regions. The first region is termed the *targeting region*, and it is complementary to the sequence of a specific RNA of interest, allowing the encoding probes to hybridize to the target RNA of interest with high efficiency and specificity. The second region is termed the *readout region*. It is comprised of one or more custom sequences that are complementary to fluorescently labeled readout probes,

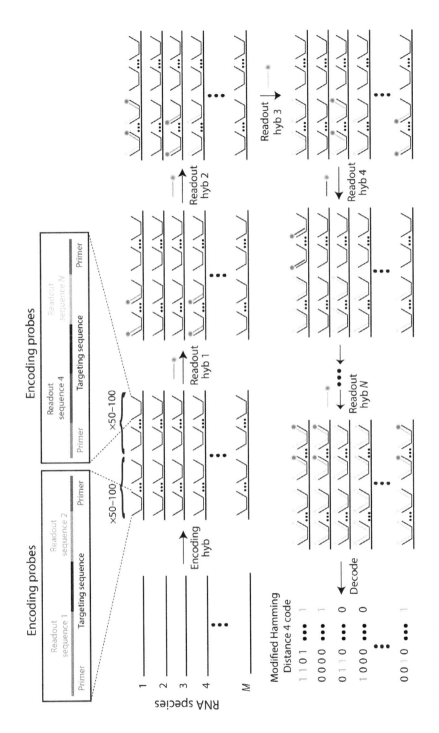

Fig. 2 See legend on next page.

again designed to bind with high efficiency and specificity to the complementary readout sequence, see Fig. 2. In the second type of hybridization, the readout probes are added to the sample and allowed to hybridized with the complementary readout sequences on the encoding probes. Binding of the encoding probes to cellular RNA is a slow process, requiring at least 12 h of hybridization. But once this slow hybridization is complete, we found that the hybridization necessary to read each readout sequence in the barcode with readout probes can be completed much more quickly, presumably because the readout sequences are not occluded by the binding of any cellular proteins or RNA. Thus, what might have been a 2-week long readout procedure can be accomplished in under 12 h for 16 rounds of readout hybridization and imaging. Fig. 2 depicts the multiple hybridization rounds required for the typical MERFISH measurement.

2.4 Performance of MERFISH

In our published proof of principle for MERFISH, we demonstrated the ability to measure 140 or 1001 RNA species simultaneously in individual cells using two different encoding schemes (Chen et al., 2015). Using an encoding scheme with 16-bit barcodes separated by a minimum HD of 4 and with a uniform Hamming Weight of 4, we demonstrated that we could identify 140 RNA species simultaneously in individual cells with accuracy and efficiency very similar to those demonstrated for smFISH. We term this encoding scheme a modified Hamming Distance 4 (MHD4) code. Fig. 3A–D

Fig. 2 Schematic depiction of the hybridization process used for MERFISH. Cellular RNAs are hybridized with a set of oligonucleotide probes, which we term *encoding probes*. These encoding probes contain a targeting sequence which directs their binding to the specific RNA. They also contain two readout sequences. For an experiment utilizing *N*-bit binary barcodes, *N* different readout sequences will be used with each bit assigned a different unique readout sequence. The specific readout sequences contained by an encoding probe to a given RNA are determined by the binary barcode assigned to that RNA: only the readout sequences assigned to bits for which this barcode contains a "1" are used. Each encoding probe also contains PCR priming regions used in its construction. To increase the signal from each copy of the RNA, multiple encoding probes, each with a different target region, are bound to the same RNA. The length of this *tile* of probes is typically between 50 and 100 probes. To identify the readout sequences contained on the encoding probes bound to each RNA, *N* rounds of hybridization and imaging are performed. Each round uses a unique, fluorescently labeled probe whose sequence is complementary to the readout sequence for that round. The binding of these fluorescent probes determines the bits which contain a "1," allowing the measurement of the specified binary code. *Modified with permission from Chen, K.H., Boettiger, A.N., Moffitt, J.R., Wang, S., & Zhuang, X. (2015). Spatially resolved, highly multiplexed RNA profiling in single cells. Science, 348, aaa6090.* (See the color plate.)

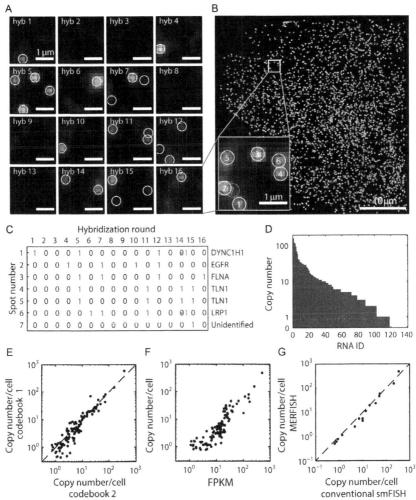

Fig. 3 Example MERFISH data for a 16-bit MHD4 code. (A) smFISH images from each of 16 rounds of hybridization of a small field of view of a Human fibroblast (IMR90) stained with encoding probes utilizing an 16-bit MHD4 code that encodes 140 RNAs. The label depicts the readout hybridization round corresponding to each Image. Circles correspond to the locations of identified fluorescent spots. (B) A single 40-μm square field of view with all measured barcodes marked. The *color* of each marker represents the measured barcode. (*Inset*) An overlay of the small section of this field of view depicted in (A) with each set of overlapping spots labeled. White circles correspond to sets of spots that represent a barcode that can be decoded into an RNA while *red* represents a set of spots for which the measured barcode does not represent an RNA. (C) The measured binary barcodes for each set of spots in the small field of view depicted in (A) with the identity of the RNA represented by that barcode. Error correction was required for two barcodes in hybridization round 14 and is represented by *red crosses*.

(*Continued*)

illustrates the use of this approach to image RNAs in human fibroblast cells (IMR90s). Control measurements in these cells revealed little technical bias, ie, the measured copy number for each RNA species did not depend on the binary barcode that had been assigned to it (Fig. 3E). Moreover, we observed excellent agreement between copy numbers as determined by MERFISH and bulk sequencing, with a Pearson correlation coefficient between the logarithmic abundances measured by these two techniques of 0.89 (Fig. 3F). Finally, we compared the copy numbers as determined by MERFISH to those determined by traditional smFISH for 15 different RNAs that span an abundance range that covers nearly three orders of magnitude (Fig. 3G). We also observed excellent correlation between these measurements and those of MERFISH; however, on average, MERFISH counted slightly smaller numbers of RNAs than that counted by traditional smFISH, roughly 80%. This slight reduction in calling rate is as expected given the measured per-bit 1-to-0 and 0-to-1 errors (Fig. 1C, magenta). The excellent agreement between MERFISH measurements and smFISH extended even to lowest abundance ranges, <1 copy per cell on average, revealing that the detection limit of MERFISH must be better than 1 RNA copy per cell.

To demonstrate the ability to further increase the multiplexing of smFISH measurements, we also tested a different encoding scheme capable of detecting but not correcting errors (Chen et al., 2015). Specifically, we used an encoding scheme with 14-bit barcodes separated by at least a HD of 2 and with a constant Hamming Weight of 4. The inability to correct errors was predicted to lower our calling rate. Indeed, we observed that the calling rate with this encoding scheme dropped to ~25%, as determined by comparing the measured copy numbers to those determined for

---

Fig. 3—Cont'd (D) The number of RNAs of each species identified in the single field of view depicted in (D). Approximately 2000 RNAs were measured in this single field of view, and in a single measurement, ~100 such fields of view containing 250,000 RNAs can be measured. (E) The average RNA copy number per cell measured with one implementation of the 16-bit MHD4 code vs the average copy number per cell for every RNA measured with another 16-bit MHD4 code in which each RNA was assigned a different barcode. The *dashed line* represents equality. (F) The average RNA copy number per cell vs the abundance as measured with bulk RNA-seq (FPKM). (G) The average copy number per cell measured via MERFISH vs that measured using conventional smFISH for 15 different RNAs. The *dashed line* represents equality. *Panels reproduced with permission from Chen, K.H., Boettiger, A.N., Moffitt, J.R., Wang, S., & Zhuang, X. (2015). Spatially resolved, highly multiplexed RNA profiling in single cells. Science, 348, aaa6090.* (See the color plate.)

individual RNAs using smFISH. However, this modified Hamming Distance 2 (MHD2) code encoding scheme still provided fairly accurate RNA measurements as evidenced by high Pearson correlation coefficients between these MERFISH measurements and bulk sequencing (0.76) as well as between these measurements and the subset of RNAs also measured via MERFISH with the higher accuracy MHD4 code (0.89). We anticipate that it should also be possible to measure ~1000 genes with our MHD4 code by using more rounds of imaging or utilizing multiple colors to readout multiple bits in each round of hybridization, for example, a 32-bit MHD4 code with a Hamming Weight of 4 can encode ~1200 RNAs.

In summary, the performance of MERFISH can be tuned to the demands of the experiment through judicious choice of the properties of the encoding scheme. If very high performance—comparable to smFISH—is required, then an encoding scheme capable of identifying and correcting errors is probably the appropriate choice. Whereas if a reduced calling rate and modest reduction in the accuracy of the measurement are tolerable, then much higher multiplexing for a fixed number of bits can also be achieved using an encoding scheme that can identify but not correct errors.

3. THE DESIGN OF OLIGONUCLEOTIDE PROBES

The first step in any MERFISH experiment is the design of the oligonucleotide probes that will be used to label individual RNA species. In our current implementation of MERFISH, each oligonucleotide encoding probe consists of three basic components as illustrated in Fig. 2. The first region is a 30-nt targeting region that is complementary to a portion of the sequence of the RNA to which it is designed to bind. The second region is a set of sequences that are called readout sequences, which were designed to be complementary and hence only bind to MERFISH readout probes and not other nucleic acid in the cell. Finally, the third region is a set of priming regions used in the construction of these probes, which will be discussed in detail in Section 4. We modified a previously developed Oligopaint approach (Beliveau et al., 2012) to make a large number of encoding probes in large quantity. In addition to the nucleotide sequences for each of these components, a *codebook*—the specific set of binary barcodes that will be used and their association with different RNA species of interest—must also be designed. In this section, we provide protocols to design these sequences and to build a codebook. Example code to perform these steps can be found at http://zhuang.harvard.edu/merfish/.

3.1 Design of Target Regions

Functionally, the goal of a target region is to direct the binding of each encoding probe to its target RNA of interest with high-binding efficiency and specificity. The central challenge in the design of target regions for MERFISH (and smFISH, in general) is to design a set of target regions where these properties are optimized for all probes under a constant set of hybridization conditions, eg, incubation temperature. To accomplish this, goal target regions are designed to cover a relatively narrow range of GC content and melting temperatures (T_M) with their target. In addition, a good target region should have limited homology to other RNAs in the transcriptome, reducing the probability that it will bind to the wrong RNA. Finally, the typical smFISH measurement does not bind each RNA with a single probe but rather tiles that RNA with multiple probes, each of which targets a different portion of the sequence of the RNA, see Fig. 2. For many smFISH measurements, the number of unique probes per RNA is often ~50; however, this number can be lowered with a corresponding reduction in the brightness of the individual RNA spots (Raj et al., 2008). For initial MERFISH work, we recommend having at least 50 encoding probes per RNA.

As described by others (Raj et al., 2008), smFISH probes can be designed online using a web interface (https://www.biosearchtech.com/support/tools/design-software/stellaris-probe-designer). In principle, probes designed with this software should work well as target regions for MERFISH. However, there is no batch processing option for such software and submitting hundreds to thousands of genes individually would be very labor intensive. Thus, we have used an alternative approach.

We design our target regions with the software package, OligoArray 2.0, which was developed for the design of microarrays (Rouillard, 2003). This software has been used previously in the design of FISH probes for DNA (Beliveau et al., 2012). In addition to the ability to batch process mRNAs, one advantage of this software is the additional stringency OligoArray applies to the design of its probes. Specifically, internal secondary structure and off-target binding are assessed not via the number of complementary bases but by the thermodynamic stability of these structures. Of course these extra calculations come at a computational cost, and calculation of target regions for a large number of genes can take days of computation on a desktop computer.

Step 1: Download and install all necessary software. OligoArray2.0 can be downloaded from http://berry.engin.umich.edu/oligoarray2_1/.

This software requires OligoArrayAux which can be downloaded from http://unafold.rna.albany.edu/?q=DINAMelt/OligoArrayAux. OligoArray2.0 will look for this software in a specific directory (C:\Program Files\ OligoArrayAux\), so it must be installed there. Finally, OligoArray2.0 also requires several legacy BLAST functions, which can be downloaded and installed from http://blast.ncbi.nlm.nih.gov/Blast.cgi?CMD=Web& PAGE_TYPE=BlastDocs&DOC_TYPE=Download.

Step 2: Download a fasta file containing the transcriptome of interest. These files can be found from a variety of repositories, such as those hosted by NCBI, UCSC, or Ensembl. For example, the human transcriptome can be downloaded from Ensembl using the cDNA link found on this page: http://www.ensembl.org/info/data/ftp/index.html.

Step 3: Create a BLAST database of the transcriptome for OligoArray2.0 to identify potential off-target binding. Instructions on how to create this database using the legacy BLAST functionality can be found here: http://blast.ncbi.nlm.nih.gov/Blast.cgi?CMD=Web&PAGE_TYPE=BlastDocs&DOC_ TYPE=Download.

Step 4: Run OligoArray2.0 on the desired transcripts. OligoArray2.0 is a java-based program and a full description of its use can be found here: http://berry.engin.umich.edu/oligoarray2_1/. When run, it must be supplied with a variety of parameters that place limits on the properties of the designed target regions, eg, the length of possible probes, the suitable GC and melting temperature ranges, as well as the maximum T_M for potential off-target sequences. For the design of our target regions, we have used the following parameters: target region length of 30 nt; the T_M of the properly hybridized probe greater than 70°C; a lower bound on the T_M of hybridization to potential off-target sequences of 72°C; no internal secondary structure with a T_M lower than 76°C; and no contiguous run of the same nucleotide longer than six. These parameter ranges were selected to balance the demands for high stringency of probe binding with the design of enough distinct target regions to label each RNA.

Step 5: Parse the OligoArray output. OligoArray2.0 generates an output file, the details of which are described here: http://berry.engin.umich.edu/ oligoarray2_1/. If a potential target region has homology to other transcripts, it will be indicated in this file. We only use potential target regions for which no potential off-targets were discovered.

Step 6: Repeat this process for all desired genes. As mentioned earlier, OligoArray2.0 is computationally intensive; thus, we recommend only using this software to design target regions for the desired subset of the transcriptome. Moreover, because computational cost does not grow linearly with the length of a transcript, we recommended that longer transcripts be split into smaller fragments and processed individually. We split transcripts into 1-kb increments.

3.2 Design of Readout Probes

There are several important considerations in the design of readout probes. First, to improve the binding efficiency of these probes, it is desirable to select probes that have similar T_M and GC content so that their hybridization properties are similar under a given hybridization condition. Second, to limit the number of potential off-target binding sites, potential sequences should be screened for homology to RNAs in the transcriptome of interest. Third, these sequences must be orthogonal, in that they should have limited homology with one another to prevent binding of one readout probe to the wrong readout sequence.

We have already screened and validated several readout probes for human samples, and we recommend the use of these sequences, which are provided in Table 1. The following steps can be taken if new or additional readout probes are required.

Step 1: Utilize existing sets of orthogonal nucleic acid sequences to design readout probes. It is important that one readout probe has little homology with a sequence of another probe to prevent potential off-target binding. Fortunately, a set of 25-mer nucleotide sequences designed to have limited cross homology exists, and we recommend using this resource to design readout sequences (Xu, Schlabach, Hannon, & Elledge, 2009). These sequences can be downloaded from http://elledgelab.med.harvard.edu/?page_id=638. We have found good performance with readout sequences of 30-nt length, which can be created from these 25-nt sequences by either concatenating portions of them or adding five random nucleotides to either end.

Step 2: Remove potential probes with homology to members of the transcriptome of interest. To remove probes with significant homology to members of the transcriptome, we create a BLAST library to the transcriptome as described in Section 3.1, Step 3. We then BLAST each potential readout probe sequence against this library and remove any probe which contains a contiguous stretch of homology longer than 14 nt. This length was selected to balance the desire for shorter regions of homology

Table 1 The Sequence of the Readouts Probes that Have Been Validated

CGCAACGCTTGGGACGGTTCCAATCGGATC	CGCGAAATCCCCGTAACGAGCGTCCCTTGC
CGAATGCTCTGGCCTCGAACGAACGATAGC	GCATGAGTTGCCTGGCGTTGCGACGACTAA
ACAAATCCGACCAGATCGGACGATCATGGG	CCGTCGTCTCCGGTCCACCGTTGCGCTTAC
CAAGTATGCAGCGCGATTGACCGTCTCGTT	GGCCAATGGCCCAGGTCCGTCACGCAATTT
GCGGGAAGCACGTGGATTAGGGCATCGACC	TTGATCGAATCGGAGCGTAGCGGAATCTGC
AAGTCGTACGCCGATGCCGCAGCAATTCACT	CGGCGCGGATCCGCTTGTCGGGAACGGATAC
CGAAACATCGGCCACGGTCCCGTTGAACTT	GCCTCGATTACGACGGATGTAATTCGGCCG
ACGAATCCACCGTCCAGCGCGTCAAACAGA	GCCCGTATTCCCGCTTGCGAGTAGGGCAAT

with the increased frequency with which such shorter regions appear in the transcriptome.

Step 3: Remove potential readout probes with significant homology to one another. While these oligos were originally designed to have limited cross homology, changing their length may introduce new regions of homology. We recommend selecting a subset of possible readout probes, building a BLAST database for these sequences, and then using BLAST to identify regions of homology, as described in Step 2. We exclude probes that contain a region of homology to another potential probe longer than 10 nt.

Step 4: Order these probes. We order probes from IDT (www.idtdna.com) tagged on the 3′ end with a Cy5. We have them synthesized on the 100 nmol synthesis scale and HPLC purified.

3.3 Design of the Codebook

Before readout sequences and target sequences can be assembled to form the sequence of encoding probes, a codebook must be designed. The first step in codebook design is the choice of an encoding scheme that is appropriate for the experimental goals. We have utilized two different encoding schemes for MERFISH. The first scheme is a constant Hamming Weight code generated by keeping all binary barcodes from an existing HD4 code known as the Extended Hamming Code that contain only four "1" bits. We call this code the MHD4 code. The 16-bit version of this encoding scheme contains 140 binary barcodes and is capable of identifying all two-bit errors and correcting any individual single-bit error. Alternatively, if the error-correcting abilities of this code are not utilized, it can also detect three-bit errors. The advantage of this encoding scheme is the very high calling rate, ~80%, and low misidentification rates that it provides, as discussed in Section 2.4. In the second encoding scheme, we utilized a constant Hamming Weight code generated from all possible combinations of four "1" bits. This code has a minimum HD of 2 between all barcodes, so we refer to it as an MHD2 code. The benefit of this encoding scheme is that the reduced HD allows a much higher level of multiplexing for a given number of a bits—a 14-bit code contains 1001 barcodes. However, the reduced HD produces a lower calling rate and a slight reduction in accuracy, again, see Section 2.4.

Here, we describe algorithms to generate both encoding schemes; however, we would encourage users to consider alternative encoding schemes, eg, versions of those provided later with fewer or more bits or schemes with different Hamming Weight or HD, in order to find a scheme that best suits their experimental needs. Again, examples of algorithms to generate these codes as well as the barcodes for the published 16-bit MHD4 and 14-bit MHD2 codes are provided at http://zhuang.harvard.edu/merfish/.

3.3.1 Generation of the MHD4 Encoding Scheme

Here, we present a method for generating sets of constant-weight-4 HD4 (MHD4) barcodes for any number of bits. As an interesting aside, we note that the task of creating the optimal constant weight coding scheme—the encoding scheme that has the maximum possible number of barcodes given a specific HD—remains an unsolved problem. Thus, the algorithm that we present here likely does not produce a constant weight HD4 code with the maximum possible number of barcodes given a specific number of bits. However, comparison of estimates of the upper bound on the number of barcodes possible in such codes (Brouwer & Etzion, 2011) and the number of barcodes generated by this algorithm suggests that it performs fairly well, typically producing ~75% of the possible number of barcodes. The 16-bit MHD4 code that we have produced is the rare exception: it contains 140 barcodes which is the current known upper limit for an MHD4 code of this length (Brouwer & Etzion, 2011).

Step 1: Generate the Extended Hamming Code that contains the desired number of bits. Details on how to generate this code are discussed many places online, and software to perform this calculation is included at http://zhuang.harvard.edu/merfish/.

Step 2: Select only barcodes with the desired Hamming Weight.

3.3.2 Generation of the MHD2 Encoding Scheme

Constant weight HD2 codes are straightforward to generate, and the algorithm that we provide here will produce the optimal code, ie, the maximum number of barcodes with four "1"s and a minimum HD of 2.

Step 1: Generate a barcode of the desired length with only four "1"s.
Step 2: Generate all possible permutations of the bits in this barcode.

3.3.3 Barcode Assignment

Once the desired barcodes have been designed, the codebook is designed by randomly associating each barcode with a specific RNA of interest. We recommend leaving 5–10% of the possible barcodes unassigned. These "blank" or "control" barcodes will serve as internal controls and will provide estimates of the frequency with which barcodes can be generated by background or spurious signals. The specific use of these "blank" measurements in discussed in Section 7.4.2.

3.4 Assembly and Screening of Encoding Probes

Once the target regions, the readout regions, and the barcodes associated with the desired encoding scheme are designed, the sequence of the encoding probes can be assembled. Each encoding probe will contain multiple readout sequences. However, given length restrictions in synthesis of these sequences, it is typically not the case that all of the readout sequences for each RNA will fit into each encoding probe. We use two readout sequences per encoding probe.

Step 1: Select the readout sequences for each encoding probe. For each possible target region, use the barcode assigned to that RNA to determine which readout sequences to use. We select the readout sequences that we use for each encoding probe randomly, without repeating a readout sequence more than once in a single encoding probe. We find that random assignment rarely produces issues with an imbalance in the usage of some readout sequences and negates concerns over potential biases that may arise from the position of the readout sequence in the encoding probe.

Step 2: Assemble the encoding probe. We place one readout sequence on either side of the targeting sequence though there is no reason why alternative arrangements could not be used, see Fig. 2. Care must be taken to insure that the proper orientation of these sequences is used: the readout sequences are the reverse complement of the readout probes designed in Section 3.2. Thus, the encoding probe sequence must contain the reverse complement of these probe sequences. Similarly, the target region portion of the encoding probe must contain the reverse complement of the corresponding sequence of the target RNA.

3.5 Design of Priming Regions

The protocol that we use to make encoding probes that contain the sequences designed earlier (Section 4) require the addition of two priming regions to each probe. The optimal regions should have similar T_M, no contiguous stretches of the same nucleotide longer than three, and relatively narrow GC content. They should also have limited homology to each other and to nonpriming regions of the encoding probes.

Step 1: Truncate an orthogonal set of 25-nt long oligonucleotides to 20-nt. To generate possible primers that have limited possibility of binding to one another, we again start with the set of orthogonal 25-mers designed by the Elledge lab as described in Step 1 of Section 3.2. We reduce the length of these priming sequences to 20-nt to decrease the overall length of the final encoding probes.

Step 2: Screen oligonucleotides for ideal primer properties. We remove any potential primer with a T_M outside of the range of 70–80°C as well as any oligonucleotide with any consecutive repeat of three or more identical nucleotides, which can create problems in the synthesis of the oligo. We also remove any oligonucleotide that does not contain a G or C in the final two nucleotides at the 3' end. The presence of this so-called GC clamp helps improve the efficiency and specificity of PCR primers.

Step 3: Screen the final set of possible primers for homology against the encoding probes. Finally, we use BLAST to identify stretches of homology within these primers and the encoding probes designed in Section 3.4. Any potential primer with a homology region of 11 nt or more is excluded.

Step 4: Add these primers to each encoding probe to form the template sequence for each oligonucleotide. The sequence of the primer that is added to the 3' end of the encoding probe should be the reverse complement of the sequence of the primer designed in Step 3. Table 2 contains an example of a template molecule for a single encoding probe.

Step 5: Screen assembled template sequences for homology to abundant RNAs. While individual components of these template molecules have been screened for homology against the transcriptome, it is possible that concatenation of the different component sequences will produce regions of homology that will allow the probe to bind to other RNAs. We recommend using BLAST to screen the set of probes designed in Step 4 for homology to abundant RNA species like rRNA and tRNA. For the human transcriptome, these sequences can be found within the ncRNA fasta file at the Ensemble website as described in Step 2 of Section 3.1. Similarly, highly abundant mRNAs can

Table 2 An Example Template Molecule for an Encoding Probe

Encoding probe template	**CGCGGGCTATATGCGAACCG** *TTAGTCGTCGCAACGCCAGGCAACTCATGC* ***TAAAGAAATTAGATAGGCTGGAAATGCTTA*** *AAATTGCGTGACGGACCTGGGCCATTGGCC* **GCGTTGTATGCCCTCCACGC**

The target region is marked with bold and italics, the readout sequences are marked with italics only, and the priming regions are marked with bold only. This encoding probe is to the VCAN RNA.

also be included in this screen if such abundance information exists. To identify regions of homology, generate a BLAST database for these sequences, and use BLAST to screen the potential encoding probes. We remove any encoding probe with homology regions longer than 14 nt.

Step 6 (Optional): Combine multiple oligo sets to fill a single pool of oligonucleotides. We typically purchase these oligonucleotides as complex oligonucleotide pools generated by array-based synthesis often from CustomArray (http://www.customarrayinc.com/). This company sells two different sizes of complex oligonucleotide pools—a pool with 12,472 unique sequences or a pool with 92,918 unique sequences. There are situations in which one set of encoding probes for one experiment does not fill an entire pool. In this case it is possible to combine multiple sets of encoding probes intended for different experiments in the same oligopool simply by assigning each a unique set of primers. However, when primers are designed for such combined pools, we recommend performing Step 3 with a BLAST database comprising all encoding probes that will be present in the single oligonucleotide pool not just those in each set of encoding probes.

4. PROBE CONSTRUCTION

Staining hundreds to thousands of RNA species each of which requires tens of unique encoding probes requires a very large number of unique oligonucleotide sequences. The sheer number of unique sequences makes traditional solid-phase oligonucleotide synthesis prohibitively expensive in most cases. For example, assuming a modest cost of $0.10/base, an encoding probe length of ~100 nt, and the need for ~50,000 unique oligos for a single MERFISH measurement, the cost of the needed oligos would be $500,000! To circumvent this astronomical cost, we have developed a high-throughput approach to generating these probes which utilizes array-derived synthesis of complex oligonucleotide pools. These array-derived complex oligonucleotide pools that contain ~100,000 custom-designed sequences can be purchased for only a few thousand dollars. The challenge to using these pools is that each individual sequence is provided in quantities far too small to be used directly for labeling. Thus, we developed an enzymatic amplification protocol to generate the encoding probes in high quantity sufficient for RNA FISH experiments using the array-derived complex oligonucleotide pools as templates.

The basic protocol involves four steps, which are described in detail in the following sections. First, the oligopool is amplified via PCR to create template molecules for in vitro transcription. Second, an in vitro transcription produces RNA from these template molecules as well as another ~100-fold amplification. Third, a reverse transcription reaction generates single-stranded DNA from this RNA. Fourth, the RNA template is removed via alkaline hydrolysis. We utilize an RNA intermediate for two reasons: (i) in vitro transcription can produce very large quantities of nucleic acid in very small volumes, reducing the amount of material that must be used, and (ii) the alkaline susceptibility of RNA allows it to be easily removed from the final DNA probes. Because this reaction involves the use of an RNA intermediate, we recommend using the higher standards of laboratory cleanliness often required for handling RNA. Specifically, all surfaces and pipettes should be cleaned daily using RNase removal solutions and separate stocks of all buffers should be kept solely for RNA work.

4.1 Required Reagents

The following protocols will require the following reagents
1. 20× EvaGreen (Biotium; 31000)
2. 2× Phusion hot start polymerase master mix (New England Biolabs; M0536S)
3. Tris–EDTA (TE) pH 8 buffer (Ambion; AM9849)
4. DNA binding buffer (Zymo Research; D4004-1-L)
5. DNA wash buffer (Zymo Research; C1016-50)
6. Oligo binding buffer (Zymo Research; D4060-1-40)
7. 100-µg capacity silicon columns (Spin-V; Zymo Research; D4003-2-48)
8. RNA binding buffer (Optional; Zymo Research; R1013-2-100)
9. RNA prep buffer (Optional; Zymo Research; R1060-2-100)
10. RNA wash buffer (Optional; Zymo Research; R1003-3-24)
11. Quick HiScribe T7 polymerase kit (New England Biolabs; E2050S)
12. RNasin plus (Promega; N2611)
13. Maxima H− reverse transcriptase (ThermoScientific; EP0751)
14. 10 mM mix of dNTPs (New England Biolabs; N0447S)
15. 0.5 M EDTA (Ambion; AM9261)
16. 1 N NaOH (VWR; JT5635-2)
17. Nuclease-free water (Ambion; AM9932)
18. 100% Ethanol (KOPTEC; VWR; 89125-186)
19. D/RNAaseFree (VWR; 47751-044)
20. 1.5-mL LoBind tubes (Eppendorf; 022431021)

21. PCR tubes

The following protocols will require the following equipments
1. Tabletop centrifuge
2. qPCR machine or thermocycler
3. 37°C incubator or water bath
4. 50°C water bath
5. 95°C water bath
6. Vacuum manifold (optional)
7. Gel electrophoresis equipment for polyacrylamide gels (optional)
8. Vacuum concentrator (optional)

4.2 Amplification of In Vitro Template

The first step in this protocol is to use PCR to amplify template molecules for the in vitro transcription. We recommend running this reaction as a limited-cycle PCR, ie, monitor the status of the reaction in real time with a qPCR machine and remove the samples immediately before the final amplification plateau. We recommend limiting the number of PCR cycles because we have found that over amplification of complex libraries can produce molecules that miss-prime on other molecules, forming long concatemers that both reduce the yield of proper encoding probes and which could produce spurious signals.

Step 1: Design the primers. This PCR will not only amplify the library, it will also add the T7 promoter to these molecules to allow in vitro transcription of these templates. The sequence of the forward primer is the same as that designed in Section 3.5. However, the sequence of the reverse primer, also designed in Section 3.5, must include a T7 promoter, TAATACGACTCACTATAGGG, at the 5′ end. Example primers can be seen in Table 3. The forward primer will also be used as the primer for reverse transcription in Section 4.4; thus, it is recommended to order this primer at a relatively large synthesis scale, such as 100 nmol or 250 nmol.

Table 3 Example Primers

Forward primer	CGCGGGCTATATGCGAACCG
Reverse primer with T7 promoter	*TAATACGACTCACTATAGGG* CGTGGAGGGCATACAACGC

These primers are compatible with the encoding probe template listed in Table 2. Note that the 5′ G in the reverse primer has been removed so as not to create a G quadruplet with the terminal GGG of the T7 promoter.

The T7 promoter ends in a G-triplet, and the presence of a G quadruplet, ie, four Gs in a row, often significantly lowers synthesis yields; thus, we recommend removing any 5′-terminal G nucleotides in the sequence of the reverse primer. The presence of the terminal G nucleotides in the T7 promoter region will serve as replacements for these nucleotides in the priming region. Resuspend forward primer to 200 μM and the reverse primer to 100 μM both in TE.

Step 2: Prepare the PCR. In a 1.7-mL Eppendorf tube, mix the following: 40 μL 20× Eva Green, 2 μL 200 μM forward primer, 4 μL 100 μM reverse primer, 1 μL of 80 ng/μL complex oligopool, 353 μL nuclease-free water, and 400 μL 2× Phusion hot start polymerase master mix. Aliquot 50 μL volumes into 16 PCR tubes.

Step 3: Amplify the template. Run the following protocol on a qPCR machine: (i) 98°C for 3 min, (ii) 98°C for 10 s, (iii) 65°C for 10 s, (iv) 72°C for 15 s, and (v) measure the fluorescence of each sample. Repeat cycle steps (ii) through (v) until the rate at which the sample amplifies decreases, which is a sign that it is approaching the final amplification plateau. Due to the complexity of these oligonucleotide pools, it is very unlikely that, once denatured, each molecule will find and rehybridize to its complement as opposed to partially hybridize with the common priming regions of a different molecule; thus, it is recommended that samples be removed after the elongation step—while the instrument is at 72°C—and before it reaches the 98°C-melting step of the next cycle. If a qPCR machine is not available, we recommend determining the appropriate number of cycles to run by quantifying the yield of small-scale PCR reactions run for different numbers of cycles.

Step 4: Purify the template. We utilize column purification to remove enzyme, nucleotides, and primers. In a 15-mL Falcon tube, mix the following: 800 μL of the PCR reaction generated in Step 3 and 4 mL of DNA binding buffer. Pull this mixture across a 100-μg capacity column using either a vacuum manifold or a centrifuge. Wash the column twice with 300 μL DNA wash buffer, spinning the column in a tabletop centrifuge at maximum speed for 30 s each time. Elute the template by adding 170 μL nuclease-free water to the column, transferring the column to a fresh 1.7-mL Eppendorf tube, and spinning at maximum speed for 30 s. Set aside 10 μL of this reaction for quality control.

Step 5 (Optional): Quality control for template reaction. Two important quality control steps can be performed at this point. First, it is useful to measure the concentration of the template with a spectrophotometer, such as the

Nanodrop. The concentration should be between 10 and 50 ng/µL. The second quality control step is gel electrophoresis and will be described in Step 5 of Section 4.4.

4.3 In Vitro Transcription

The second step of this protocol is a high-yield in vitro transcription reaction that further amplifies the template molecules created in Section 4.2 as well as converts them into RNA.

Step 1: In vitro transcription. In a fresh 1.7-mL Eppendorf tube, mix the following: 160 µL of the in vitro template created in Section 4.2, 176 µL of nuclease-free water, 250 µL of the NTP buffer mix provided with the Quick HiScribe T7 polymerase kit, 25 µL of RNasin Plus, and 25 µL T7 polymerase (from the same HiScribe kit). Incubate the reaction in a 37°C incubator for 12–16 h. Often the reaction is complete after 6–8 h, but it is typically convenient to leave this reaction overnight. Remove 20 µL for quality control.

Step 2 (Optional): Quality control for the in vitro transcription. To confirm that the in vitro transcription was successful, purify the reaction and measure its concentration with a spectrophotometer. To purify, mix 20 µL of the in vitro reaction with 30 µL nuclease-free water, 100 µL RNA binding buffer, and 150 µL 100% ethanol. Pass across a 100-µg capacity spin column in a tabletop centrifuge. Wash this column once with 400 µL RNA prep buffer, and then twice with 200 µL RNA wash buffer, each time with a 30-s spin at the maximum speed of the tabletop centrifuge. Elute the RNA with 100 µL nuclease-free water. If successful, the concentration of the in vitro transcription should be between 0.5 and 2 µg/µL. Purified RNA can also be run on a gel as described in Step 5 of Section 4.4.

4.4 Reverse Transcription and Purification

In this step of the protocol, the large quantities of RNA produced by the high-yield in vitro transcription are converted to single-stranded DNA using a reverse transcription reaction. This RNA template is then removed via alkaline hydrolysis, and the final encoding probes are purified and concentrated.

Step 1: Reverse transcription. To the unpurified in vitro transcription created in Section 4.3, add the following and mix well: 200 µL 10 mM dNTP mix, 120 µL 200 µM forward primer, 240 µL 5× Maxima buffer, 24 µL RNasin Plus, and 24 µL Maxima H− reverse transcriptase. Incubate in a 50°C water bath for 1 h. It is important to use a water bath, not an air incubator, to insure that the temperature of the sample rises to 50°C quickly.

Step 2: Alkaline hydrolysis. Split the above reaction into two 1.7-mL Eppendorf tubes and add the following to each: 300 μL 0.5 M EDTA and 300 μL 1 N NaOH. Incubate in a 95°C water bath for 15 min.

Step 3: Purification of ssDNA probe. Combine the two aliquots above into a single 50-mL Falcon tube and add the following: 4.8 mL Oligo binding buffer and 19.2 mL 100% ethanol. Mix well and split equally between eight 100-μg capacity spin columns. Pull the sample across the columns with a vacuum manifold or via centrifugation. Wash the columns once with 750 μL DNA wash buffer. Elute the columns using 100 μL of nuclease-free water. Combine eluates and set aside 10 μL for quality control.

Step 4: Concentration of probe. Use a vacuum concentrator to dry the samples. This process could take several hours. Resuspend the dried pellet in 24 μL of nuclease-free water, or if desired, the hybridization buffer described in Section 5.1. Store probe at −20°C and avoid unnecessary freeze-thaw cycles. If a vacuum concentrator is not available, it is also possible to concentrate the probe using ethanol precipitation.

Step 5 (Optional): Quality control of in vitro template, RNA, and probe. We recommend running the in vitro template, the RNA, and the final probe on a 15% TBE-urea polyacrylamide gel to identify both RNase contamination and low conversion of the reverse transcription primer to full-length probe. Large smearing both in the RNA band and the probe band can indicate RNase contamination. Failure to efficiently convert the reverse transcription primer into probe is revealed by a bright band running at the 20-nt length corresponding to the primer. Increasing the amount of RNA template in the reverse transcription often improves the fraction of primer converted into probe. We routinely obtain ~75% or greater incorporation of the reverse transcription primer into probe with the above protocol.

5. SAMPLE PREPARATION AND STAINING

The preparation and staining of samples for MERFISH follows closely the typical protocols used for smFISH (Raj et al., 2008). However, there are a few places in which we have modified these protocols to optimize MERFISH staining. Again, RNase contamination can destroy samples, so care should be taken to work in an RNase-free environment.

5.1 Required Materials

The following protocol will require these reagents and buffers
1. Fixation buffer (4% PFA in 1 × PBS)
   a. 1 mL 32% paraformaldehyde (PFA; Electron Microscopy Sciences; 15714)
   b. 0.8 mL 10× phosphate buffered saline (PBS; Ambion AM9625)
   c. 6.2 mL nuclease-free water
   d. Store at −20°C in single-use aliquots
2. Permeabilization buffer (0.5% v/v Triton X-100 in 1 × PBS)
   a. 5 mL 10×PBS
   b. 45 mL nuclease-free water
   c. 250 µL Triton X-100 (Sigma; T8787)
   d. Store at room temperature
3. 2 × SSC buffer
   a. 5 mL 20 × saline sodium citrate (SSC; Ambion; AM9763)
   b. 45 mL nuclease-free water
   c. Store at room temperature
4. 1 × PBS buffer
   a. 5 mL 10 × PBS
   b. 45 mL nuclease-free water
   c. Store at room temperature
5. 20% w/v Dextran sulfate
   a. In an RNase-free bottle (such as an empty nuclease-free water bottle), mix the following
   b. 100 g dextran sulfate (Fisher Scientific; bp1585100)
   c. 300 mL nuclease-free water
   d. Stir with a stir bar cleaned with D/RNaseFree while heating gently on a hot plate (37°C) until dissolved
   e. Add nuclease-free water to produce a 500 mL total volume
   f. Store at room temperature
6. Encoding probe hybridization buffer
   a. 300 µL 100% deionized formamide (Ambion; AM9342)
   b. 500 µL 20% w/v dextran sulfate (above)
   c. 100 µL 20 × SSC
   d. 90 µL nuclease-free water
   e. 1 mg yeast tRNA (Life Technologies; 15401-011)
   f. 10 µL 200 mM vandyl ribonucleoside complex (VRC; New England Biolabs; S1402S)
   g. Store at −20°C

7. Encoding probe wash buffer
   a. 15 mL 100% deionized formamide
   b. 5 mL 20 × SSC
   c. 29.5 mL nuclease-free water
   d. 0.5 mL 200 mM VRC
   e. Store at 4 °C in the dark and use within a few days
8. 40-mm diameter, No. 1.5 coverslips (Bioptechs; 40-1313-0319)
9. 60-mm diameter, Cell culture petri dish (Corning; 353802)
10. $1'' \times 3''$ microscope slides
11. Parafilm (VWR; 52858-076)
12. Orange FluoSphere carboxylate-modified beads, 0.1 μm diameter (ThermoScientific; F-8800)

The following protocol requires the following equipments
1. Equipment for cell culture
2. 37°C incubator
3. 47°C incubator
4. Laboratory rocker

5.2 Fixation and Permeabilization of Cells

The first step in most FISH protocols is to fix the cells and then to permeabilize the membrane using either an overnight incubation in 70% ethanol or a brief exposure to surfactant. We prefer the latter approach since this protocol requires less time; however, we have had reasonable performance with both approaches.

Step 1: Culture cells on a suitable coverglass. The culture substrate for cells should be the same coverglass that will be used for imaging. We utilize 40-mm diameter No. 1.5 coverslips from Bioptechs since these coverslips fit within the flow chamber that we use for imaging (discussed in Section 6). We typically culture cells in sterile, 60-mm-diameter petri dishes with one coverslip placed in the bottom of each dish.

Step 2: Fix cells. Allow the fixation buffer described in Section 5.1 to warm to room temperature. Aspirate culture medium from cells then gently decant 2–3 mL of fixation buffer into the petri dish that holds the coverslip. Cover and rock gently for 15 min at room temperature. Aspirate the fixation buffer and gently decant 2–3 mL of 1 × PBS into the petri dish to wash the cells. Immediately aspirate this buffer. Repeat this 1 × PBS wash for a total of three times.

Step 3: Permeabilize the cells. Aspirate any residual 1 × PBS from the petri dish and decant 2–3 mL permeabilization buffer into the petri dish.

Incubate the cells at room temperature with gentle rocking for 2 min. Aspirate out the permeabilization buffer and wash the cells three times with 1 × PBS as described in Step 2.

5.3 Hybridization of Encoding Probes

Step 1: Exchange buffer on cells. Aspirate any residual 1×PBS and gently decant 2–3 mL of room temperature encoding probe wash buffer into each petri dish. Rock the sample at room temperature for 5 min. This step insures that any residual buffer left on the coverslip will not significantly dilute the formamide in the hybridization buffer in the next step.

Step 2: Add hybridization buffer with probes. Place a layer of fresh parafilm on a 1″ × 3″ microscope slide. Dilute the encoding probes made in Section 4 into encoding probe hybridization buffer to the desired concentration, typically somewhere between 10 and 200 μM depending on the number of unique encoding probes in the probe set. Add 30 µL of this probe solution to the parafilm surface. Aspirate any residual encoding probe wash buffer from the coverslip and gently place the coverslip, cell-side-down, onto the droplet of encoding probes. Press gently to create a thin layer of encoding probe. The probe concentration needs to be optimized for each probe set: higher probe concentrations produce brighter spots but a higher level of nonspecifically bound probe background. We recommend titrating the concentration of probe across a ~40-fold concentration range (5–200 μM) to identify the concentration that produces the clearest signal relative to the background. We find that the optimal probe concentration only needs to be established once per probe set.

Step 3: Incubation. Create a humidity chamber by pouring nuclease-free water into the base of an old pipette tip box. Place the sample within this box and seal with parafilm. The sample will be sitting on the upper plastic level and should not come into contact with the water reservoir below. Place this box into a 37 °C incubation chamber and incubate for at least 12 h. In some cases, we find that longer incubations (~36 h) increase the quality of staining. Again, we recommend varying the incubation time for each probe set to identify the optimal incubation time.

Step 4: Wash away residual encoding probes. Preheat 6 mL of encoding probe wash buffer per sample to 47 °C in a water bath. Slowly peel the coverslip off of the microscope slide taking care not to crack the coverslip. If it appears to be stuck, immerse the assembly in a layer of encoding probe wash buffer for 5 min to loosen. Place the coverslip, cell-side-up, in a fresh

60-mm-diameter petri dish and add 3 mL of preheated encoding probe wash buffer. Place the petri dish in a 47°C incubator for 30 min. Aspirate the encoding probe wash buffer, add 3 mL of fresh encoding probe wash buffer, and repeat the 30-min, 47°C incubation. Aspirate the wash buffer and wash the sample in $2\times$ SSC by adding 3 mL of $2\times$ SSC and then immediately aspirating this buffer. Repeat this SSC wash for a total of three times.

Step 5: Addition of fiducial beads. Prepare a $100\times$ solution of fluorescent beads by adding 10 μL of 0.1-nm-diameter Orange Fluospheres beads to 10 mL $2\times$ SSC. Vortex to mix. This stock solution can be stored at 4°C in the dark for several months. Further dilute this stock solution 1–100 in $2\times$ SSC to create 3 mL of bead solution for each sample. Aspirate the $2\times$ SSC from the samples and add the bead solution to each petri dish. Gently rock at room temperature for 15 min. We choose orange-colored beads (561 nm excitation) because we prefer to use the red imaging channel (641 nm excitation) for our smFISH images; however, the color of these beads could be changed to suit different imaging demands. We aim to have ~20 beads per field of view; thus, if different beads are used, the concentration should be adjusted to produce a similar final density of beads on the sample.

Step 6: Postfixation. To fix the fiducial beads in place and to prevent subtle changes in the shape of the cell induced by the repeated washing that will occur in Section 6, we perform an additional fixation of the samples. Aspirate the bead solution and add 3 mL of fixation buffer prewarmed to room temperature to each petri dish. Gently rock the sample at room temperature for 10 min. Aspirate the fixation buffer and wash the sample three times with $2\times$ SSC. If the sample was prepared in an RNase-free environment, it can now be stored in $2\times$ SSC at 4°C for several days. We do not recommend prolonged storage.

6. MERFISH IMAGING

The basic imaging process for MERFISH involves several fluid-handling steps, eg, introduction of hybridization buffers and wash and imaging buffers, in conjunction with the collection of smFISH images and the photobleaching of the sample. While it would be possible to perform these fluid exchange and imaging steps manually, we strongly recommend the use of an integrated and automated fluid exchange and imaging approach. Here, we provide protocols for how to construct a fluid-handling system that should be compatible with a variety of microscopes. We also discuss the basic protocols for hybridizing readout probes to samples on the microscope, imaging the sample, and bleaching residual signal.

6.1 Required Materials

The following protocols will require these reagents
1. Readout probe hybridization buffer
   a. 25 mL 20% w/v dextran sulfate (Section 5.1)
   b. 5 mL 20× SSC
   c. 5 mL 100% deionized formamide
   d. 14.5 mL nuclease-free water
   e. 0.5 mL 200 mM VRC
   f. Make fresh each day
2. Imaging buffer master mix
   a. In an RNAse-free bottle (such as an empty nuclease-free water bottle) add the following
   b. 50 mL 20× SSC
   c. 25 mL 1 M Tris–HCl pH 8 (Ambion; AM9856)
   d. 325 mL nuclease-free water
   e. 50 g glucose (Sigma; G8270)
   f. A D/RNaseFree-treated stir bar
   g. Stir until glucose is dissolved
   h. Add nuclease-free water to 500 mL
   i. Store at room temperature for up to a few weeks
3. Trolox solution
   a. 20 mg (±)-6-hydroxy-2,5,7,8-tetramethylchromane-2-carboxylic acid (Trolox; Sigma; 238813)
   b. 200 µL methanol (Sigma; 322415)
   c. Make fresh each day
4. Imaging buffer
   a. 40 mL imaging buffer master mix
   b. 200 µL Trolox solution
   c. 200 µL catalase (Sigma; C30)
   d. 42 mg glucose oxidase (Sigma; G2133)
   e. Make fresh and use immediately
5. Photobleaching buffer
   a. 5 mL 20× SSC
   b. 44.5 mL nuclease-free water
   c. 0.5 mL 200 mM VRC

    d. Make fresh each day
6. Readout probe wash buffer
    a. 5 mL 2×SSC
    b. 10 mL 100% deionized formamide
    c. 34.5 mL nuclease-free water
    d. 0.5 mL 200 mM VRC
    e. Make fresh each day
7. Light mineral oil (Sigma; M5904)

Components for an automated fluidics system
1. Flow/imaging chamber (Bioptechs; FCS2)
2. 5″-long 20-gauge needles (Hamilton; 7750-11)
3. Luer to ¼-28 female fitting adaptor (IDEX; P-655)
4. 1/16″ barb to ¼-28 female fitting adapter (IDEX; P-646)
5. ¼-28 fitting (IDEX; XP-202)
6. Clear ETFE tubing, 0.02″ inner diameter (McMasterCarr; 5583K52)
7. Clear PVC tubing, 1/16″ inner diameter (McMasterCarr; 5233K51)
8. 0.38-mm inner diameter peristaltic tubing (Pulse Instrumentation; 116-0549-04)
9. Peristaltic pump (Gilson; Minipuls 3)
10. Computer-controlled valve (Hamilton; MVP; 36798)
11. 8-way valve (Hamilton; HVXM 8-5; 36766)
12. 5-min epoxy (VWR; 300050-778)
13. 25-gauge needles (VWR; BD305125)
14. 18-gauge needles (VWR; BD305196)
15. USB to RS-232 converter (Keyspan; USA-19HS)

The following protocol requires the following equipments
1. FCS2 flow system (Bioptechs; FCS2)
2. Objective heater (Bioptechs; 150803)
3. Automated flow system (see Section 6.2)
4. Microscope capable of smFISH imaging (see Section 6.3)

6.2 Assembly and Operation of Flow System

Each round of MERFISH hybridization and imaging requires the controlled exchange of four different buffers at precise intervals. For this reason, integration of a fully automated fluid-handling system with the microscope control software can significantly facilitate the collection of MERFISH data. Here, we describe the construction and operation of the automated fluid system that we have used. This system is constructed around a peristaltic pump

(Gilson; Minipuls 3) which controls the flow rate of different buffers through the flow system and a series of valves (Hamilton; MVP) that control which buffer is being pulled across the sample at any given time. The components required for the flow system are listed in Section 6.1, and a schematic diagram of the flow system required for 16 rounds of hybridization is included in Fig. 4. Once assembled, the components of this flow system do not need to be replaced between measurements, unless otherwise specified.

Step 1: Assemble buffer reservoirs. Using an 18-gauge needle, create a hole in the cap of one 15-mL Falcon tube for each hybridization round. In the example depicted in Fig. 4, 16 such tubes would be needed. Insert a 5″-long 20-gauge needle into each hole. Slide the needle so that it is ~1 mm from the bottom of the 15-mL Falcon tube and use 5-min epoxy to secure it. Using the same protocol, insert and secure a needle into the cap of three 50-mL Falcon tubes (one each for the imaging buffer, wash buffer, and bleaching buffer). Connect a luer to ¼-28 fitting adapter to the end of each needle.

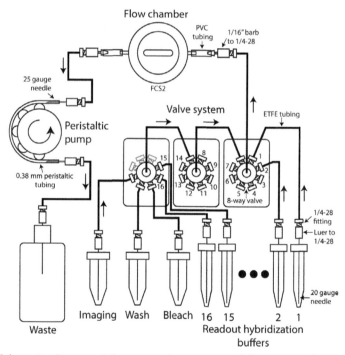

Fig. 4 Schematic diagram of the setup of an automated flow system for a 16-round MERFISH measurement. *Arrows* mark the local flow direction. For clarity, only 4 of the 16 tubes and flow lines required for the different hybridization buffers are depicted. The sample is contained within the FCS2 flow chamber.

Step 2: Create a waste bottle. Using the same approach as described in Step 1, insert a needle into the top of a 1-L plastic bottle which will serve as waste collection. The collected waste from a MERFISH measurement will contain formamide and, thus, must be discarded as toxic waste. Choose a plastic bottle appropriate for this requirement.

Step 3: Assemble tubing for flow lines. Cut the ETFE tubing for each desired flow line using a fresh razor blade. Take care to make the end as flat as possible. The length of each tubing section should be as short as possible to reduce the dead volume within each flow line but long enough to allow easy manipulation of the component to which it is attached, such as a Falcon tube. Typical lengths are ~8″. Assemble a ¼-28 fitting on each end of each tubing section.

Step 4: Assemble the flow lines. Insert the 8-way valve into each MVP valve system following the manufacturer's instructions. The valve system is going to be run as a daisy chain, so connect fluidics lines such that the first valve addresses the first seven hybridization buffers and the contents of the second valve system, the second valve system addresses the next seven hybridization buffers and the contents of the third valve, and the third valve addresses the remaining two hybridization buffers (assuming 16 rounds of hybridization) and the imaging, wash, and bleaching buffers, see Fig. 4 for a flow diagram. This layout insures that when common buffers are flown across the sample, ie, imaging, wash, or bleach buffers, residual hybridization buffer from other rounds are not accidentally introduce into the sample.

Step 5: Connect flow lines to the FCS2 chamber. Flow connections are made to the FCS2 chamber via two metal tubes cast into this chamber. Cut two short sections of PVC tubing and gently slide these onto the metal tubes. Insert a 1/16″ barb to ¼-28 female adapter into each PVC tube to allow flow lines to be connected to the flow chamber.

Step 6: Connect flow lines to the peristaltic tubing. Insert a 25-gauge needle into each end of the peristaltic tubing. Connect these needles to the ETFE tubing flow lines using luer to ¼-28 adapters

Step 7: Setup computer control. Both the peristaltic pump and the valve units can be controlled via serial communication. Most computers no longer contain serial ports, so we recommend using USB to serial converters such as a USB to RS-232 adapter as well as a RS-422 to RS-232 converter for the peristaltic pump. Custom software can be written to control these pumps via serial command; however, we have written a python-based graphical user-interface program to control a pump and several valves. This software—named Kilroy in honor of the World-War-II-era character—is open source and can be found here: https://github.com/zhuanglab/storm-control. It can communicate with other software using TCP/IP, and, thus, it should be possible to integrate this fluidics control software with a wide variety of scripting languages available on many commercial microscopes.

6.3 Microscope Requirements

The physical requirements for a microscope for MERFISH measurements are essentially identical to the requirements for smFISH measurements. Even with many probes per RNA, the signal from individual RNAs is often relatively dim; thus, high numerical aperture objectives and sensitive cameras such as electron-multiplying CCD or scientific CMOS are often needed. MERFISH requires relatively high laser powers to bleach the sample after each round of imaging; thus, the microscope should be able to illuminate the sample with at least 100 mW of light in the wavelength used to image the labeled RNAs, as measured at the back focal plane of the objective. Only a few mW of light is needed for illumination of the fluorescent fiducial beads. If less illumination light is available in the color channel used for smFISH, it should still be possible to perform MERFISH measurements. The user will simply need to devote more time to photobleaching each region of the sample. Finally, to achieve any reasonable level of automation and throughput with MERFISH, a motorized sample stage will be required as well as some form of automatic focus system.

For published MERFISH measurements, we used a 1.45 NA, 100× oil immersion Olympus objective, and illuminated our sample with ~200 mW of 641-nm light and ~20 mW of 561-nm light, as measured at the back focal plane of the objective, using solid state lasers (MPB communications; VFL-P500-642 and Coherent, 561-200CWCDRH). The 641-nm laser was used to excite our Cy5-labeled readout probes while the 561-nm laser was used to excite the fiducial beads. A custom dichroic (Chroma, zy405/488/561/647/752RP-UF1) and a custom notch filter (Chroma, ZET405/488/561/647-656/752m) were used to couple this light into the objective and filter the emission. These custom optics were used so that our home-built microscope had the capability of imaging at 750 nm. This color band was not used in our published MERFISH work (Chen et al., 2015), and thus stock dichroics and emission filters would also have worked. For example, the Chroma 89016bs dichroic in combination with the Chroma ZET561/10× and ZET642NF notch filters would also work. The fluorescent signal from the sample was imaged onto an EMCDD camera (Andor;

iXon-897) through a QuadView (Photometrics), which used several stock dichroics (Chroma, T560lpxr; T650lpxr, and 750dcxxr) and emission filters (Chroma; ET525/50 m, WT59550m-2f, ET700/75 m, HQ770lp) to image several different color channels onto different quadrants of the camera. The 640-nm channel, corresponding to our RNA signal, and the 561-nm channel, corresponding to our fiducial beads, were excited and imaged simultaneously with this setup. The 750-nm channel allows future work to include probes of this color but again was not used in our published MERFISH work (Chen et al., 2015). However, it would also be possible to forego the use of a QuadView and image each color channel one at a time, using the selective excitation of each laser to discriminate the color channels corresponding to Cy5 and the fiducial beads. That being said, it is crucial that no offset be introduced between the image of the smFISH signal and the image of the fiducial beads; thus, the same dichroic filter cube should be used for both color channels. We collected images corresponding to a 40×40 μm field of view with a pixel size of 167 nm in the sample plane. Our sample position was controlled with a motorized stage (Marzhauser) and our focus was maintained with a home-built autofocus system consisting of an objective nanopositioner (Mad City Labs, Nano-F) whose position was locked to the position of an IR laser spot (940 nm) reflected off of the sample–coverslip interface and imaged with a CMOS camera (Thorlabs, uc480).

6.4 MERFISH Imaging Protocol

The general MERFISH imaging protocol involves the hybridization of each readout probe, a brief wash step to remove some of the nonspecifically bound probe, imaging of the sample, and then photobleaching of the sample, before hybridization of the next readout probe.

Step 1: Prepare the microscope and the flow system. Preheat the objective using the Bioptechs objective heater to 37.0°C for at least 2 h before the start of the measurement. This increased temperature is required to favor the proper hybridization of readout probes to the readout sequences. Room temperature is also suitable if the concentration of formamide is increased to 30% and 40% v/v in the readout probe hybridization buffer and the readout probe wash buffer, respectively. Prepare a 5 mL aliquot of readout hybridization buffer as described in Section 6.1 for each round of hybridization and dilute a single readout probe (designed in Section 3.2) into each buffer to a final concentration of 10 nM. Prepare the imaging, wash, and bleaching buffers as described in Section 6.1. Load all buffers into the flow system. The imaging buffer is O_2 sensitive, so add ~5–10 mL of light mineral oil to the top of this buffer by decanting it into the 50-mL Falcon tube holding this buffer. This oil layer prevents the diffusion of O_2 into this buffer during the course of the measurement. Connect the input flow line to the FCS2 channel directly to the exit line, bypassing the FCS2 system, and prime the flow system by flowing enough of each buffer to completely fill its specific flow line. Flush the system with bleaching buffer to ensure that no hybridization buffers will flow onto the sample once it is added. Inspect each connection to see if bubbles are forming. The presence of bubbles indicates a loose or faulty connection. Tighten the connector or replace the tubing. If flow rates appear to be lower than expected, replace the peristaltic tubing. We recommend replacing the peristaltic tubing every few measurements.

Step 2: Load the sample. Assemble the 40-mm coverslip containing the sample as prepared in Section 5 into the FCS2 flow system. Flow bleaching buffer through the tubes to fill the chamber with liquid. Inspect the flow chamber to insure that no air bubbles remain in the chamber. If bubbles are present, gentle agitation of the flow chamber, via tapping on the coverslip, can often dislodge them.

Step 3: Hybridize the first readout probe. Flow ~2 mL of the first readout hybridization buffer across the sample at a flow rate of 0.5 mL/min. This volume should be sufficient to fill the dead volume of the flow system and chamber as well as to exchange the volume of the chamber many times over. Stop the flow and incubate the sample in the first hybridization buffer for 15 min. If the quality of individual stains is low, increasing this hybridization time can sometimes improve the quality of the stain.

Step 4: Wash the sample. Flow ~2 mL of the readout wash buffer across the sample, again at a flow rate of 0.5 mL/min. Incubate the sample for 5 min.

Step 5 (Optional): Inspect the quality of the sample and select regions of interest. After the first readout probe is hybridized to the sample, it is often useful to pause the automated imaging and fluid-handling program and visually inspect the quality of the sample. For new samples, it is often necessary to adjust the excitation light intensity so as to produce bright signal but not saturate the camera. Similarly, it can be convenient to examine the sample at this point to identify specific regions or cells to image. Before examining the sample, be sure to flow imaging buffer across the sample as described in Step 6 later. The brief exposure to light at this stage typically is insufficient to produce any significant photobleaching as long as the sample is immersed in imaging buffer.

Step 6: Image the sample. Flow ~2 mL of imaging buffer across the sample at 0.5 mL/min. When flow has stopped, collect an image in both the color channel corresponding to the labeled readout probes (eg, 641 nm) and the color channel corresponding to the fiducial beads (eg, 561 nm) for all desired fields of view. We utilize ~50 mW at the microscope back port to excite our Cy5-labeled readout probes, which corresponds to an average power density of ~1 kW/cm^2, and we use ~5 mW at the microscope back port to excite the fiducial beads, which corresponds to an average power density of 100 W/cm^2.

Step 7: Bleach the signal from each region. Flow ~2 mL of the bleaching buffer across the sample at 0.5 mL/min. When flow has stopped, return to each field of view, turn up the illumination to the maximum value for the color channel corresponding to the labeled readout probes. Illuminate each sample for a sufficient time to completely photobleach the signal. For our measurements, this time was typically 3 s. In our measurements, we utilize 200 mW of 641-nm laser power at the microscope back port, which corresponds to a power density of ~4 kW/cm^2. However, the bleaching rate will depend strongly on the illumination properties of each microscope, and the optimal bleaching time should be determined empirically for each microscope.

Step 8: Repeat Steps 3–7 for the remaining hybridization rounds.

Step 9: Cleanup the system. When the experiment is complete, replace all buffers with nuclease-free water and flow 5 mL through each flow line. Store the flow lines and buffer reservoirs filled with deionized water to prevent the crystallization of salts within the flow system. If the mineral oil used to protect the imaging buffer has been accidentally introduced into the flow system, wash the affected lines with 5–10 mL of isopropanol and then nuclease-free water.

7. MERFISH DATA ANALYSIS

The basic decoding of MERFISH data—the identification of specific RNA species for each spot in the sample—consists of three basic steps. First, fluorescent spots must be identified in all images. Second, slight changes in the stage position between images of the same region from different hybridization rounds must be corrected. Finally, sets of fluorescent spots that occupy the same location in the sample must be decoded into binary barcodes, which, in turn, are decoded into the specific RNA species. After these RNAs have been decoded, there are a series of simple calculations that can often be performed on the measured barcodes to determine important parameters associated with the performance of the measurements, such as estimates of the binary errors made at each bit or the relative confidence that can be ascribed to the counts for each RNA species. In this section, we provide protocols for performing these basic computational tasks. Example software and data can be found at http://zhuang.harvard.edu/merfish/.

7.1 Identification of Fluorescent Spots

There are a variety of software packages and approaches to identify the location and brightness of fluorescent spots in images. In principle, any of these approaches should work reasonably well for this initial stage of the analysis of MERFISH data. However, some of the more advanced algorithms for the identification of fluorescent spots have been developed in the context of localization-based super-resolution microscopy. In these measurements, there is a desire to be able to properly localize molecules that are close enough together that their fluorescent spots partially overlap. Depending on the local density of RNAs within the cell, the signal from different RNAs does occasionally overlap, necessitating the use of these more advanced spot finders. The algorithm we use—3D-daoSTORM—was developed in our laboratory (Babcock, Sigal, & Zhuang, 2012) and is an extension of the daoPHOT algorithm originally developed for the analysis of astronomy data (Holden, Uphoff, & Kapanidis, 2011). 3D-daoSTORM is open source and can be found with documentation here: https://github.com/ZhuangLab/storm-analysis. In this section, we describe the basic steps for the use of this software for finding fluorescent spots in MERFISH data.

Step 1: Data conversion. The first challenge associated with any analysis software will be to insure that the image data is in an appropriate format. Our data is typically stored in a custom file format. While conversion to this format is possible, we instead recommend converting data to the more commonly used tif format, which the 3D daoSTORM algorithm can process. Many microscopes can already save images as tif format, so conversion may not be necessary.

Step 2: Determine the appropriate intensity threshold for spot finding. The daoSTORM algorithm requires that a threshold be selected to determine whether a spot is bright enough to fit or not. It is useful to first run spot-finding software for a range of thresholds to determine which value performs the best at discriminating between bright spots and low-intensity spots, which are likely due to off-target binding of encoding or readout probes. The brightness of spots can differ between different hybridization rounds, so it is often worthwhile to explore different thresholds for different readout probes. In practice, it may be difficult to determine the optimal threshold for discriminating background spots and spots that correspond to RNAs by simple visual inspection. Thus, we recommend analyzing data with a range of thresholds and using an iterative approach to select the threshold that produces the best decoding of MERFISH data. This process is discussed in detail in Section 7.5.

Step 3: Batch analyze the data. Once an appropriate threshold has been selected, use this software to identify the location of potential molecules in both the smFISH images and the images of fiducial beads.

7.2 Correction of Image Offsets

Every time the microscopy stage returns to a given location there will be a slight offset, often no more than a single pixel or two. Nonetheless, this small offset can significantly degrade the ability to align fluorescent spots within images of the same sample stained with different readout probes. For this reason, in each round of hybridization, we collect an image of small fluorescent beads stuck to the surface of the sample. Because the location of these beads are fixed with respect to the sample, apparent differences between the position of these beads between different rounds of hybridization can be used to correct these small-stage offsets. We perform this correction by using the positions of these beads to create affine transformations that map the position of each spot in each round of hybridization back to the coordinate system defined by the image collected in the first round of hybridization.

Step 1: Find fiducial bead centroids. We utilize the same spot-fitting approach described in Section 7.1 to determine the location of the fiducial beads in the images of these beads corresponding to each round of hybridization.

Step 2: Create the affine transformations. For each field of view, load the locations of the fiducial beads in each round of hybridization. To use these beads as control points in the construction of affine transformation, each bead in one image must be associated with beads in another image. This can be accomplished by associating each bead with its nearest neighbor in the other image. However, this simple approach will fail if the offset between images is substantially larger than the average distance between beads within a single field of view since the nearest neighbor between images would likely correspond to a different bead. In this case, an initial crude offset can be found by using image cross correlation either with the original bead images or with a lower resolution image created from the 2D histogram of the bead positions. This crude offset can be used to correct the positions of these beads to sufficient accuracy to allow simple association of nearest neighbors, which then allows the construction of the final affine transformation. Using this approach, we can reregister different frames with a residual error of roughly 20 nm, much better than a single pixel.

Step 3: Correct the location of fluorescent spots. Once the affine transformations are calculated, apply these transformations to the locations of all found fluorescent spots in each smFISH image and save the revised locations.

7.3 Decoding Barcodes

Once all spots have been transformed to the same coordinate system, spots in different rounds of hybridization that occupy the same or similar physical locations within the sample should be associated to create the specific binary barcode associated with the putative RNA at that location in the cell. We have explored several different approaches to perform this association. Here, we present the algorithm that has performed best in our hands. One advantage of this algorithm is that it requires only one parameter: the maximum distance between spots in different rounds of hybridization. Fig. 3A–C illustrates the decoding of an example MERFISH data set.

Step 1: Create a list of all found spots. For a single field of view, create a list of the location of all spots in all rounds of hybridization. This algorithm will create an N-bit binary barcode for each of these spots.

Step 2: Construct a barcode for each found spot. For each spot in the list created in Step 1, compute the distance to the nearest spot in each round of hybridization. This will create a vector of N distances. To convert this vector into a binary barcode, compare the ith distance to the maximum allowed distance (a parameter of this algorithm). If this distance is less than or equal to this maximum distance, assign a "1" to the ith bit of the barcode for this spot. Otherwise, assign a "0" to that bit. We use a maximum distance of 160 nm, ie, one camera pixel. Practically, we found that a smaller maximum distance between spots tends to discard more real barcodes than false barcodes, despite the fact that we can resolve the centroid of each RNA spot to better than 160 nm. This observation may be due to the fact that the position of the centroid for the observed spot for a single RNA varies on average by ~100 nm between different rounds of hybridization—an effect that may be due to the finite cellular volume occupied by each RNA.

Step 3: Remove redundant barcodes. This algorithm creates a barcode for all spots in all rounds of hybridization, ignoring the fact that spots in different rounds will be combined to form individual barcodes. Thus, each barcode may be replicated in the results returned by Step 2. Redundant barcodes will share the same set of spots, and, thus, the average centroid of the spots that form a barcode will be unique for each barcode. Thus, to remove these redundant barcodes, we compute the average centroid of all spots that comprise each barcode and remove extra copies of barcodes that share the same centroid. We perform this step because it is computationally faster to create redundant barcodes and then remove them in this fashion than to identify such barcodes as they are created.

Step 4: Associate each binary barcode with the RNA it encodes. Using the codebook designed in Section 3.3, create a lookup table that associates each encoded RNA with all binary barcodes that encode it. If the encoding scheme is not capable of error correction, then there will only be one binary barcode for each RNA. If, on the other hand, the encoding scheme allows error correction, then each RNA will have multiple binary barcodes that correspond to that word. For the purposes of calculating several properties associated with the basic performance of MERFISH, it is useful to create two lookup tables: one that corresponds to the correct binary barcodes and one that corresponds to all binary barcodes that can be matched to an RNA after error correction. The use of two lookup tables will allow each decoded RNA to be marked as an exact match, ie, no error correction was applied, or an error corrected match.

7.4 Calculate MERFISH Performance

The basic performance of a MERFISH measurement can be quickly assessed from several quantities that can be calculated from the barcodes decoded in Section 7.3. Here, we describe the calculation of these quantities and their interpretation.

7.4.1 The Per-Bit Error Rate

If the encoding scheme has the capacity to correct errors, then it will also have the ability to determine the bit at which the corrected errors occurred. Using this ability, it is possible to compute the average "1" → "0" or "0" → "1" error rates associated with each round of hybridization. These quantities can be very useful in diagnosing the performance of a MERFISH run. For example, a high "1" → "0" error rate for one round of hybridization may indicate poor staining in that round or perhaps an inappropriately high threshold for spot fitting. Alternatively, a high "0" → "1" error rate may suggest that the spot-fitting threshold is too low for that round of hybridization. To calculate this quantity, compute the number of times each RNA species was measured correctly N_C and the number of times it was measured with a word that contains a single-bit error at each possible position N_i. Fig. 5A illustrates these counts for two example barcodes. These two quantities can then be used to estimate the probability of flipping the ith bit via $N_i/(N_i+N_C)$ (see the Materials and Methods of Chen et al., 2015 for a derivation of this expression.) Once the probability of flipping each bit is calculated for each RNA, an estimate of the bit-flipping probability for each round is calculated from the average across all genes, see Fig. 5B for an example of these per-bit error rates.

7.4.2 Background Counts and the Confidence Ratio

As discussed in Section 3.3.3, we recommend leaving a few percent of the possible barcodes unassigned. Measurements of these "blank" or "control" barcodes can then serve as a simple measure of the background count rate in MERFISH and can be used to assign a qualitative degree of confidence to the observed counts for different RNAs.

Perhaps the most obvious use of these "blank" barcodes is to set a lower limit on the number of counts that can be trusted. For example, if an RNA-encoding barcode is counted fewer times than a "blank" barcode, then it might be reasonable to assume that the counts for this RNA are dominated by background counts and should not be trusted. However, this use of the "blank" barcodes provides an overly cautious measure of the rate of background counts because it does not properly take into consideration the way in which different binary barcodes are related to one another. For example, imagine that one "blank" word is counted more frequently than a barcode representing a low-abundance RNA. If these two barcodes are separated by a large HD, ie, these barcodes have very few "1" bits in common, then it is possible that the errors that produced observations of this "blank" barcode may not be relevant for the measurement of the barcode associated with the low-abundance RNA.

To partially capture the relationship between different barcodes and improve estimates of the accuracy of different barcode measurements, we have developed a metric that we term the confidence ratio. This ratio is defined as the number of measured binary barcodes that exactly match

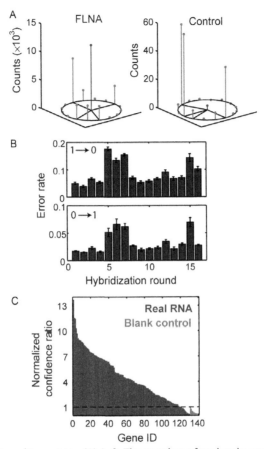

Fig. 5 MERFISH quality metrics. (A) *Left*: The number of molecule counts whose code exactly match that of *FLNA* (*blue* (*dark gray* in the print version)) and the number of molecule counts whose code differ in one bit from the *FLNA* barcode (*red* (*gray* in the print version)). (*Right*) as in (*left*) except for a barcode that was left intentionally unused to serve as a misidentification control (*Control*). Lines connecting the central exact counts to barcodes that contain a single-bit error denote "1" → "0" errors. (B) The average rate at which a "1" → "0" (*top*) or "0" → "1" (*bottom*) error occurs at each bit. These error rates are derived from the ratios of the counts to the correct barcode (A, *blue* (*dark gray* in the print version)) relative to the counts to the barcodes that differ in a single-bit (A, *red* (*gray* in the print version)). (C) The confidence ratio for each used barcode (*Real RNA, blue* (*dark gray* in the print version)) from a 16-bit, MHD4 code measurement normalized to the largest confidence ratio observed for the "blank" barcodes (*Blank control, red* (*gray* in the print version)). *All data are reproduced with permission from Chen, K.H., Boettiger, A.N., Moffitt, J.R., Wang, S., & Zhuang, X. (2015). Spatially resolved, highly multiplexed RNA profiling in single cells. Science, 348, aaa6090.*

the binary barcode for a given RNA over the sum of this number and the number of counts for all binary words that differ from this barcode in only one bit, ie, the ratio of the blue bars to the sum of the blue bar and all red bars in Fig. 5A. For an error-correcting encoding scheme, this quantity is simply the ratio of the number of times an RNA was observed without the application of error correction to the number that was observed with error correction applied. However, error correction is not required to compute the confidence ratio. All that is required is that all barcodes are separated by at least an HD of 2.

To illustrate the usefulness of this ratio, consider the following scenario. Imagine that one specific hybridization round is particularly problematic, and the "1"→"0" error rate for that round is higher than others (as Fig. 5B illustrates, there can be variation between error rates in different rounds of hybridization). Then RNAs encoded with binary barcodes that contain a "1" in that bit will tend to have more single-bit errors. This will decrease the confidence ratio for these words by generating a larger proportion of counts for these single-bit errors. However, RNAs encoded with binary words that do not contain a "1" in that bit will not see an increase in the number of single-bit errors in this round, and the confidence ratio for these RNAs will not be lowered.

Finally, a word of caution: the confidence ratio does not imply a quantitative level of confidence in the counts for an RNA. For example, a ratio of 0.3 does not imply that 30% of the measured counts are correct. The proper interpretation of this ratio is more qualitative: one should trust the counts for an RNA with a confidence ratio of 0.5 more than those for a value of 0.3. To provide a heuristic cut-off on the confidence ratio, we use the measured confidence ratios for the "blank" barcodes. If an RNA-encoding barcode has a confidence ratio larger than the largest confidence ratio observed for the "blank" barcodes then we generally trust the counts associated with this RNA, see Fig. 5C. Of course, this cut-off is still somewhat arbitrary, and we only recommend using it as a rule of thumb. The confidence ratio is one of the simplest quantities one could compute in an effort to exploit the natural geometric connectivity of the different measured binary barcodes. In the future, we expect that more complicated calculations could be performed to exploit this connectivity and further improve both the accuracy of the measured counts with MERFISH and our metrics of the performance of a given measurement.

7.5 (Optional) Iterative Identification of Optimal Thresholds

As discussed in Section 7.1, we observe that different rounds of hybridization produce somewhat different levels of background and spot brightness. Thus, it is typically advisable to tune the spot-finding threshold for each round of hybridization. To further improve this process, it is possible to automate this threshold search by using some of the performance metrics discussed in Section 7.4 to determine the optimal threshold for each hybridization round.

Step 1: Run the spot-fitting algorithm for a range of spot-brightness thresholds. We analyze each MERFISH data set with 10–20 thresholds spanning a range set by examining these data by hand.

Step 2: Select a starting set of thresholds and decode barcodes using these found spots. The initial thresholds can be selected at random or one can start at one limit—ie, all thresholds are the lowest or highest possible values. With these found spots, run the decoding algorithm discussed in Section 7.3.

Step 3: Compute a quality metric. For the MHD4 code, where it is possible to calculate the per-bit error rates, we calculate the geometric mean of all "1"→"0" and "0"→"1" error rates. For the MHD2 code that does not have the ability to correct errors and calculate these per-bit error rates, we utilize the ratio of the number of barcodes with four "1" bits to those measured with three or five "1" bits, ie, the average confidence ratio.

Step 4: Iterate. Repeat Steps 2 and 3 for a variety of threshold combinations and select the threshold combination that maximizes the selected quality metric. Typically, we change our threshold combinations during each iteration round by selecting one hybridization round, fixing the threshold for all other rounds, and screening through all possible threshold values for this round. Once all thresholds have been screened for this round, we select the threshold value for that round that produces the highest quality metric and then fix this threshold value for this hybridization round for subsequent iterations. We then repeat this process for each subsequent hybridization round. Once we have screened the thresholds for all hybridization rounds in this fashion, we repeat this process, often for a total of three times. We find that both quality metrics that we have used typically converge with only a few iterations through all hybridization rounds. More complicated search strategies are also possible.

8. SUMMARY

smFISH is a powerful technique because it allows the quantitative measurement of the exact copy number and spatial distribution of individual mRNAs within single cells in intact tissues. MERFISH extends these powerful abilities to the transcriptome scale by massively multiplexing smFISH measurements by encoding individual RNA species with error-robust barcodes and reading out the barcodes via a series of smFISH measurements. We anticipate this method to be relatively easily adopted by other laboratories and the techniques described in this chapter should facilitate these efforts as will the supporting protocols and code that can be found at http://zhuang.harvard.edu/merfish/.

ACKNOWLEDGMENTS

We thank Guiping Wang, Kok Hao Chen, Siyuan Wang, Alistair Boettiger, and Junjie Hao for comments on the manuscript. This work was in part supported by the National Institutes of Health. J.R.M. was funded in part by a Helen Hey Whitney Postdoctoral Fellowship. X.Z. is a Howard Hughes Medical Institute Investigator.

REFERENCES

Babcock, H., Sigal, Y. M., & Zhuang, X. (2012). A high-density 3D localization algorithm for stochastic optical reconstruction microscopy. *Optical Nanoscopy, 1*, 6.

Balagopal, V., & Parker, R. (2009). Polysomes, P bodies and stress granules: States and fates of eukaryotic mRNAs. *Current Opinion in Cell Biology, 21*, 403–408.

Batish, M., Raj, A., & Tyagi, S. (2011). RNA detection and visualization. In J. E. Gerst (Ed.), *Methods in molecular biology*. Totowa, NJ: Humana Press.

Beliveau, B. J., Joyce, E. F., Apostolopoulos, N., Yilmaz, F., Fonseka, C. Y., McCole, R. B., et al. (2012). Versatile design and synthesis platform for visualizing genomes with Oligopaint FISH probes. *Proceedings of the National Academy of Sciences of the United States of America, 109*, 21301–21306.

Besse, F., & Ephrussi, A. (2008). Translational control of localized mRNAs: Restricting protein synthesis in space and time. *Nature Reviews. Molecular Cell Biology, 9*, 971–980.

Brouwer, A. E., & Etzion, T. (2011). Some new distance-4 constant weight codes. *Advances in Mathematics of Communications, 5*, 417–424.

Buxbaum, A. R., Haimovich, G., & Singer, R. H. (2014). In the right place at the right time: Visualizing and understanding mRNA localization. *Nature, 16*, 95–109.

Chen, K. H., Boettiger, A. N., Moffitt, J. R., Wang, S., & Zhuang, X. (2015). Spatially resolved, highly multiplexed RNA profiling in single cells. *Science, 348*, aaa6090.

Femino, A. M., Fay, F. S., Fogarty, K., & Singer, R. H. (1998). Visualization of single RNA transcripts in situ. *Science, 280*, 585–590.

Holden, S. J., Uphoff, S., & Kapanidis, A. N. (2011). DAOSTORM: An algorithm for high-density super-resolution microscopy. *Nature Methods, 8*, 279–280.

Holt, C. E., & Schuman, E. M. (2013). The central dogma decentralized: New perspectives on RNA function and local translation in neurons. *Neuron, 80*, 648–657.

Itzkovitz, S., & van Oudenaarden, A. (2011). Validating transcripts with probes and imaging technology. *Nature Methods*, *8*, S12–S19.

Jakt, L. M., Moriwaki, S., & Nishikawa, S. (2013). A continuum of transcriptional identities visualized by combinatorial fluorescent in situ hybridization. *Development*, *140*, 216–225.

Larson, D. R., Singer, R. H., & Zenklusen, D. (2009). A single molecule view of gene expression. *Trends in Cell Biology*, *19*, 630–637.

Lécuyer, E., Yoshida, H., & Krause, H. M. (2009). Global implications of mRNA localization pathways in cellular organization. *Current Opinion in Cell Biology*, *21*, 409–415.

Levesque, M. J., & Raj, A. (2013). Single-chromosome transcriptional profiling reveals chromosomal gene expression regulation. *Nature Methods*, *10*, 246–248.

Levsky, J. M., Shenoy, S. M., Pezo, R. C., & Singer, R. H. (2002). Single-cell gene expression profiling. *Science*, *297*, 836–840.

Lubeck, E., & Cai, L. (2012). Single-cell systems biology by super-resolution imaging and combinatorial labeling. *Nature Methods*, *9*, 743–748.

Lubeck, E., Coskun, A. F., Zhiyentayev, T., Ahmad, M., & Cai, L. (2014). Single-cell in situ RNA profiling by sequential hybridization. *Nature Methods*, *11*, 360–361.

Martin, K. C., & Ephrussi, A. (2009). mRNA localization: Gene expression in the spatial dimension. *Cell*, *136*, 719–730.

Moon, T. (2005). *Error correction coding: Mathematical methods and algorithms*. New York, NY: Wiley.

Munsky, B., Neuert, G., & van Oudenaarden, A. (2012). Using gene expression noise to understand gene regulation. *Science*, *336*, 183–187.

Padovan-Merhar, O., & Raj, A. (2013). Using variability in gene expression as a tool for studying gene regulation. *Wiley Interdisciplinary Reviews Systems. Biology and Medicine*, *5*, 751–759.

Raj, A., & van Oudenaarden, A. (2008). Nature, nurture, or chance: Stochastic gene expression and its consequences. *Cell*, *135*, 216–226.

Raj, A., van den Bogaard, P., Rifkin, S. A., van Oudenaarden, A., & Tyagi, S. (2008). Imaging individual mRNA molecules using multiple singly labeled probes. *Nature Methods*, *5*, 877–879.

Rodriguez, A. J., Czaplinski, K., Condeelis, J. S., & Singer, R. H. (2008). Mechanisms and cellular roles of local protein synthesis in mammalian cells. *Current Opinion in Cell Biology*, *20*, 144–149.

Rouillard, J.-M. (2003). OligoArray 2.0: Design of oligonucleotide probes for DNA microarrays using a thermodynamic approach. *Nucleic Acids Research*, *31*, 3057–3062.

Sanchez, A., & Golding, I. (2013). Genetic determinants and cellular constraints in noisy gene expression. *Science*, *342*, 1188–1193.

Shaffer, S. M., Wu, M.-T., Levesque, M. J., & Raj, A. (2013). Turbo FISH: A method for rapid single molecule RNA FISH. *PLoS One*, *8*, e75120.

St Johnston, D. (2005). Moving messages: The intracellular localization of mRNAs. *Nature Reviews. Molecular Cell Biology*, *6*, 363–375.

Xu, Q., Schlabach, M. R., Hannon, G. J., & Elledge, S. J. (2009). Design of 240,000 orthogonal 25mer DNA barcode probes. *Proceedings of the National Academy of Sciences of the United States of America*, *106*, 2289–2294.

Youk, H., Raj, A., & Van Oudenaarden, A. (2010). Imaging single mRNA molecules in yeast (Elsevier Inc.). *Methods in Enzymology*, *470*, 429–446.

Zenklusen, D., & Singer, R. H. (2010). Analyzing mRNA expression using single mRNA resolution fluorescent in situ hybridization. *Methods in Enzymology*, *470*, 641–659.

CHAPTER TWO

Imaging Single mRNA Dynamics in Live Neurons and Brains

H.C. Moon*, H.Y. Park*,[1]
*Seoul National University, Seoul, South Korea
[1]Corresponding author: e-mail address: hyeyoon.park@snu.ac.kr

Contents

1. Introduction 52
2. Neuron Culture Imaging 55
    2.1 Materials 55
    2.2 Protocols 56
3. Brain Slice Imaging 61
    3.1 Materials 61
    3.2 Protocols 61
Acknowledgments 63
References 63

Abstract

RNA is a key player in the process of gene expression. Whereas fluorescence in situ hybridization allows single mRNA imaging in fixed cells, the MS2-GFP labeling technique enables the observation of mRNA dynamics in living cells. Recently, two genetically engineered mouse models have been developed for the application of the MS2-GFP system in live animals. First, the Actb-MBS mouse was generated by knocking in 24 repeats of the MS2 stem-loop sequence in the 3′ untranslated region of the β-actin gene. Second, the MCP mouse was made to express the NLS-HA-MCP-GFP transgene in all cell types. By crossing Actb-MBS and MCP mice, a double homozygous mouse line, MCP × MBS, was established to visualize endogenous β-actin mRNA labeled with multiple green fluorescent proteins. By imaging hippocampal neurons or brain slices from MCP × MBS mice, the dynamics of mRNA, such as transcription, transport, and localization, can be studied at single mRNA resolution. In this chapter, we explain the basics of MCP × MBS mice and describe methods for utilizing these animals.

1. INTRODUCTION

Cells respond to their environment in various ways. One way in which cells respond to stimuli is by controlling production of specific proteins, which can be done by first controlling the RNA level. Therefore, quantifying RNA levels has been an important means to determine changes in gene expression in response to specific stimuli. Quantitative reverse transcriptase polymerase chain reaction and northern blots are widely used to measure the average RNA level of a whole population of cells (Gibson, Heid, & Williams, 1996; Vandesompele et al., 2002). However, these conventional methods average out single-cell aspects such as intrinsic noise and extrinsic noise in gene expression (Elowitz, Levine, Siggia, & Swain, 2002) and cannot detect heterogeneous populations of cells. Moreover, biochemical techniques involve breaking up tissue architecture and thus cannot manifest the complex interactions of cells in their microenvironment.

To distinguish different states of single cells and probe their spatial organization, it is necessary to directly image the cells in their native environment. Fluorescence in situ hybridization has provided tremendous insights into RNA expression in fixed cells and tissues (Singer & Ward, 1982; Zenklusen, Larson, & Singer, 2008). The MS2-GFP system has been one of the most widely used methods of imaging RNA dynamics in living cells. The MS2-GFP labeling technique employs the high affinity ($K_d \sim 5$ nM) binding between MS2 bacteriophage coat protein (MCP) and the stem-loop structure of the MS2 binding site (MBS) (Bertrand et al., 1998). The MBS sequence is typically inserted into the 3′ untranslated region (3′ UTR) of the gene of interest because the 3′ UTR is usually longer than the 5′ untranslated region (5′ UTR). A transgene that encodes MCP fused with GFP (MCP-GFP) is cotransfected with an MBS-tagged reporter construct to label the reporter RNA with GFP.

The Singer laboratory has recently developed two genetically engineered mouse models for live tissue imaging of mRNA at single-molecule resolution. The Actb-MBS mouse is a knock-in mouse whose β-actin gene is modified in a way that a 1.2-kbp sequence containing 24 repeats of MBS stem-loops is inserted 441 bp downstream of the β-actin stop codon (Lionnet et al., 2011). Any primary cells can be cultured from Actb-MBS mice and transfected with MCP-GFP for live-cell imaging of β-actin mRNA. Because MCP-GFP dimerizes and attaches to an MBS stem-loop, a total of 48 GFP molecules are localized in a small volume, and β-actin mRNA can be observed as a single bright spot at high magnification.

A transgenic mouse that expresses MCP-GFP allows circumvention of transfection procedure, which in turn facilitates observing single mRNA dynamics in live mice. The MCP mouse was generated by lentiviral transgenesis to express NLS-HA-MCP-GFP under the human ubiquitin C promoter (Park et al., 2014). The nuclear localization sequence (NLS) is a peptide sequence that attaches to importin α, which enables sequestration of MCP-GFP in the nucleus and decreases the background of free MCP-GFP in the cytoplasm. High concentrations of MCP-GFP in the nucleus also allow for immediate labeling of β-actin mRNA with GFP upon transcription. An HA tag from human influenza hemagglutinin is inserted for the purpose of identifying the list of proteins attached to β-actin mRNA with pull-down experiments.

Through mating Actb-MBS and MCP mice, a hybrid mouse line, MCP × MBS, was established and bred to double homozygosity. Double homozygous MCP × MBS mice have both β-actin alleles tagged with 24 MBS stem-loops and express NLS-HA-MCP-GFP in all cell types. Thus, all of the endogenous β-actin mRNAs are labeled with multiple GFPs in vivo (Park et al., 2014). Therefore, it is possible to observe the dynamics of mRNA in many different types of primary cells and tissues from MCP × MBS mice (Fig. 1).

With high spatial and temporal resolution, we can observe localization of mRNA, which is a prerequisite for local protein synthesis (Aakalu, Smith, Nguyen, Jiang, & Schuman, 2001; Kang & Schuman, 1996; Martin et al., 1997). In neurons, local protein synthesis is one of the important mechanisms contributing to synaptic plasticity. For example, the synaptic tagging model suggests that mRNA is captured locally near the synapse, and with appropriate stimulation, such as receptor binding of the neurotransmitter, mRNA is translated to make long-lasting changes in synaptic strength (Frey & Morris, 1997, 1998; Morris, 2006; Reymann & Frey, 2007). Live-cell imaging of mRNA also makes it possible to observe the burst of transcription in real time (Golding, Paulsson, Zawilski, & Cox, 2005; Larson, Zenklusen, Wu, Chao, & Singer, 2011; Muramoto et al., 2012). If many RNAs are transcribed in a short period of time, which is usually referred to as a burst of transcription, multiple MCP-GFP-tagged mRNAs are

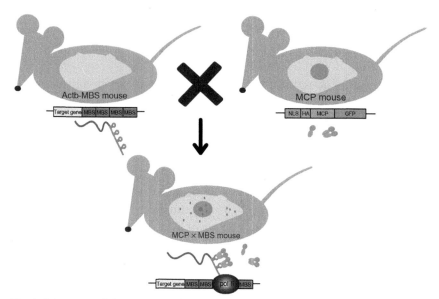

Fig. 1 Schematic of the MCP × MBS mouse system. Inside each mouse body is a schematic of a fibroblast from the mouse. The nucleus of MCP mouse cells is filled with free MCP-GFP. Actb-MBS mouse cells do not fluoresce, but β-actin mRNAs have 24 MBS stem-loops. In MCP × MBS mouse cells, β-actin mRNAs are tagged with multiple MCP-GFP dimers and appear as bright single spots. Inside the nucleus, a burst of transcription can be observed as a large bright spot.

localized in a relatively small area. Thus, transcription sites appear as bright foci that appear slightly larger than MCP-GFP-tagged single mRNA. Analyzing the correlation between the effect of a stimulus and the resulting transcriptional dynamics is expected to provide interesting insights into the mechanism of transcription.

Thus far, mRNA dynamics has been mostly studied in two-dimensional (2D) cultures on a flat surface, which is different from the three-dimensional (3D) structure of the brain in vivo. Lacking the normal cell–cell interaction and having intrinsically different microenvironments, mRNAs in 2D culture may exhibit different properties from their behavior in real tissue environments. By using acute brain slices from MCP × MBS double homozygous mice, we have begun to uncover the nature of mRNA in its native tissue environment. In this protocol, we describe how to image GFP-labeled endogenous β-actin mRNA in live neurons and acute brain slices prepared from MCP × MBS mice (Park et al., 2014).

2. NEURON CULTURE IMAGING
2.1 Materials
Ultrapure water (18 MΩ cm equivalent) is used for all procedures.
1. 10× boric acid buffer (BAB)
   Final concentration: 50 mM boric acid (Sigma-Aldrich, cat. No. B6768; MW 61.83), 12.5 mM sodium tetraborate decahydrate (Sigma-Aldrich, cat. No. B9876; MW 381.37), pH 8.5.
2. 10× Poly-D-Lysine (PDL) solution
   Final concentration: 2 mg/mL PDL (Sigma-Aldrich, cat. No. P7886; MW 30,000–70,000) in 10× BAB.
   It is recommended to make a 10× PDL solution and store aliquots at −20 °C.
3. Neural dissection solution (NDS)
   Final concentration: 1× Hank's balanced salt solution (Gibco, cat. No. 14185) and 10 mM HEPES (Gibco, cat. No. 15630-080) in water.
   Prepare ~50 mL of NDS, and keep it on ice. NDS is used for dissection of hippocampi.
4. Plating medium (PM)
   Final concentration: 10% FBS heat inactivated (Gibco, cat. No. 10082-147), 1× Glutamax (Gibco, cat. No. 35050-061), and 0.1 mg/mL Primocin (Invivogen, cat. No. ant-pm-1) in Neurobasal A medium (Gibco, cat. No. 10888-022).
   Prepare ~20 mL of PM. Keep this solution at 4 °C. PM, which includes heat inactivated FBS, is used for plating trypsinized neuron cells on glass-bottom culture dishes.
5. B27 medium
   Final concentration: 1× B27 (Gibco, cat. No. 17504-044), 1× Glutamax (Gibco, cat. No. 35050-061), and 0.1 mg/mL primocin (Invivogen, cat. No. ant-pm-1) in Neurobasal A medium (Gibco, cat. No. 10888-022).
   B27 medium, which includes B27 supplement instead of FBS, is used in the last step after plating. For each glass-bottom culture dish, 1.8 mL of B27 is used. Make an appropriate amount based on your need. Equilibrate B27 medium in the incubator that maintains 5% CO_2 and 37 °C for 2–4 h before use. Note that B27 medium evaporates during equilibration in the incubator. Phenol-red free neurobasal A medium (Gibco, cat. No. 12349-015) can also be used with the

advantage of less background noise than phenol-red included neurobasal A medium.

6. HEPES-buffered saline (HBS)

    Final concentration: 119 mM NaCl, 5 mM KCl, 2 mM CaCl$_2$, 2 mM MgCl$_2$, 30 mM D-glucose (Sigma-Aldrich, cat. No. G7021), 20 mM HEPES (Gibco, cat. No. 15630-080) at pH 7.4.

    If a CO_2 incubator is not available on the microscope, media must be changed to HBS to maintain the pH of the neuron culture during imaging. Approximately 2 mL of HBS is used for one glass-bottom culture dish. It is recommended to block the exposure of light by wrapping the HBS-containing bottle with aluminum foil. Make fresh HBS once a week.

7. Dissection tools

    Prepare a dissection stereoscope, two fine-tipped straight forceps (No. 5), one curved forceps, small dissection scissors, a small spatula, No. 10 blade, plastic transfer pipets, and a hemocytometer.

8. Microscopy equipment

    Live-cell imaging of single mRNA requires an inverted microscope (Olympus IX73 or similar), 488 nm laser or high power LED light source, electron multiplying charge-coupled device (EMCCD; Andor iXon or similar), 100× or 150× high magnification objective, and temperature-controlled imaging chamber.

2.2 Protocols

2.2.1 PDL Coating for Neuron Imaging

Cells are usually cultured on polystyrene tissue culture dishes. However, imaging single mRNA requires a high magnification and high numerical aperture (NA) objective lens. Therefore, it is essential to culture neurons on glass coverslips for live-cell imaging of mRNA. To cause a hydrophobic glass surface to become hydrophilic, PDL, a highly positively charged synthetic amino acid chain, is used.

1. The day prior to culture, thaw a vial of 10 × PDL and make a 1:10 dilution (final: 0.2 mg/mL) in water.
2. Filter the diluted PDL solution with a syringe filter (0.45 μm pore).
3. For each glass-bottom culture dish, add 200 μL of filtered PDL solution to the glass section. Incubate dishes at 37°C overnight. Incomplete coating may lead to aggregation of neurons. Do not let the PDL dry.

4. Remove the PDL solution and rinse the glass-bottom culture dish with sterile water before seeding neurons. Remove remaining water.
5. Incubate dishes in incubator for several minutes before seeding neuron cells.

2.2.2 Dissection

Sacrifice 1–3-day-old pups through a method approved by the Institutional Animal Care and Use Committee (IACUC). It is important to dissect and seed neurons as quickly as possible because neurons are very sensitive compared to other types of cells. Note that 3–4 pups will yield ~16 glass-bottom culture dishes at a concentration of 425,000 cells/mL. There are various methods of dissecting hippocampi, and we recommend a protocol published with an excellent video tutorial (Seibenhener & Wooten, 2012).

1. Prepare one curved forceps, two sharp forceps, dissection scissors, one small spatula, No. 10 blade, and pipets (Fig. 2). Place Kimwipes inside a beaker, and fill the beaker with 70% ethanol. Kimwipes are used to prevent damaging sharp forceps tips. Sterilize the tools with 70% ethanol before dissection.
2. Fill a 15 mL conical tube with approximately 5 mL of NDS to keep dissected hippocampi. Place NDS-filled 15 mL conical tubes in an ice bucket. Place filter paper on the bottom of a 50 mm Petri dish and fill it with ~5 mL NDS. Fill a 150 mm Petri dish with ice, and place it on the stereoscope stage. Place an NDS-filled 50 mm Petri dish on the ice-filled 150 mm Petri dish to keep NDS cold during dissection (Fig. 3).
3. Decapitate a pup. Hold the nose part of the head with a curved forceps and cut a T-shape with a No. 10 blade. Peel off the skin and carefully cut the skull with a No. 10 blade. With a sharp forceps, tear off the skull

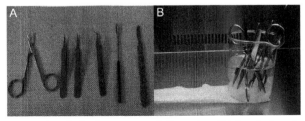

Fig. 2 Preparation of tools for hippocampal dissection. (A) Surgical tools used during dissection. (B) Sterile preparation of the surgical tools.

Fig. 3 Setup for dissection.

and scoop out the brain with a small spatula. Place the brain in the NDS-filled 50 mm Petri dish.
4. Dissect the brain under a stereoscope. Cut the hindbrain with a blade.
5. Split the brain in half with a blade (right and left hemispheres).
6. From this point, use two sharp forceps for dissection. Remove (dig out) the midbrain from a hemisphere.
7. Remove meninges (membranes) and blood vessels carefully. Remember the position of the hippocampus and pinch the interlayer between the hippocampus and cortex.
8. Remove the remaining meninges from the hippocampus.
9. Collect the dissected hippocampus with a wide-bore plastic pipet, and put it in the ice-cold NDS-filled 15 mL falcon tube.
10. Repeat dissection for all pups.

2.2.3 Seeding Neuron Cells

It is important to note that dissection and seeding steps must be performed as quickly as possible for viability of neurons. PM should be warmed to 37°C before use. It is recommended to warm PM several minutes before dissecting the hippocampi from all pups.
1. Warm up the PM to 37°C.
2. Bring hippocampi in the NDS-filled 15 mL conical tube to the clean bench. Leave it until hippocampi settle to the bottom of the tube.
3. Carefully aspirate off NDS until ~2 mL of NDS is left.
4. Add 200 μL of 10× trypsin and incubate in a 37°C water bath for 15 min.
5. Remove trypsin by aspirating it off carefully and add 3 mL of PM to stop trypsin activity by FBS. Triturate with a 5 mL pipet 20–30 times until brain tissues are dissociated. At the end of triturating, most of the

large chunks should be crushed. Wait for ~3 min until undigested tissues settle to the bottom. Optionally, cell strainer can be used for filtering out undigested brain tissues.

6. Determine the cell density of solution by using a hemocytometer. If the cell density is too low, you can triturate again with 1 mL pipet.
7. Resuspend the cells to 425,000–950,000 cells/mL by adding an appropriate amount of PM. We have observed that extremely high density of cells can result in aggregation of neurons in culture, while too low plating density cannot produce viable neuron culture.
8. Seed the cells on the glass part of the PDL-coated glass-bottom dish and incubate them in a cell incubator that maintains 5% CO_2 and 37°C.
9. Incubate the cells for 4 h until cells attach to the bottom.
10. Prepare B27 medium and equilibrate it by incubating in the cell incubator for 2–4 h.
11. After 4 h of plating, carefully add 1.8 mL of equilibrated B27 medium to each glass-bottom culture dish. Incubate the cells until they are used in experiments.

2.2.4 Maintaining Neuron Culture

Every 7 days after seeding, gently add 300 μL of equilibrated B27 medium to the cells. Fig. 4 shows typical images of hippocampal neurons with good morphology and density at different stages. Because neurons are vulnerable to environmental changes, it is recommended to refrain from taking neurons out of the incubator often.

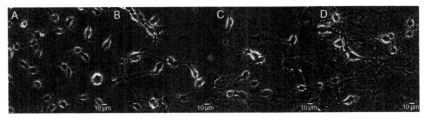

Fig. 4 Images of a cultured neuron with the density of 650,000 cells/mL. (A) 24 h after seeding. Neurites are starting to grow. Still, a lot of extra substances can be observed on the culture. These extra substances disappear as the time passes. (B) 3 Days in vitro (DIV). Relatively long neurites are observed. (C) 7 DIV. (D) 15 DIV. Neural network becomes more complex as the neurons differentiate and make synapses.

2.2.5 Imaging mRNA in Cultured Neurons

If a CO_2 incubator is available on the microscope, neuron culture with B27 medium can be imaged right away. Even though phenol-red free B27 medium exhibits less background than phenol-red included B27 medium, HBS provides the most high quality image. However, HEPES which is included in HBS is known to have phototoxicity and is not good for long-term imaging over several hours. Depending on your experiment, choose appropriate media that will be used during image acquisition.

1. If using B27 medium during image acquisition, just take image without extra steps. If using HBS, warm HBS to 37°C in a water bath.
2. Carefully remove the B27 medium from the dish and add HBS medium as slowly as possible.
3. Use a 150× high magnification, high NA oil immersion objective lens (Olympus, U Apochromat 150× TIRF objective, NA=1.45) and EMCCD (Andor, iXon Ultra DU-897U) to collect as much signal as possible. It is also possible to observe single mRNAs with 100× oil immersion objective lens (Olympus, UPlanSApo 100× Objective, NA=1.40). Turn on the LED (Lumencor SOLA SE), and use an appropriate filter set for GFP (Chroma #49002). An open source microscopy software such as Micro-Manager can be used for image acquisition. Generally, it is desirable to use low excitation power and high EMCCD gain to minimize photobleaching. Depending on the experiment, pulsed illumination might be useful for reducing phototoxicity. We typically use 100 µW excitation power on the sample, 200 ms exposure time, and 90% EMCCD gain (Fig. 5).

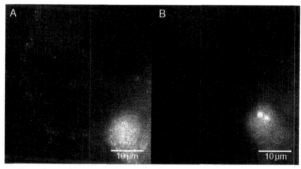

Fig. 5 Images of a cultured neuron (DIV 2). Two images were taken on the same cell but at different z positions. Image conditions: exposure time 200 ms, LED power 5%, and EMCCD gain 900. (A) β-actin mRNAs in the neurite. (B) Two strong transcription sites inside the nucleus.

3. BRAIN SLICE IMAGING

3.1 Materials

Ultrapure water (18 MΩ cm equivalent) is used for all procedures.

1. Artificial cerebrospinal fluid (ACSF).

    Final concentration: 119 mM NaCl, 2.5 mM KCl, 1.3 mM MgSO$_4$ (Sigma-Aldrich, cat. No. M7506; MW 120.37), 2.5 mM CaCl$_2$, 1.0 mM NaH$_2$PO$_4$ (Sigma-Aldrich, cat. No. 71507; MW 155.99), 26.4 mM NaHCO$_3$ (Sigma-Aldrich, cat. No. S5761; MW 84.01), 11 mM D-glucose (Sigma-Aldrich, cat. No. G7021).

    Prepare ACSF and oxygenate it by bubbling carbogen gas (95% O$_2$/5% CO$_2$) for ∼30 min on ice.

2. Dissection tools and materials.

    Isoflurane (anesthetic), carbogen gas (95% O$_2$/5% CO$_2$), vibratome (Leica VT1000S or similar), large scissors, small surgical scissors, rongeur (Fine Science Tools Inc., cat. No. 16220-14 or similar), mini hippocampal tool, spatulas, blunt tweezers, filter papers (Whatman, grade 50, 9 cm or similar), plastic spoon, small paint brush, No. 10 blades, and vetbond.

3. Prechamber for holding slices (Warner Instruments, cat. No. 65-0076 or similar).

4. Microscopy equipment.

    Upright microscope, two-photon or confocal laser-scanning system, water immersion objective (Olympus XLPLN25XWMP2 25×/1.05 NA or similar), peristaltic pump, recording chamber (Warner Instruments, cat. No. 64-0236 or similar), platform for recording chamber heater (Warner Instruments, cat. No. 64-0284 or similar), in-line solution heater (Warner Instruments, cat. No. 64-0102 or similar), heater controller (Warner Instruments, cat. No. 64-0101 or similar), and slice hold-down harp (Warner Instruments, cat. No. 64-0251 or similar).

3.2 Protocols

3.2.1 Preparation of Acute Brain Slices

To image single mRNA molecules in live brain tissues, it is desirable to use mice that are younger than 1 month of age. Because the endogenous fluorescent pigment lipofuscin accumulates in the cytoplasm of neurons with age (Schnell, Staines, & Wessendorf, 1999), it becomes difficult to distinguish mRNA from the autofluorescent spots after ∼1 month of age.

1. Euthanize a mouse through a method approved by the IACUC. A recommended method is to anesthetize a mouse with isoflurane and decapitate it with sharp scissors.
2. After decapitation, immediately make an incision along the midline of the scalp with a blade. Flip the skin and cut the skull along the midline from the neck to the eyes with small surgical scissors.
3. Use a rongeur to open the skull on both sides. Quickly remove the brain and put it in oxygenated ice-cold ACSF for 5 min.
4. Fill a 100 mm Petri dish with ice, place filter paper on the cover of the Petri dish, and apply oxygenated ice-cold ACSF to wet the filter paper thoroughly.
5. Place the brain on the wet filter paper. With a blade, remove the hindbrain and a quarter of the frontal lobes. Cut the remaining brain into two hemispheres.
6. Apply a drop of vetbond on the cutting stage of the vibratome, and place each hemisphere with front side down.
7. Fill the cutting chamber with oxygenated ice-cold ACSF until the hemispheres are fully immersed.
8. Cut 350–500 µm thick coronal sections at slow speed ($\sim$0.2 mm/s) and high vibration frequency ($\sim$80 Hz).
9. Collect slices with a small paint brush and transfer them into a Petri dish filled with oxygenated ice-cold ACSF.
10. Transfer brain slices into a prechamber for recovery in oxygenated ACSF at room temperature for approximately 2 h.

3.2.2 Imaging mRNA in Brain Slices

To image mRNA in live brain tissues, either confocal or multiphoton laser-scanning microscopy must be used. Because cells on the surface of brain slices are mostly damaged, it is desirable to image cells that are at least $\sim$30 µm deep from the surface. With a confocal microscope, it is possible to detect individual mRNP particles down to $\sim$60 µm deep in brain tissues. Two-photon excitation with near-infrared light allows imaging of mRNP particles down to $\sim$300 µm deep inside live brain tissues. In addition, two-photon microscopy is advantageous for tissue imaging because of less photobleaching and phototoxicity.

1. Transfer a brain slice into the recording chamber maintained at 32°C and perfused with warm oxygenated ACSF flowing at $\sim$1 mL/min. Secure the position of the slice by using a slice hold-down harp.

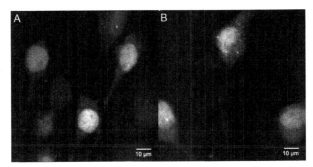

Fig. 6 Images of acute brain slices acquired by two-photon microscopy. (A) Hippocampus. (B) Cortex.

2. For two-photon microscopy, use an XLPlan N 25 × 1.05 NA water immersion lens (Olympus) with 880 nm excitation wavelength to detect single mRNP particles labeled with GFP.
3. Choose a region of interest with a low magnification overview and zoom in to meet the Nyquist sampling criteria. When using an 880 nm excitation beam and a water immersion objective with 1.05 NA, the pixel size should be ∼190 nm, and the z-step size should be ∼640 nm (Zipfel, Williams, & Webb, 2003).
4. Take time-lapse z-stack images with appropriate imaging parameters, time interval, and duration for the mRNA dynamics being examined. We typically use 10–30 mW two-photon excitation power on the sample using a mode-locked Ti:Sapphire laser at 880 nm wavelength with 100 fs pulse duration and 80 MHz repetition rate. Fig. 6 was acquired using 30 mW excitation power, 4.4 µs pixel dwell time, and 5 × zoom with a 25 × objective.

ACKNOWLEDGMENTS

This work was supported by the Seoul National University Research Grant in 2015 and Basic Science Research Program through the National Research Foundation of Korea (NRF) funded by the Ministry of Science, ICT & Future Planning (2015R1C1A1A02036674).

REFERENCES

Aakalu, G., Smith, W. B., Nguyen, N., Jiang, C., & Schuman, E. M. (2001). Dynamic visualization of local protein synthesis in hippocampal neurons. *Neuron*, *30*(2), 489–502.

Bertrand, E., Chartrand, P., Schaefer, M., Shenoy, S. M., Singer, R. H., & Long, R. M. (1998). Localization of ASH1 mRNA particles in living yeast. *Molecular Cell*, *2*(4), 437–445.

Elowitz, M. B., Levine, A. J., Siggia, E. D., & Swain, P. S. (2002). Stochastic gene expression in a single cell. *Science*, *297*(5584), 1183–1186.

Frey, U., & Morris, R. G. M. (1997). Synaptic tagging and long-term potentiation. *Nature*, *385*(6616), 533–536.

Frey, U., & Morris, R. G. M. (1998). Synaptic tagging: Implications for late maintenance of hippocampal long-term potentiation. *Trends in Neurosciences*, *21*(5), 181–188.

Gibson, U. E. M., Heid, C. A., & Williams, P. M. (1996). A novel method for real time quantitative RT PCR. *Genome Research*, *6*(10), 995–1001.

Golding, I., Paulsson, J., Zawilski, S. M., & Cox, E. C. (2005). Real-time kinetics of gene activity in individual bacteria. *Cell*, *123*(6), 1025–1036.

Kang, H., & Schuman, E. M. (1996). A requirement for local protein synthesis in neurotrophin-induced hippocampal synaptic plasticity. *Science*, *273*(5280), 1402–1406.

Larson, D. R., Zenklusen, D., Wu, B., Chao, J. A., & Singer, R. H. (2011). Real-time observation of transcription initiation and elongation on an endogenous yeast gene. *Science*, *332*(6028), 475–478. http://dx.doi.org/10.1126/science.1202142.

Lionnet, T., Czaplinski, K., Darzacq, X., Shav-Tal, Y., Wells, A. L., Chao, J. A., ... Singer, R. H. (2011). A transgenic mouse for in vivo detection of endogenous labeled mRNA. *Nature Methods*, *8*(2), 165–170. http://dx.doi.org/10.1038/nmeth.1551.

Martin, K. C., Casadio, A., Zhu, H., Yaping, E., Rose, J. C., Chen, M., ... Kandel, E. R. (1997). Synapse-specific, long-term facilitation of aplysia sensory to motor synapses: A function for local protein synthesis in memory storage. *Cell*, *91*(7), 927–938.

Morris, R. G. M. (2006). Elements of a neurobiological theory of hippocampal function: The role of synaptic plasticity, synaptic tagging and schemas. *European Journal of Neuroscience*, *23*(11), 2829–2846.

Muramoto, T., Cannon, D., Gierlinski, M., Corrigan, A., Barton, G. J., & Chubb, J. R. (2012). Live imaging of nascent RNA dynamics reveals distinct types of transcriptional pulse regulation. *Proceedings of the National Academy of Sciences of the United States of America*, *109*(19), 7350–7355.

Park, H. Y., Lim, H., Yoon, Y. J., Follenzi, A., Nwokafor, C., Lopez-Jones, M., ... Singer, R. H. (2014). Visualization of dynamics of single endogenous mRNA labeled in live mouse. *Science*, *343*(6169), 422–424. http://dx.doi.org/10.1126/science.1239200.

Reymann, K. G., & Frey, J. U. (2007). The late maintenance of hippocampal LTP: Requirements, phases, 'synaptic tagging', 'late-associativity' and implications. *Neuropharmacology*, *52*(1), 24–40. http://dx.doi.org/10.1016/j.neuropharm.2006.07.026.

Schnell, S. A., Staines, W. A., & Wessendorf, M. W. (1999). Reduction of lipofuscin-like autofluorescence in fluorescently labeled tissue. *The Journal of Histochemistry and Cytochemistry*, *47*(6), 719–730.

Seibenhener, M. L., & Wooten, M. W. (2012). Isolation and culture of hippocampal neurons from prenatal mice. *Journal of Visualized Experiments*, (65), e3634.

Singer, R. H., & Ward, D. C. (1982). Actin gene expression visualized in chicken muscle tissue culture by using in situ hybridization with a biotinated nucleotide analog. *Proceedings of the National Academy of Sciences of the United States of America*, *79*(23), 7331–7335.

Vandesompele, J., De Preter, K., Pattyn, F., Poppe, B., Van Roy, N., De Paepe, A., & Speleman, F. (2002). Accurate normalization of real-time quantitative RT-PCR data by geometric averaging of multiple internal control genes. *Genome Biology*, *3*(7), research0034.1–0034.11.

Zenklusen, D., Larson, D. R., & Singer, R. H. (2008). Single-RNA counting reveals alternative modes of gene expression in yeast. *Nature Structural & Molecular Biology*, *15*(12), 1263–1271. http://dx.doi.org/10.1038/nsmb.1514.

Zipfel, W. R., Williams, R. M., & Webb, W. W. (2003). Nonlinear magic: Multiphoton microscopy in the biosciences. *Nature Biotechnology*, *21*(11), 1368–1376.

CHAPTER THREE

Monitoring of RNA Dynamics in Living Cells Using PUM-HD and Fluorescent Protein Reconstitution Technique

H. Yoshimura, T. Ozawa[1]
School of Science, The University of Tokyo, Tokyo, Japan
[1]Corresponding author: e-mail address: ozawa@chem.s.u-tokyo.ac.jp

Contents

1. Introduction	66
1.1 RNA Imaging in Living Cells	66
1.2 An Ideal Approach for RNA Labeling in Living Cells	67
1.3 Previous Approaches to Visualize RNA in Cultured Cells	68
2. Principle of PUM-HD-Based RNA Probes	70
2.1 RNA-Binding Protein Domain, PUM-HD	70
2.2 Fluorescent Protein Reconstitution Method	72
3. Development of PUM-HD-Based Probes	73
3.1 Design Principle	73
3.2 Materials	75
3.3 Methods	75
4. Microscopy Setup and Visualization of Single-Molecule RNA in Living Cells	76
4.1 Entire Design of the Optics of the Microscope System	77
4.2 The Detail Design of Excitation Optical System	77
4.3 Methods	79
5. Example of RNA Visualization by Using PUM-HD-Based RNA Probes	80
5.1 Monitoring Localization of Mitochondrial mRNA, ND6 mRNA	80
5.2 Monitoring Localization and Dynamics of Single β-Actin mRNA Molecule in Living Cells	81
6. Conclusion	82
Acknowledgment	84
References	84

Abstract

Fluorescence live-cell RNA imaging to monitor the intracellular localization and dynamics of the target RNA is a challenging subject. One of the difficulties to achieve this is to establish a precise method to enable a fluorescent labeling to the target RNA in living cells. Technologies to reduce the background fluorescence and to detect the RNA with high sensitivity are also necessary to visualize and analyze the intracellular localization and dynamic of the target RNA precisely. Especially in monitoring single-molecule motion, a special setup of a microscope system is required. Such technical problems make the live-cell RNA imaging to be a difficult subject. We recently developed a methodology to label and to visualize a target RNA in living cells with low background fluorescence by using a probe that is based on an RNA-binding protein domain PUM-HD (pumilio homology domain) and a fluorescent protein reconstitution method. A noteworthy property of PUM-HD to apply RNA probes is that this protein domain can be modified to recognize a particular 8-base RNA sequence by inducing tailor-made designed mutagenesis. The fluorescent protein reconstitution method allows us to detect the target RNA with high signal-to-noise ratio. Using the probe based on PUM-HD, a fluorescent protein reconstitution method, and a homebuilt fluorescent microscope system, we succeeded in single-molecule observation of a target RNA in living cells. In this chapter, the techniques to establish the probe and to observe the motion of single-molecule RNA are described.

1. INTRODUCTION
1.1 RNA Imaging in Living Cells

RNA was, in past, mostly considered as just a "messenger" of genetic information from chromosome DNA to generate proteins. Recently, intracellular localization and dynamics of RNA, even mRNA, are recognized to be implicated in controlling a variety of physiological functions (Little, Sinsimer, Lee, Wieschaus, & Gavis, 2015; Martin & Ephrussi, 2009). For example, localization of an mRNA in a specific region induces a local protein synthesis in the site that requires the protein immediately. One of the well-investigated examples for mRNA localization is β-actin mRNA; this mRNA is concentrated in the subcellular regions that require β-actin protein intensively such as pseudopodia of fibroblast, and growth cones and spines of neuron (Farina, Huttelmaier, Musunuru, Darnell, & Singer, 2003; Lin & Holt, 2007; Shav-Tal et al., 2004; Tiruchinapalli et al., 2003). These regions require abundant β-actin proteins to construct newly synthesized actin cytoskeleton networks immediately in responding to external stimulations. In addition to the β-actin mRNA, various mRNAs are known to localize in specific subcellular regions, implicating in regulation of different cell functions. In addition to mRNAs, other RNA categories such as functional RNAs or noncoding RNAs are recent topics of importance of RNA in physiological events (Quinodoz & Guttman,

2014). These RNAs have their original physiological functions like protein enzymes. Therefore, it is natural that the intracellular localization of these RNAs is important for them to express the precise functions.

Despite that much knowledge has been reported on the relation between intracellular RNA localization and expression of cellular functions, detail understanding of functions and mechanisms of RNA localization and dynamics demands novel methodologies to monitor target mRNAs in living cells. One of the major reasons why the mechanisms and functions of intracellular RNA localization have not been fully understood is lack of methodological approaches to monitor the localization and dynamics of target RNAs in living cells. Many researches to date have been performed investigations on RNAs mostly based on biochemical experimental approaches, in which information of subcellular localization and dynamics of target RNAs were diminished or not detailed. A live-cell imaging approach is one of the promising methods to trace the time-course of localization and dynamics of target molecule in living cells.

The authors recently succeeded in visualizing intact endogenous RNA molecules in living cells using a newly developed protein-based RNA probe (Yamada, Yoshimura, Inaguma, & Ozawa, 2011; Yoshimura, Inaguma, Yamada, & Ozawa, 2012). In this chapter, we introduce the design of the RNA probe, which consists of mutants of an RNA-binding protein domain PUM-HD (pumilio homology domain) and a pair of split fragments of a fluorescent protein. Next, we describe a microscope setup to visualize the target RNA molecules labeled with the probe. We then show examples of monitoring on the localization and dynamics of endogenous RNAs in living cells.

1.2 An Ideal Approach for RNA Labeling in Living Cells

Methodologies to visualize RNAs in living cells have some different requirements from those to visualize proteins in living cells. Unlike the labeling of proteins in living cells, RNAs cannot be conjugated with a fluorescent tag. A protein fused with a fluorescent protein can be generated in living cells by using genetic engineering technique and subjected to fluorescence microscopic observation. In case of RNAs, to generate a genetically encoded RNA fused with a fluorescent tag in living cells is impossible. Therefore, some other approaches to enable a fluorescent labeling with a probe molecule to the target RNA are required. The probe molecule should have following properties: high specificity to the target RNA, high affinity to the target RNA, and design flexibility to be applied to visualization of various RNAs. In addition, ideally the probe can be induced into cultured cells with high efficiency through a simple procedure, be used in living cells, and have high detectability in fluorescence microscope observation. Many researchers have made intensive efforts to develop RNA probes that possess such properties.

1.3 Previous Approaches to Visualize RNA in Cultured Cells

Spatial information of RNAs has been usually obtained through bioimaging on chemically fixed cells with labeling the target RNA through in situ hybridization techniques. However, this approach cannot yield information of target RNA dynamics because the sample cells are chemically fixed ones, in which RNAs do not allow any motions. Moreover, the fixation process in the in situ hybridization possibly causes damages to the sample cells, perturbing the intracellular distribution of RNAs. On the other hand, fluorescence live-cell imaging is a preferable approach to analyze dynamics and subcellular localization of target molecules in intact living cells (Tyagi, 2009; Weil, Parton, & Davis, 2010). Using this approach, time-course of RNA localization changes caused by stimulation-induced dynamics and single-molecule motion of the target RNA will be able to be monitored. However, fluorescence live-cell imaging approaches for RNAs have not fully established. A major issue that hampers to develop methods of RNA imaging in living cells is in difficulties of RNA labeling with sufficient selectivity and low background fluorescence. In particular, endogenous RNAs cannot be fused directly with fluorescent proteins in living cells. Development of RNA visualization probes that bind selectively to the target RNA and emit a fluorescent signal only when they bind to the target RNA is desirable for RNA live-cell imaging.

Much efforts have been made to develop probes to visualize RNAs in living cells. A typical approach, called a molecular beacon method, is based on a probe that consists of an oligonucleotide and synthetic fluorophores (Bratu, Cha, Mhlanga, Kramer, & Tyagi, 2003; Tyagi, 2009). The oligonucleotide probes have a stem–loop structure, the loop region which has complementary sequences to the target mRNAs. Both termini of the oligonucleotide are conjugated with fluorophores or a pair of a fluorophore and a quencher. The fluorophore is quenched or causes FRET without the target RNA because the stem–loop structure makes the both termini to be close to each other. In the presence of the target RNA, the loop region hybridized on the target RNA and the stem–loop structure is dissolved. These result in termination of the quenching or FRET of the fluorophore and make it possible to detect the target RNA using fluorescence microscopy. This method, however, necessitates injection of the probe into individual cells. In addition, oligonucleotides tend to be concentrated into particular intracellular compartments like the nucleus, preventing precise labeling of the target RNA without hampering its generous localization features. The molecular beacon method is based on resonance energy transfer of fluorescence and substantial background fluorescence is unavoidable and therefore accumulation of many probe molecules on individual target RNA molecules is necessary to make a sufficient signal-to-noise ratio for high-sensitive imaging such as single-molecule observations (Raj, van den Bogaard, Rifkin, van Oudenaarden, & Tyagi, 2008; Vargas et al., 2011). Although the molecular beacon method has high selectivity to the target RNA based on the Watson–Crick interaction between the probe and the target RNA, the molecular beacon method is not suitable to visualize real-time single-molecule dynamics of RNAs in living cells.

Protein-based probes make it easier to prepare the sample cells that harbor the probe molecules because an expression vector which codes the amino acid sequence of the probe can be induced to many cultured cells at a time by a transfection method. However, designs of protein-based probes generally require additional tag sequences on the target mRNA, which act as the recognition site of the probe. A typical example of protein-based RNA probes is the MS2 system (Bertrand et al., 1998; Darzacq et al., 2007; Dictenberg, 2012). In this system, the probes composed of a fluorescent protein domain and an MS2-coat protein domain that specifically binds to an RNA with a particular sequence called MS2 sequence. The target RNA is conjugated with an additional tag consisting of a repeat of MS2 sequences. The probes concentrated on the MS2 repeat tag on the target RNA to visualize it as a fluorescence spot in fluorescence microscope images. The MS2 system has enabled to monitor various RNAs in living subjects. However, the target RNA needs be modified artificially to attach the additional MS2 repeat tag, and therefore, the target RNA is not endogenous one for the sample cell. The observation results of this artificial RNA are possibly different from that of the endogenous one and might cause misunderstanding of the localization and dynamics of the original target RNA in living cells.

Split fluorescent protein reconstitution technologies provide a methodology to satisfy some requirements for live-cell RNA imaging probes

(Ozawa, Yoshimura, & Kim, 2013; Yoshimura & Ozawa, 2014). Split fragments of a fluorescent protein fused with target proteins or protein domains can be generated in living cells in a genetically encoded manner. Fusing a pair of split fluorescent protein fragments with RNA-binding proteins is a possible strategy in design of an RNA probe to visualize RNAs in living cells. An RNA probe based on fluorescent protein reconstitution technique was developed with using eIF4A as the RNA-binding moiety of the probe (Valencia-Burton, McCullough, Cantor, & Broude, 2007). The eIF4A was separated into two globular domains. Each domain was fused with N- and C-terminal fragments of EGFP. In the presence of an RNA aptamer, which includes the sequences to bind to eIF4A, the domains of eIF4A bind to the RNA, bringing the EGFP fragments mutually close and induce the reconstitution of the fragments. The pair of probes was applied for visualization of an RNA aptamer-tagged mRNA and a 5S ribosomal RNA. The method using eIF4A can be applied to detect RNAs which have the target sequence of eIF4A. Conjugation of the aptamer to the target RNA is necessary to visualize the target RNA in living cells, but the RNA is no longer endogenous one as is the case of the MS2 system.

2. PRINCIPLE OF PUM-HD-BASED RNA PROBES

The ideal approach for RNA monitoring is based on a protein-based probe that is designed to recognize an endogenous target RNA selectively. In addition, the probe should emit signal only upon binding the target RNA. Therefore, design of the RNA-recognizing module and the signal emitting module is critical to the performance of the probe. An RNA-binding protein human pumilio 1 provides an outstanding tool for the RNA-recognizing module, and split fluorescent protein reconstitution technique is an excellent approach to make signal of binding the probe to the target RNA molecule. In this section, we introduce these tools to produce RNA probes for live-cell imaging.

2.1 RNA-Binding Protein Domain, PUM-HD

We focused on PUM-HD of human PUMILIO1 to generate protein-based RNA probes for live-cell imaging (Ozawa, Natori, Sato, & Umezawa, 2007; Yamada et al., 2011; Yoshimura et al., 2012). PUMILIO1 is an RNA-binding protein that associates with 3′UTR of mRNAs to modulate gene translation in cooperation with other proteins such as eIF4 (Vardy & Orr-Weaver, 2007). PUM-HD is an RNA-binding domain of human

PUMILIO1. In 2002, Hall and coworkers reported a crystal structure of PUM-HD (Fig. 1A) (Wang, McLachlan, Zamore, & Hall, 2002). Based on this report, PUM-HD was revealed to consist of eight repeated motifs (repeats 1–8) that recognize an RNA sequence of UGUAYAUA (Y is C or U). In each motif, three amino acids form hydrogen bonds and Van der Waals force specifically to a single base of the RNA, resulting in the selective 8-base RNA recognition (Fig. 1B and C). A noteworthy property of PUM-HD is that PUM-HD can be modified to recognize other 8-base RNA sequences than the original one by inducing tailor-made substitution of particular amino acids involving RNA recognition based on the crystal structure (Cheong & Hall, 2006). Although there is no motif in the wild-type PUM-HD to recognize a cytosine base specifically, a study with a screening on random mutagenesis approach has generated a motif that recognizes a cytosine base selectively (Filipovska, Razif, Nygard, & Rackham, 2011). An important limitation of PUM-HD mutant design is difference of affinity between the mutant and the target RNA depending on the recognition sequence. In particular, the three bases at the 5′-terminus of the RNA should be conserved to be UGU to maintain high affinity between the PUM-HD mutant and the RNA. The dissociation constant of a PUM-HD mutant and the cognate RNA is nanomolar or subnanomolar in case that the

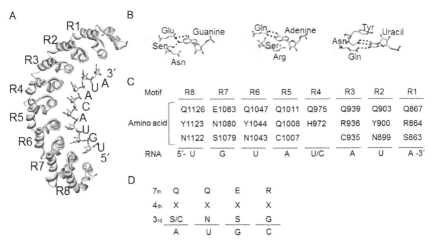

Fig. 1 The structure and RNA recognition of PUM-HD. (A) Crystal structure of PUM-HD. (B) Enlarged structure of the RNA-base-recognizing portions. (C) Combination of RNA-recognizing amino acids in each repeat motif in the wild-type PUM-HD. (D) The universal code of RNA-recognizing amino acids (3rd, 4th, and 7th position in the second helix) and recognized RNA bases. Any amino acid is acceptable at the position indicated by X.

5′-terminus of the RNA is UGU, whereas the dissociation constant is tens or hundreds nanomolar when the 5′-terminus is not UGU.

2.2 Fluorescent Protein Reconstitution Method

In addition to the usage of PUM-HD, the use of a fluorescent protein reconstitution technique is an important point in the design of PUM-HD-based probes. In development of RNA probes, establishing a technique to detect target RNA molecules with high sensitivity and accuracy is an important requirement in addition to the specific labeling to the target RNA. In case of RNA probes, fluorescence from excess probes that is not binding to the target RNA possibly hampers to monitor the target RNA. Therefore, a technique to obtain fluorescence signals from the probes only after binding to the target RNAs is designed. The fluorescent protein reconstitution technique is a possible approach to diminish background fluorescence signal from probe molecules that do not attach on the target RNA (Ozawa et al., 2013; Yoshimura & Ozawa, 2014).

Split fluorescent protein reconstitution approach is based on the reaction between fragments of a fluorescent protein to recover original fluorescence, the reaction of which occurs when the two fragments come close to each other (Magliery et al., 2005; Ozawa, Nogami, Sato, Ohya, & Umezawa, 2000; Yoshimura & Ozawa, 2014). This approach is generally used to detect protein–protein interactions and localization changes of a protein of interest. In this approach, fluorescent protein is dissected at a particular site into N- and C-terminal fragments, which are fused to proteins A and B, respectively. When the protein A interacts with the protein B, the two fragments of the split fluorescent protein come up to each other to fold into the native ternary structure, leading to the recovery of fluorescence. The reconstruction of a fluorescent protein is an irreversible process, and the minimal affinity between the proteins A and B required for the reconstruction is 500 μM to 1 mM. The establishment of the split GFP reconstitution technique has led to the development of different dissection sites and application of fluorescent proteins with different fluorescence properties.

Although the approach on reconstitution of fluorescent protein fragments eliminates background fluorescence and improves the signal-to-background ratio in the obtained images, this approach has some limitations. A major limitation is reaction time for reconstitution. The reconstitution reaction of the pair of EGFP fragments, from coming close to emitting fluorescence, takes tens of minutes, which might prevent precise analysis of dynamics of the target molecules before completing the reconstitution. Another limitation is the requirement to induce two expression plasmids that code two probe molecules into the sample cells. Different expression levels of the probes in respective cells possibly cause different fluorescence intensities even in the same cell line. Balanced expression of the two probes is not controllable in individual cells by using general techniques of plasmid transfection methods. This situation may mislead to wrong results especially in a single cell analysis. A possible method to avoid these limitations is to use a full-length fluorescent protein, because it produces fluorescence immediately after the maturation of the protein (Yoshimura et al., 2012). In addition, only a single expression vector is necessary to generate the probe in target cells. Of course, excess probe molecules based on full-length fluorescent protein cause fatal background fluorescence that critically inhibits RNA visualization. Therefore, use of full-length fluorescent protein-based probe requires more strict control of expression level of the probe. Thus, though there is an option to use full-length fluorescent protein-based probes, split fluorescent protein reconstitution approach is the first choice to be adopted in the RNA probe designs.

3. DEVELOPMENT OF PUM-HD-BASED PROBES
3.1 Design Principle
3.1.1 Design of Probe Construction

Based on the principle of selective RNA recognition by PUM-HD mutants and split fluorescent protein reconstitution technology, we generated a structural backbone of an RNA probe to visualize a target RNA in living cells (Ozawa et al., 2007; Yamada et al., 2011). The probe consists of two subunits: one is a fusion protein of the N-terminal fluorescent protein fragment and a PUM-HD mutant that binds to a particular 8-base sequence in the target RNA. The other is composed of the C-terminal fluorescent protein fragment and another PUM-HD mutant to recognize a different 8-base portion in the target RNA. Upon binding the probe pair to a target RNA molecule at a time, the fragments of fluorescent protein come close to each other, inducing the reconstitution reaction to recover its fluorescence. In this design, the probe totally recognizes a 16-base RNA sequence. The number of RNA sequence patterns that the probe can distinguish is 4^{16} (about 4.3×10^9), which is larger than the number of genes. Therefore, recognizing a specific 16-base RNA sequence theoretically achieves sufficient selectivity to detect the target RNA selectively. In case of our previous β-actin visualization, the target sites of the two PUM-HD mutants locate in the 3' UTR, so that binding of the probe molecules to the RNA did not inhibit translation of the β-actin gene. In addition, fluorescence of the probe is recovered only the case in which both subunits of the probe bind on a target RNA molecule at a time. Therefore, errors from the probe based on nonspecific binding to other RNAs are controlled enough to be negligible.

3.1.2 Design of PUM-HD Mutant to Recognize the Target RNA Sequence

The following rules are available to engineer the PUM-HD specificity to a particular RNA sequence. As mentioned earlier, PUM-HD consists of eight tandem repeats, named R1 to R8, that run antiparallel to the recognized RNA; namely, nucleotides from $5'-U_1$ to A_8-3' are recognized individually by the repeats from R8 to R1, respectively. Each repeat structure has three helices. The third, fourth, and seventh amino acids in the second helix interact with an RNA base to recognize. Ser or Cys at the third amino acid, and Gln at the seventh one are necessary for recognition of adenine, Asn at the third and Gln at the seventh for uracil, and Ser at the third or Glu at the seventh for guanine (Cheong & Hall, 2006). Although no motifs recognize a cytosine base selectively, Filipovska et al. found that a mutated motif with Gly at the third position and Arg at the seventh position binds selectively to a cytosine (Filipovska et al., 2011). Based on this principle, mutant PUM-HDs that recognize a particular 8-base RNA sequence can be designed (Fig. 1D).

Contribution of each repeat to the affinity of PUM-HD to the target RNA sequence is not equivalent. The initial three bases of an RNA ($5'-U_1G_2U_3-3'$ for wild-type PUM-HD) strongly affect the total affinity of PUM-HD mutant to the RNA: PUM-HD mutant that recognizes RNA sequence without $5'-U_1G_2U_3-3'$ has generally about two-digit lower affinity than those to recognize RNA with $5'-U_1G_2U_3-3'$. The R4 in the wild-type PUM-HD recognizes purine base but its recognition is not strict. The following sequence, $5'-U_5A_6U_7A_8-3'$, is relatively variable. In case of designing a β-actin mRNA probe, we introduced two amino acid substitutions (C935E, Q939E) into the wild-type PUM-HD in order to generate a mutant capable of specifically recognizing the sequence of 5'-UGUACGUA-3'. To generate the other mutant PUM-HD addressing the sequence of UGUGCUGU, we introduced seven amino acid substitutions (S863N, N899S, Q903E, C935E, Q939E, C1007S, Q1011E). These PUM-HD mutants, named mPUM3 and mPUM4, respectively, were connected with N- and C-terminal fragments of EGFP, named GN-mPUM3 and mPUM4-GC.

3.2 Materials
3.2.1 Cell Culture and Transfection
1. Dulbecco's modified Eagle's medium (DMEM, Gibco, CA) supplemented with 10% fetal bovine serum (FBS; Gibco).
2. Trypsin solution (0.025%) (Gibco).
3. OPTI-MEM (Gibco).
4. Lipofectamine LTX (Invitrogen, CA).
5. Phosphate-buffered saline (PBS). Autoclave the before storage at room temperature.

3.2.2 Immunoprecipitation and Reverse Transcription PCR Analysis
1. Recombinant RNase inhibitor (Takara, Japan)
2. Anti-GFP antibody (Roche Diagnostics, Basel, Switzerland)
3. Protein G sepharose (GE Healthcare, U.K.)
4. TRIzol (Invitrogen)
5. Lysis buffer (150 mM NaCl, 5 mM EDTA, 50 mM NaF, 0.5% NP-40, 10 mM Tris–HCl, pH 7.4)
6. Superscript III (Invitrogen)

3.2.3 Fluorescence Imaging with TIRF Microscope System
1. Hanks' balanced salt solution (HBSS, Gibco).
2. Inverted fluorescence microscope, IX81 (Olympus, Japan), equipped with a 100 × 1.49 NA oil immersion objective, and a homebuilt excitation optical system (see Section 4).
3. EM-CCD camera (ImagEM, Hamamatsu, Japan).
4. Aquacosmos software.

3.3 Methods
3.3.1 Immunoprecipitation and Reverse Transcription PCR
1. Seed cells in 6 cm culture dishes with DMEM including 10% FBS and culture the cells up to 80–90% confluence.
2. Transfect the probe expression plasmids into the cells using Lipofectamine LTX.
3. Incubate the cells at 37°C for 2 days in an atmosphere of 5% CO_2.
4. Wash the cells twice with PBS and collect with a scraper.

5. Centrifuge the cell suspension at 4000 rpm for 1 min and keep the cell precipitate.
6. Add 400 μL of the lysis buffer to the cells and rotated for 1 h at 4°C.
7. Centrifuge the lysate at 15,000 × g for 5 min and keep the supernatant.
8. Add 1 μL of the recombinant RNase inhibitor to the supernatant and then add 300 μL of the lysis buffer.
9. Add anti-GFP antibody into the solution.
10. Incubate the solution for 2 h at 4°C on a rotator.
11. Mix 30 μL of Protein G sepharose beads to the lysate to adsorb the immune precipitates.
12. Rotate the solution for 1 h at 4°C.
13. Wash the beads three times with the lysis buffer.
14. Add 200 μL of TRIzol reagent and 40 μL of chloroform.
15. Centrifuge the mixture at 15,000 × g for 15 min and keep the water layer.
16. Conduct RNA precipitation with isopropyl alcohol and 70% ethanol.
17. Convert the RNA into cDNA using a Superscript III and Oligo dT.
18. Run PCR using the cDNA as a template.

3.3.2 Cell Preparation for Fluorescence Imaging
1. Seed cells onto a 3.5 cm glass-bottom dish 24 h before transfection and incubate the cells at 37°C in 5% CO_2 atmosphere.
2. Mix the probe expression plasmids.
3. Transfect the mixed plasmids using Lipofectamine LTX as following the manufacture's protocol.
4. Incubate the cells for 10 h.
5. Rinse the cells twice with HBSS before the observation.

4. MICROSCOPY SETUP AND VISUALIZATION OF SINGLE-MOLECULE RNA IN LIVING CELLS

The RNA probes based on PUM-HD and a fluorescent protein reconstitution method can be used for live-cell RNA imaging on not only conventional fluorescence microscopy but also real-time single-molecule fluorescence imaging. In this section, we show a microscope setup that was used in single-molecule RNA imaging with the PUM-HD-based probe.

4.1 Entire Design of the Optics of the Microscope System

Microscope systems for single-molecule imaging are generally constructed based on total internalization reflection fluorescence (TIRF) microscopy technique (Kusumi, Tsunoyama, Hirosawa, Kasai, & Fujiwara, 2014; Pitchiaya, Heinicke, Custer, & Walter, 2014; Tokunaga, Kitamura, Saito, Iwane, & Yanagida, 1997; Toomre & Bewersdorf, 2010). Our previous single-molecule RNA imaging studies were also performed on a TIRF microscope-based system. Our system equips homebuilt excitation optics to induce laser lights into the microscope and to illuminate the sample, and detection system in which fluorescence lights from different colored fluorophores are split and lead into respective EM-CCD cameras. Using this system, simultaneous dual color fluorescence observation of the target RNA and a related protein labeled with another fluorescent tag can be performed.

The most important apparatus for single-molecule imaging is the excitation optics. Because the single-molecule fluorescence is very weak, the excitation light power has to be stronger than that in the conventional epi fluorescence microscopy using a mercury lamp. Laser light is often used as the light source for single fluorescence excitation. In addition, the incidence angle of the excitation laser light to the sample has to be modulated. The illumination of TIRF microscopy, however, excites fluorophore in a very thin region with the depth of about 150 nm from the coverslip. In this excitation condition, most of the intracellular molecules that do not localize on the plasma membrane, including RNAs, are not illuminated and therefore are invisible through complete TIRF microscopy. Highly inclined illumination, which has large incidence angle but not in total internal reflection condition, is useful to observe intracellular molecules with low background signal (Tokunaga, Imamoto, & Sakata-Sogawa, 2008).

4.2 The Detail Design of Excitation Optical System

The excitation optical setup of our microscope system for observation of single-molecule RNA consists of laser apparatuses, mechanical shutters, beam expanders, neutral density (ND) filters, a laser combiner, a full mirror, and a collector lens (Fig. 2A). The lasers of 488 nm for GFP and 561 nm for red fluorescent proteins (RFPs) or other red fluorescence dyes are put on a vibration removal table, and the mechanical shutters are placed just in front of the laser to switch the laser injection to the optical setup. The beam expander is composed of two convex lenses with the focus distance of 10 and 200 mm, respectively, so that the diameter of the laser light magnified

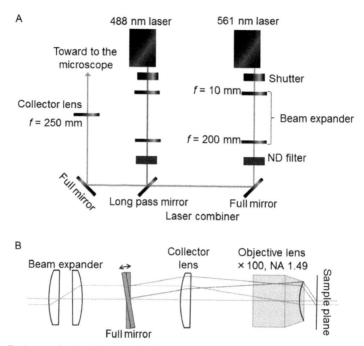

Fig. 2 Design and principle of optical setup in the excitation light path system. (A) The arrangement of optical components in the excitation system. (B) Principle to generate inclined illumination to the sample plane. The *blue* (*light gray* in the print version) *lines* indicate the light path on the optical axis. The *red* (*gray* in the print version) ones represent that off but parallel to the optical axis, generating inclined illumination on the sample plane.

20 times. To modulate the intensity, the laser lights pass through ND filters and combined by a dichroic mirror device that is used as a laser combiner. The laser light is reflected on a full mirror and induced to the collection lens. The position arrangement of these materials is important for generating and fine-tuning TIRF or highly inclined illumination. The collector lens, of which focus distance in our setup is 250 mm, is placed at 250 mm away from the back-focal plane of the objective lens on the microscope. Given that the laser light is originally a parallel light, the light path through the collector lens to be focused on the back-focal plane of the objective lens. Then the light injected from the objective lens as a parallel light. The full mirror is set at the distance of 250 mm from the collector lens, the same as the focus distance of this lens. In this design, modulation of the angle of this full mirror makes the laser light after the collector lens parallel to but off the optical axis. Then the light is injected to the objective lens and refracted to make TIRF or highly inclined illumination (Fig. 2B).

In addition of the design to make TIRF and inclined illuminations, the setup to illuminate the whole field to be observed is also important. This issue is related to laser diameter, the focus distances of expander lenses, collector lens, objective lens, and the size of the CCD device in the camera. The diagonal length of the CCD device in a typical EM-CCD camera (512 pixel × 512 pixel, such as ImagEM from Hamamatsu and iXon from Andor) is about 11.6 mm. In an observation with a microscope of 200 magnifications, a 58 μm diameter region in the sample is imaged by this CCD device, and therefore the excitation light has to illuminate whole of this region; otherwise the edges of the view are lost in the obtained image. The laser light diameter is 0.7 mm in many laser apparatuses that suit for microscope in bioscience. In our setup, the expander is a combination of two convex lenses with the focus distance of 10 and 200 mm, respectively. Therefore, the laser diameter is expanded to be 14 mm. Then the expanded laser light is focused by the collector lens and injected by objective lens. The diameter of injected light is determined by the ratio of the focus distances of the collector and objective lenses. The focus distance of the objective lens is different on the manufacturers even if the magnification is the same, but the distance is generally a few millimeter in case of 100 magnification lenses. If the focus distances are 2 and 250 mm on the objective and the collector lenses, the laser light diameter of 14 mm gets to be 112 μm after injection from the objective lens, respectively, which is enough to illuminate the whole region of the observation with 200 magnifications.

4.3 Methods

1. Turn on the excitation lasers.
2. Modulate the angle of the full mirror to have the laser light outgo vertically from the objective.
3. Modulate the distance between the beam expander lenses to be the summation of the focal distances of the two lenses, and the expanded laser light become a parallel beam.
4. Place a sample on the microscope stage.
5. Focus initially at the surface of the coverslip.
6. Tilt the incident angle of the excitation laser light to make TIRF or oblique illumination.
7. Find a cell expressing the probes at appropriate level.
8. Modulate the focus to an appropriate position for observation of the target RNA.
9. Capture images or movies of fluorescent-labeled RNA in living cells.

5. EXAMPLE OF RNA VISUALIZATION BY USING PUM-HD-BASED RNA PROBES

5.1 Monitoring Localization of Mitochondrial mRNA, ND6 mRNA

The first trial to use a PUM-HD-based RNA probe was visualization of NADH dehydrogenase 6 (ND6) mRNA that is transcribed from the mitochondrial genome (Ozawa et al., 2007). To label ND6 mRNA, we selected two target sites (5′-UGAUGGUU-3′ and 5′-UGAUAUA-3′) of 8-base sequences in ND6 mRNA and prepared two PUM-HD mutants (mPUM1 and mPUM2) that recognize the target sites in the ND6 mRNA, respectively. The dissociation constants of mPUM1 and mPUM2 to the respective target sequence were 163 and 92 nM, which is sufficient to label ND6 mRNA. Then, mPUM1 and mPUM2 were fused to N- and C-terminal fragments of EGFP or a yellow fluorescent protein Venus, respectively. These probe subunits were additionally connected to three repeats of mitochondrial localization signal sequences to make them localize in mitochondria.

In the observation of cultured cells expressing the probe, fluorescence from the probe was detected especially on double-strand DNA in mitochondria. On the other hand, a probe variant that lack mPUM1 or mPUM2 did not emit fluorescence. This result indicates that EGFP reconstitution and emission of fluorescence occurred only when both probe molecules have an ability to bind ND6 mRNA by the function of mPUM1 and mPUM2 to have the split EGFP fragments come close to each other. Next, we performed fluorescence recovery after photobleaching (FRAP) experiment to analyze the dynamics of ND6 mRNA in living cells. The fluorescence of the probe in a region of 4 μm diameter in a mitochondrion of a probe-expressing cell was bleached by irradiation of 490 nm light, and recovery of fluorescence was monitored for 30 min. The fluorescence was recovered to be about 20% intensity of that before the bleaching in the case of the ND6 mRNA probe, whereas EGFP as a negative control recovered the fluorescence to be 85% intensity of that before the bleaching (Fig. 3). This result indicates that the ND6 mRNA was static and did not diffuse freely in mitochondria.

Intracellular DNA is reportedly decomposed by oxidation stresses. We observed mitochondrial DNA and ND6 mRNA in cultured cells treated with hydrogen peroxide. As was the FRAP experiment described earlier,

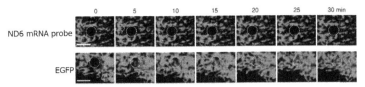

Fig. 3 FRAP experiment of the ND6 mRNA probe and EGFP in mitochondria. The fluorescence of the ND6 mRNA probe in the bleached region did not fully recover, whereas that of EGFP almost recovered within 30 min. This result indicates restriction of ND6 mRNA in mitochondria. The *circles* of *white dotted lines* indicate the photobleaching region with the diameter of 4 μm. Scale bar: 5 μm. (See the color plate.)

the ND6 mRNA probe was bleached by photoirradiation, and the cells were treated with hydrogen peroxide immediately after the photobleaching. Then the recovery of fluorescence of the ND6 mRNA probe and the fluorescence of a DNA marker DAPI was monitored. In this observation, we found a decrease in DAPI fluorescence gradually and recovering the fluorescence of ND6 mRNA probe in the photobleaching region. This result implies that degradation of mitochondrial DNA by hydrogen peroxide treatment released ND6 mRNA to be able to diffuse in mitochondria.

5.2 Monitoring Localization and Dynamics of Single β-Actin mRNA Molecule in Living Cells

Using the PUM-HD-based probe, we also performed single-molecule live-cell observation of β-actin mRNA, of which information on localization and dynamics had been accumulated in previous reports (Yamada et al., 2011). First, we designed PUM-HD mutants that recognize two 8-base portions of β-actin mRNA 3' UTR. The selected sequences are 5'-UGUACGUA-3' and 5'-UGUGCUGU-3', and based on these sequences, we prepared mutants of PUM-HD with amino acid substitution of Q939E and C935S to recognize 5'-UGUACGUA-3' and Q1011E, C1007S, Q939E, C935N, Q903E, N889S, and S863N to recognize 5'-UGUGCUGU-3'. The former was fused with the N-terminal fragment of EGFP and the latter was connected with the C-terminal one.

This probe was expressed in NIH3T3 cells and observed using a TIRF microscope equipped with 488 nm laser for excitation of the sample. Because β-actin mRNA is a cytosolic RNA, the incidence angle of the excitation light was modulated to be smaller than that to generate complete TIRF illumination. Detection of single-molecule β-actin mRNA was performed with an EM-CCD camera (ImagEM, Hamamatsu) at the frame rate of 30 Hz. In this observation, fluorescent spots that showed diffusion motion in the cytosol were visualized. These fluorescent spots bleached in single-step manner and their intensity distribution was approximated with a single Gaussian function, indicating that these spots represent single-molecule-reconstituted EGFP. The fluorescent spots of the probe were overlapped with those of β-actin mRNA stained with an in situ hybridization method in chemically fixed cells. These results indicate that the probe precisely labels the target β-actin mRNA and makes it visualized in single-molecule sensitivity under a fluorescence microscopy.

Next, the ability of the probe to monitor the single-molecule dynamics of β-actin mRNA in living cells was estimated. NIH3T3 cells were cultured in a medium containing no serum to make the cells sensitive to external signals and then were subjected to single-molecule fluorescence observation after stimulation with serum. In this result, the distribution of β-actin mRNA was homogeneous before the stimulation, whereas after the stimulation β-actin mRNA localized in a cell peripheral region, which exhibited expansion with making a lamellipodium (Fig. 4A). Furthermore, we performed simultaneous observation of β-actin mRNA and microtubules that were labeled with RFPs. In this observation, a dynamics of β-actin mRNA moving linearly along a microtubule was visualized (Fig. 4B). The velocity of the motion was 1.78 μm/s in average, which is comparable with that of motor proteins moving on microtubules. Such dynamics of β-actin mRNA is consistent with previous reports in which this mRNA is transported to particular intracellular regions after extracellular stimulation is induced to the cell. These results demonstrated that the RNA probe based on RNA recognition by PUM-HD and a split EGFP reconstitution technique are applicable to single-molecule analysis of the target RNA dynamics in living cells.

6. CONCLUSION

Live-cell RNA imaging is a desirable technique to understand the mechanism of physiological phenomena in which RNA is implicated. An approach using PUM-HD mutants and the split fluorescent protein complementation method are promising candidates to visualize RNA in living cells. The advantage of this methodology is especially present in design flexibility to match the target RNA sequence. PUM-HD can be modulated to recognize a particular 8-base RNA sequence in the RNA of interest in a tailor-made manner. This property of PUM-HD makes it possible to label

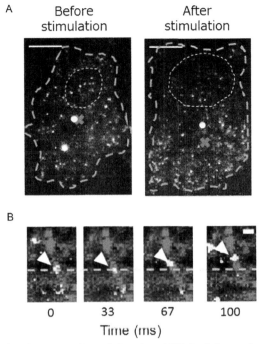

Fig. 4 Single-molecule monitoring of β-actin mRNA in living cells. (A) Localization alteration of β-actin mRNA upon serum stimulation. The *blue* and *white broken lines* represent the edges of the cell and the nucleus, respectively. The *yellow spots* indicate the centroid of the cell, and the *red crosses* show the median points of all fluorescent spots of β-actin mRNA probes. Scale bar: 8 μm. (B) Directed motion of a single β-actin mRNA molecule along a microtubule. The *green* and *red signals* represent β-actin mRNA probes and microtubules, respectively. The β-actin mRNA indicated by the *white arrow head* moved directionally along a microtubule. The *blue broken line* indicates the initial position of the spot. Scale bar: 800 nm. (See the color plate.)

intact endogenous RNAs of interest in living cells. Also, detectability based on reconstituted fluorescent proteins is another strong point in this probe design. Fluorescent protein reconstitution technique reduces background fluorescence drastically and realizes single-molecule detection by using TIRF or inclined illumination fluorescence microscopy. This technology based on PUM-HD mutants and split fluorescent protein reconstitution was applied also for live-cell imaging of the genomic RNA of tobacco mosaic virus in plant cells in addition to our researches on ND6 mRNA and β-actin mRNA (Tilsner et al., 2009). Thus, this methodology is expected to become a general approach to monitor localization and dynamics of RNAs of interest in living cells. Single-molecule fluorescent imaging requires some special microscope apparatuses and techniques, but the principle is based on the theory of conventional fluorescent microscopy. Therefore, constructing a precise excitation optical system and adopting high-sensitivity cameras allow us to monitor the single-molecule dynamics of RNA in living samples. The total design of this experimental system would provide a substantial progress in the field of RNA research.

ACKNOWLEDGMENT
This work was supported in part by Japan Society for the Promotion of Science.

REFERENCES
Bertrand, E., Chartrand, P., Schaefer, M., Shenoy, S. M., Singer, R. H., & Long, R. M. (1998). Localization of ASH1 mRNA particles in living yeast. *Molecular Cell, 2*, 437–445.

Bratu, D. P., Cha, B. J., Mhlanga, M. M., Kramer, F. R., & Tyagi, S. (2003). Visualizing the distribution and transport of mRNAs in living cells. *Proceedings of the National Academy of Sciences of the United States of America, 100*, 13308–13313.

Cheong, C. G., & Hall, T. M. (2006). Engineering RNA sequence specificity of Pumilio repeats. *Proceedings of the National Academy of Sciences of the United States of America, 103*, 13635–13639.

Darzacq, X., Shav-Tal, Y., de Turris, V., Brody, Y., Shenoy, S. M., Phair, R. D., et al. (2007). In vivo dynamics of RNA polymerase II transcription. *Nature Structural & Molecular Biology, 14*, 796–806.

Dictenberg, J. (2012). Genetic encoding of fluorescent RNA ensures a bright future for visualizing nucleic acid dynamics. *Trends in Biotechnology, 30*, 621–626.

Farina, K. L., Huttelmaier, S., Musunuru, K., Darnell, R., & Singer, R. H. (2003). Two ZBP1 KH domains facilitate beta-actin mRNA localization, granule formation, and cytoskeletal attachment. *The Journal of Cell Biology, 160*, 77–87.

Filipovska, A., Razif, M. F., Nygard, K. K., & Rackham, O. (2011). A universal code for RNA recognition by PUF proteins. *Nature Chemical Biology, 7*, 425–427.

Kusumi, A., Tsunoyama, T. A., Hirosawa, K. M., Kasai, R. S., & Fujiwara, T. K. (2014). Tracking single molecules at work in living cells. *Nature Chemical Biology, 10*, 524–532.

Lin, A. C., & Holt, C. E. (2007). Local translation and directional steering in axons. *The EMBO Journal, 26*, 3729–3736.

Little, S. C., Sinsimer, K. S., Lee, J. J., Wieschaus, E. F., & Gavis, E. R. (2015). Independent and coordinate trafficking of single Drosophila germ plasm mRNAs. *Nature Cell Biology, 17*, 558–568.

Magliery, T. J., Wilson, C. G., Pan, W., Mishler, D., Ghosh, I., Hamilton, A. D., et al. (2005). Detecting protein-protein interactions with a green fluorescent protein fragment reassembly trap: Scope and mechanism. *Journal of the American Chemical Society, 127*, 146–157.

Martin, K. C., & Ephrussi, A. (2009). mRNA localization: Gene expression in the spatial dimension. *Cell, 136*, 719–730.

Ozawa, T., Natori, Y., Sato, M., & Umezawa, Y. (2007). Imaging dynamics of endogenous mitochondrial RNA in single living cells. *Nature Methods, 4*, 413–419.

Ozawa, T., Nogami, S., Sato, M., Ohya, Y., & Umezawa, Y. (2000). A fluorescent indicator for detecting protein-protein interactions in vivo based on protein splicing. *Analytical Chemistry, 72*, 5151–5157.

Ozawa, T., Yoshimura, H., & Kim, S. B. (2013). Advances in fluorescence and bioluminescence imaging. *Analytical Chemistry*, *85*, 590–609.

Pitchiaya, S., Heinicke, L. A., Custer, T. C., & Walter, N. G. (2014). Single molecule fluorescence approaches shed light on intracellular RNAs. *Chemical Reviews*, *114*, 3224–3265.

Quinodoz, S., & Guttman, M. (2014). Long noncoding RNAs: An emerging link between gene regulation and nuclear organization. *Trends in Cell Biology*, *24*, 651–663.

Raj, A., van den Bogaard, P., Rifkin, S. A., van Oudenaarden, A., & Tyagi, S. (2008). Imaging individual mRNA molecules using multiple singly labeled probes. *Nature Methods*, *5*, 877–879.

Shav-Tal, Y., Darzacq, X., Shenoy, S. M., Fusco, D., Janicki, S. M., Spector, D. L., et al. (2004). Dynamics of single mRNPs in nuclei of living cells. *Science*, *304*, 1797–1800.

Tilsner, J., Linnik, O., Christensen, N. M., Bell, K., Roberts, I. M., Lacomme, C., et al. (2009). Live-cell imaging of viral RNA genomes using a Pumilio-based reporter. *The Plant Journal*, *57*, 758–770.

Tiruchinapalli, D. M., Oleynikov, Y., Kelic, S., Shenoy, S. M., Hartley, A., Stanton, P. K., et al. (2003). Activity-dependent trafficking and dynamic localization of zipcode binding protein 1 and beta-actin mRNA in dendrites and spines of hippocampal neurons. *The Journal of Neuroscience*, *23*, 3251–3261.

Tokunaga, M., Imamoto, N., & Sakata-Sogawa, K. (2008). Highly inclined thin illumination enables clear single-molecule imaging in cells. *Nature Methods*, *5*, 159–161.

Tokunaga, M., Kitamura, K., Saito, K., Iwane, A. H., & Yanagida, T. (1997). Single molecule imaging of fluorophores and enzymatic reactions achieved by objective-type total internal reflection fluorescence microscopy. *Biochemical and Biophysical Research Communications*, *235*, 47–53.

Toomre, D., & Bewersdorf, J. (2010). A new wave of cellular imaging. *Annual Review of Cell and Developmental Biology*, *26*, 285–314.

Tyagi, S. (2009). Imaging intracellular RNA distribution and dynamics in living cells. *Nature Methods*, *6*, 331–338.

Valencia-Burton, M., McCullough, R. M., Cantor, C. R., & Broude, N. E. (2007). RNA visualization in live bacterial cells using fluorescent protein complementation. *Nature Methods*, *4*, 421–427.

Vardy, L., & Orr-Weaver, T. L. (2007). Regulating translation of maternal messages: Multiple repression mechanisms. *Trends in Cell Biology*, *17*, 547–554.

Vargas, D. Y., Shah, K., Batish, M., Levandoski, M., Sinha, S., Marras, S. A., et al. (2011). Single-molecule imaging of transcriptionally coupled and uncoupled splicing. *Cell*, *147*, 1054–1065.

Wang, X., McLachlan, J., Zamore, P. D., & Hall, T. M. (2002). Modular recognition of RNA by a human pumilio-homology domain. *Cell*, *110*, 501–512.

Weil, T. T., Parton, R. M., & Davis, I. (2010). Making the message clear: Visualizing mRNA localization. *Trends in Cell Biology*, *20*, 380–390.

Yamada, T., Yoshimura, H., Inaguma, A., & Ozawa, T. (2011). Visualization of nonengineered single mRNAs in living cells using genetically encoded fluorescent probes. *Analytical Chemistry*, *83*, 5708–5714.

Yoshimura, H., Inaguma, A., Yamada, T., & Ozawa, T. (2012). Fluorescent probes for imaging endogenous beta-actin mRNA in living cells using fluorescent protein-tagged pumilio. *ACS Chemical Biology*, *7*, 999–1005.

Yoshimura, H., & Ozawa, T. (2014). Methods of split reporter reconstitution for the analysis of biomolecules. *Chemical Record*, *14*, 492–501.

CHAPTER FOUR

Applications of Hairpin DNA-Functionalized Gold Nanoparticles for Imaging mRNA in Living Cells

S.R. Jackson*, A.C. Wong*, A.R. Travis*, I.E. Catrina[†], D.P. Bratu[†,‡], D.W. Wright*, A. Jayagopal[§,1]

*Vanderbilt University, Nashville, TN, United States
[†]Hunter College, City University of New York, New York, NY, United States
[‡]Program in Molecular, Cellular, and Developmental Biology, and Program in Biochemistry, The Graduate Center, City University of New York, New York, NY, United States
[§]Pharma Research and Early Development (pRED), F. Hoffmann-La Roche Ltd., Basel, Switzerland
[1]Corresponding author: e-mail address: ash.jayagopal@gmail.com

Contents

1. Introduction   88
   1.1 Mechanism of Hairpin DNA-Functionalized Gold Nanoparticles   88
   1.2 Optical Imaging of Matrix Metalloproteinases in Breast Cancer   89
   1.3 Strategy for hAuNP-Guided Imaging of MMP Subtypes   91
2. Protocol for hAuNP-Guided Imaging of mRNA in Living Cells   92
   2.1 Materials and Instrumentation   92
   2.2 DNA Hairpin and Oligonucleotide Synthesis   93
   2.3 Synthesis and Characterization of hAuNP   94
   2.4 Cell Culture Studies   95
   2.5 Flow Cytometric Analysis   95
   2.6 Confocal Microscopy   95
3. Key Results   96
   3.1 Design of Hairpin Sequence   96
   3.2 Characterization of hAuNPs   96
   3.3 Flow Cytometric Analysis of hAuNP Uptake in Live Breast Cancer Cells   97
   3.4 Confocal Microscopy of hAuNP in Live Breast Cancer Cell Lines   99
4. Conclusions   99
Acknowledgments   101
References   101

Abstract

Molecular imaging agents are useful for imaging molecular processes in living systems in order to elucidate the function of molecular mediators in health and disease. Here, we demonstrate a technique for the synthesis, characterization, and application of hairpin DNA-functionalized gold nanoparticles (hAuNPs) as fluorescent hybridization probes for imaging mRNA expression and spatiotemporal dynamics in living cells. These imaging probes feature gold colloids linked to fluorophores via engineered oligonucleotides to resemble a molecular beacon in which the gold colloid serves as the fluorescence quencher in a fluorescence resonance energy transfer system. Target-specific hybridization of the hairpin oligonucleotide enables fluorescence de-quenching and subsequent emission with high signal to noise ratios. hAuNPs exhibit high specificity without adverse toxicity or the need for transfection reagents. Furthermore, tunability of hAuNP emission profiles by selection of spectrally distinct fluorophores enables multiplexed mRNA imaging applications. Therefore, hAuNPs are promising tools for imaging gene expression in living cells. As a representative application of this technology, we discuss the design and applications of hAuNP targeted against distinct matrix metalloproteinase enzymes for the multiplexed detection of mRNA expression in live breast cancer cells using flow cytometry and fluorescence microscopy.

1. INTRODUCTION

1.1 Mechanism of Hairpin DNA-Functionalized Gold Nanoparticles

Hairpin DNA-functionalized gold nanoparticles (hAuNPs) are constructed of a 15 nm core gold nanoparticle with covalently attached DNA hairpin beacons via gold–thiol bond formation (Jayagopal, Halfpenny, Perez, & Wright, 2010). hAuNPs are taken up by cells nonspecifically and emit fluorescence upon hybridization with an intracellular mRNA target. Each DNA hairpin is constructed of a 5′ hexane thiol linker covalently attached to a 10-thymine extension, followed by a ∼30 base hairpin recognition sequence and a 3′ fluorophore (Fig. 1). The first and last five bases of the recognition sequence are complementary resulting in a five-base duplex stem. While in the closed position, this construct utilizes the fluorescence quenching ability of the gold nanoparticle to quench the fluorescent signal, within the cytoplasm, binding of the complementary mRNA sequence to the hairpin DNA (hDNA) causes the 3′ fluorophore to extend beyond the quenching range of the gold nanoparticle resulting in emission of a fluorescent signal. hAuNPs improve upon currently used technologies for studying RNA trafficking due to their efficient internalization within live cells without requiring transfection reagents, improved resistance to DNAse degradation, low cytotoxicity, and high specificity and sensitivity toward the target mRNA sequence (Jayagopal et al., 2010). hAuNPs are water soluble, unlike many molecular beacons which feature poorly soluble fluorescence quenchers like QSY-21, and tunability of hAuNP fluorescence emission from the visible to

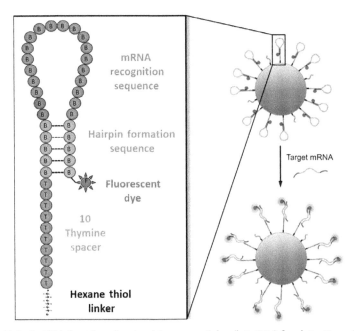

Fig. 1 Hairpin DNA-functionalized gold nanoparticles (hAuNPs) for detection of mRNA in living cells. Hairpin DNA is coupled to the gold colloid surface via a 5′ hexane thiol, thereby quenching fluorescence emission of the 3′ fluorescent dye. The hairpin formation sequences maintain the hairpin conformation through complementary hydrogen bonding. Hybridization between the mRNA recognition sequence and the target DNA or RNA sequence results in opening of the hairpin DNA and fluorescence emission.

near-infrared (NIR) wavelengths is possible attributable to the wide fluorescence quenching range of gold colloids (Shi et al., 2015; Xue et al., 2012). Gold nanoparticles, furthermore, can be readily made in the laboratory to the required specifications, without specialized equipment using published protocols (Chen, Huang, Zhao, Lu, & Tian, 2012; Feng et al., 2006).

1.2 Optical Imaging of Matrix Metalloproteinases in Breast Cancer

As a representative application of hAuNP technology, we describe in this report a protocol for assaying distinct matrix metalloproteinase (MMP) expression in breast cancer cell lines of varying aggressiveness. In the era of personalized medicine for cancer therapy, the ability to detect molecular expression and activity using highly sensitive techniques and distinguish tumor-to-tumor heterogeneity represents an urgent need. One of the defining characteristics of malignancy is the ability of cancer cells to metastasize and invade distant organ sites. Proteolysis of the extracellular matrix (ECM) is required to accommodate angiogenesis, invasion, and metastasis of tumor cells. The proteases responsible for degrading most of the components of the ECM are MMPs, a family of more than 20 extracellular, zinc-dependent endopeptidases (Sternlicht & Werb, 2001).

In the tumor microenvironment, host- and tumor-derived MMPs are often dysregulated leading to uncontrolled degradation of the ECM. For this reason, MMPs are potential biomarkers for elucidating disease type, disease activity, and response to therapy. The recognition of the role of MMPs in tumor growth, migration, and invasion has guided the development of methods to detect and image tumors *in vivo* including magnetic resonance imaging (Lepage et al., 2007; Schellenberger et al., 2008), single-photon emission computed tomography (Medina et al., 2005), and positron emission tomography (Sprague, Li, Liang, Achilefu, & Anderson, 2006). However, optical imaging has been among the most extensively validated techniques, particularly with activatable fluorescence imaging probes that rely on fluorescence resonance energy transfer (FRET) between fluorophores and spectra-matched quenchers (Yang, Hong, Zhang, & Cai, 2009; Zhu, Niu, Fang, & Chen, 2010).

Optical imaging is a noninvasive technique that uses light to probe tissue, making it a distinct alternative to other imaging modalities that require strong magnetic fields or radioactive tracers to image tissues (Weissleder & Pittet, 2008). Fluorophore-labeled substrates have been designed that are quenched due to the proximity of the fluorophores (self-quenching), or utilize fluorophore–quencher pairs to enable proteolytically based FRET detection of MMPs (McIntyre et al., 2004; McIntyre & Matrisian, 2003; Tsien, 2005). Optical detection and imaging of in vivo protease activity were first demonstrated by Weissleder et al. in mouse xenograft tumors using a linear poly-lysine–polyethylene glycol copolymer with cleavable sequences carrying a NIR fluorophore, Cy5.5 (Weissleder, Tung, Mahmood, & Bogdanov, 1999). Due to the proximity of the fluorophores, self-quenching prevented almost any detectable fluorescent signal in the nonactivated state. Upon proteolytic cleavage of the poly-lysine peptide linker, Cy5.5 molecules were released from the carrier, and a NIR fluorescence signal was optically detected. This construct was later modified to include a proteolytic beacon for MMP2 which was able to image HT-1080 human fibrosarcoma tumors in mice (Bremer, Bredow, Mahmood, Weissleder, & Tung, 2001; Bremer, Tung, & Weissleder, 2001). In addition, probes have been synthesized which are composed of activatable cell-penetrating peptides (ACPPs) containing an MMP recognition site (Jiang et al., 2004; Olson et al., 2009). These ACPPs consist of a fluorophore-tagged cationic peptide covalently attached to a short stretch of acidic residues via a cleavable linker. The acidic residues prevent cellular uptake of the cationic peptide, but when the ACPPs reach tumor surfaces expressing MMPs, the linker on ACPPs is cleaved, releasing the acidic inhibitory domain, and the cationic peptides are free to carry the fluorophore into cells.

These types of activatable probes have now become commercially available and are utilized by many research groups interested in imaging tumor-associated MMPs. However, the majority of these studies are based on proteolytic substrates, which are often cleaved by multiple MMPs, thus lacking the ability to assess and efficiently determine specific protease activity. In addition, MMPs are inhibited by various means, including binding to tissue inhibitors of metalloproteinases (Brew, Dinakarpandian, & Nagase, 2000), which would give inconsistent results if measuring gene expression through proteolytic activity. Therefore, additional types of MMP imaging probes are warranted to enable more accurate detection of MMP subtypes.

1.3 Strategy for hAuNP-Guided Imaging of MMP Subtypes

In order to enable specific imaging of MMP subtype expression in cancer cells, we describe in this report a method for developing hAuNP targeted toward specific MMP mRNA transcripts. hAuNPs are used here to image the expression of three specific MMPs in three breast cancer cells of varying invasiveness. By using hAuNPs to target an mRNA sequence instead of utilizing a proteolytic substrate, we are able to assess the expression levels of specific MMPs. Furthermore, by engineering the recognition sequence on the hairpin probe, we enable the detection of three different MMPs. Finally, the observed expression of specific MMP mRNA using hAuNPs is compared to previously published levels of mRNA detected by qRT-PCR to validate the use of hAuNPs as qualitative mRNA expression probes in living cells (Hegedus, Cho, Xie, & Eliceiri, 2008).

These types of hAuNP-guided imaging applications may allow for generation of phenotypic "fingerprints" for different types of cancers and their invasiveness, ultimately leading to more effective personalized therapies. The approach can be broadly applied to imaging mRNA and miRNA expression in a number of diverse cell types for studying the roles of specific RNA expression in health and disease, as well as the importance of RNA trafficking, as reported by coauthors (Bratu, 2003, 2006; Bratu, Catrina, & Marras, 2011; Catrina, Marras, & Bratu, 2012).

2. PROTOCOL FOR hAuNP-GUIDED IMAGING OF mRNA IN LIVING CELLS

2.1 Materials and Instrumentation

(1) DA-MB-231 human mammary adenocarcinoma cell line (ATCC #HT-26)
(2) MCF-7 human mammary adenocarcinoma cell line (ATCC #HT-22)
(3) MCF-10A human mammary fibrocystic disease cell line (ATCC #CRL-10317)
(4) High glucose DMEM media, 1% Penicillin–Streptomycin, and 0.25% Trypsin–EDTA (Life Technologies)
(5) Mammary Epithelial Basal/Growth Medium (MEBM and MEGM, Lonza Biosciences)
(6) 0.25% Trypsin–EDTA (Life Technologies Inc.)
(7) Chambered coverslip μ-Slides (ibidi GmbH)
(8) 15 nm gold nanoparticles (Ted Pella, Inc.)
(9) Custom synthesized oligonucleotides with 5′ thiol linker and 3′ fluorophore (Sigma Life Sciences; Tables 1 and 2)
(10) Zetasizer Nano ZS Dynamic Light Scattering System (Malvern Instruments)

Table 1 Hairpin DNA (hDNA) Sequences

Target	Sequence (5′–3′)
GAPDH	5′-(SH)-TTT TTT TTT TGC ACG **AGG TTT TTC TAG ACG GCA GG**C GTG C-(TAMRA)-3′
MMP13	5′-(SH)-TTT TTT TTT TGC AGC **TTG ACG CGA ACA ATA CGG TTA A**GC TGC-(Cy5)-3′
MMP14	5′-(SH)-TTT TTT TTT TGC AGC **GCT CTT CTC CTC TTT TCC GGT TTT** GCT GC-(Cy5)-3′
MMP26	5′(SH)-TTT TTT TTT TGC AGC **TAG TGT GCT TAT TCC ACT TGC A**GC TGC-(Cy5)-3′
MMP26 control (scrambled)	5′(SH)-TTT TTT TTT TGC AGC **GTT ATC TTA CTC GTT CGA CAG T**GC TGC-(Cy5)-3′

Table 2 Target Sequences Used for hAuNP Specificity and Sensitivity Studies

Complement	Sequence (5'-3')
GAPDH	5'-TTC CTG CCG TCT AGA AAA ACC TTT-3'
MMP13	5'-TTT TAA CCG TAT TGT TCG CGT CAA TT-3'
MMP14	5'-TTA AAA CCG GAA AAG AGG AGA AGA GCT T-3'
MMP26	5'-TTT GCA AGT GGA ATA AGC ACA CTA TT-3'
MMP26 scrambled	5'-TTA CTG TCG AAC GAG TAA GAT AAC TT-3'

(11) Nanodrop microvolume absorbance spectrophotometer ND-1000 (Thermo Fisher Scientific, Inc.)
(12) Biotek Synergy Mx Multimode Fluorescence/Absorbance Spectrophotometer with Take3 microvolume analysis plate (BioTek Instruments, Inc.)
(13) Dithiothreitol (DTT) (Sigma-Aldrich Inc.)
(14) BD LSR II Flow Cytometer (Becton-Dickinson Life Sciences Inc.)
(15) Zeiss LSM 710 META confocal microscope

2.2 DNA Hairpin and Oligonucleotide Synthesis

hDNA used for coupling to the AuNP surface was synthesized with the following sequences (Table 1): GAPDH, 5'-(SH)-TTT TTT TTT TGC ACG **AGG TTT TTC TAG ACG GCA GG**C GTG C-(TAMRA)-3'; MMP13, 5'-(SH)-TTT TTT TTT TGC AGC **TTG ACG CGA ACA ATA CGG TTA A**GC TGC-(Cy5)-3'; MMP14, 5'-(SH)-TTT TTT TTT TGC AGC **GCT CTT CTC CTC TTT TCC GGT TTT** GCT GC-(Cy5)-3'; MMP26, 5'(SH)-TTT TTT TTT TGC AGC **TAG TGT GCT TAT TCC ACT TGC A**GC TGC-(Cy5)-3'; and MMP26scram, 5'(SH)-TTT TTT TTT TGC AGC **GTT ATC TTA CTC GTT CGA CAG T**GC TGC-(Cy5)-3'. Bolded regions indicate sequences encoding specifically for target mRNAs (except for negative control MMP26scram, which has a scrambled, nonspecific hybridization region that has no target in the human transcriptome), and regions in parentheses indicate modifications of 5' thiol and 3' fluorophore conjugations.

Sequences for complementary targets for evaluation of hAuNP hybridization specificity were as follows (Table 2): GAPDHcomp, 5'-TTC CTG CCG TCT AGA AAA ACC TTT-3'; MMP13comp, 5'-TTT TAA CCG TAT TGT TCG CGT CAA TT-3'; MMP14comp, 5'-TTA AAA CCG GAA AAG AGG AGA AGA GCT T-3'; MMP26comp, 5'-TTT GCA

AGT GGA ATA AGC ACA CTA TT-3′; and MMP26scramcomp, 5′-TTA CTG TCG AAC GAG TAA GAT AAC TT-3′. A spacing ligand for coupling to the AuNP surface was T10, 5′-(SH)-TTT TTT TTT T-3′.

2.3 Synthesis and Characterization of hAuNP

The protocol for synthesis and characterization of hAuNP is described later:
(1) DNA concentration was verified by UV absorbance at 260 nm using a Nanodrop microvolume spectrophotometer.
(2) Prior to coupling of hDNA to AuNP, lyophilized DNA was resuspended in 100 nM DTT in 0.1 mM phosphate buffer (pH 8.3) and incubated for 1 h at room temperature (RT) to reduce residual 5′ disulfide bonds.
(3) A solution of 1 nM 15 nm AuNPs was prepared in nuclease-free water and adjusted to contain 0.1% Tween-20.
(4) T10 DNA was added to the solution to make 100 nM T10 DNA and incubated for 4 h at RT while protected from light.
(5) hDNA was added to the solution to give 300 nM hDNA and incubated for an additional 4 h.
(6) The solution was buffered to 10 mM phosphate buffer (pH 7.0) and adjusted to contain 0.1 M NaCl. The hAuNPs were then incubated for an additional 4 h at RT. The following steps involve gradually increasing salt concentration to stabilize hAuNP without precipitating the nanoparticles.
(7) The NaCl concentration was then adjusted to 0.2 M and the hAuNPs were incubated for an additional 4 h at RT.
(8) The NaCl concentration was finally adjusted to 0.3 M and incubated for 4 h at RT. The solution was then centrifuged at $21,100 \times g$ for 20 min and the hAuNPs were resuspended in PBS.
(9) The process of centrifugation and resuspension was repeated two more times.
(10) hAuNP concentration was measured by absorbance at 520 nm, using the extinction coefficient for 15 nm AuNP $\varepsilon_{520} = 3.64 \times 10^8\ M^{-1}\ cm^{-1}$.
(11) To determine sensitivity and specificity of hAuNPs to synthetic targets, increasing concentrations of complementary or mismatch target DNA were added to 1 nM hAuNP solutions in PBS and measured for fluorescence emission intensity using a BioTek Synergy Mx microplate reader.

(12) For measurement of hydrodynamic diameter, 0.1 nM solutions of AuNPs or hAuNPs were resuspended in nuclease-free PBS and measured using dynamic light scattering (DLS) on a Malvern Nano ZS.

2.4 Cell Culture Studies

All cells were cultured as monolayers in a humidified atmosphere of 95% air and 5% CO_2 at 37°C. MCF-7 and MDA-MB-231 cells were maintained in high glucose DMEM with L-glutamine supplemented with 10% FBS and 1% Penicillin/Streptomycin MCF-10A cells were maintained in MEBM supplemented with 100 ng/mL cholera toxin and Lonza CC-4136 but with 1% final Penicillin/Streptomycin rather than the supplied GA-1000 aliquot.

MCF-10A cells are a fibrocystic benign breast cell line which exhibit little invasiveness in culture or *in vivo*. MCF-7 cells are an invasive breast carcinoma cell line, and MDA-MB-231 are considered to be highly invasive breast carcinoma cell lines. Analysis of hAuNP in these three cell lines therefore allowed for detecting relative differences in MMP mRNA expression according to aggressiveness of the cell type.

2.5 Flow Cytometric Analysis

For each experiment, cells at 80–90% confluence were incubated with 0.5 nM hAuNPs for 4 h, rinsed three times with Ca^{2+} and Mg^{2+} free DPBS to remove unbound and noninternalized hAuNPs, and detached using trypsin. Cells were assayed using a BD LSR II flow cytometer. All experiments consisted of $\geq$10,000 events.

2.6 Confocal Microscopy

Twelve hours prior to the addition of hAuNPs, chambered coverslip μ-Slides from ibidi were prepared. A culture of each cell type was lifted with .25% trypsin–EDTA and then plated onto collagen-IV coated slides in their respective media. The slides were then placed in an incubator at 37°C until the addition of hAuNPs. Cells were cultured to 10% confluency for experiments. hAuNPs were added to a 0.5 nM concentration in solution. After 4 h the media and suspended hAuNPs were aspirated away, and the chambered slides were washed thrice with Ca^{2+}- and Mg^{2+}-free DPBS. After the final wash, all cells were kept in phenol red free, high glucose DMEM supplemented with 10% FBS while being imaged.

3. KEY RESULTS
3.1 Design of Hairpin Sequence

In order to evaluate the effectiveness of hAuNPs as qualitative mRNA expression probes, we chose three individual MMP mRNAs to target: MMP13, MMP14, and MMP26. These three MMPs were selected to screen for high abundance (MMP14), moderate abundance (MMP13), and little-or-no abundance (MMP26) expression targets in MDA-MB-231 invasive breast cancer cells (Hegedus et al., 2008). By comparison, we expected lower abundance of MMP mRNA relative to the MDA-MB-231 cells when analyzing the less invasive MCF-7 and relatively noninvasive MCF10A cells. We chose to use the housekeeping gene glyceraldehyde 3-phosphate dehydrogenase (GAPDH) as the target for a positive control probe. DNA hairpins with a scrambled version of the MMP26 recognition loop (MMP26scram) were used to create negative control hAuNPs.

Theoretically, any sequence within a target mRNA can be chosen as a site for recognition loop binding. The most straightforward approach for probe sequence selection is to search the literature for priming sites used successfully in quantitative reverse transcription PCR (qRT-PCR) of the biomarker of interest. However, target accessibility is primarily a consequence of complex secondary and tertiary intramolecular structures, which are not easy to predict and can mask many regions used for primer binding. To help predict sites readily available for binding, we have employed RNAstructure, a program available from the Mathews Lab at the University of Rochester Medical Center, to predict the folding of the target mRNA sequence. Use of this program for molecular beacon design has been detailed elsewhere by coauthors (Bratu et al., 2011). After the sequence is folded, a section of 18–25 bases is chosen from an easily accessible region. Using public databases and sequence comparison tools, it is confirmed that the chosen probe sequence is unique to the target biomarker and has little or no homology with other cellular RNA sequences. The full probe sequences we selected for this study are shown in Table 1.

3.2 Characterization of hAuNPs

AuNPs heavily loaded with DNA strands possess strong interparticle electrostatic repulsion which protects the AuNPs from being aggregating in salt solution. However, heavily loading the particle with hDNA causes steric interference between hairpins, inhibiting a complementary sequence from binding to a hairpin or preventing a hairpin from properly opening after binding. To remedy this effect, we employed the use of spacing ligands composed of 10 sequential thymine bases with a hexane-thiol linker at the 5′ end. Indeed, hAuNPs synthesized with a 3:1 ratio of hairpins:spacing ligands have provided the highest fluorescence intensity and signal to noise ratio in response to excess complement in our experience (unpublished data).

The number of hairpins loaded on each particle was determined by subjecting a 10 nM solution of hAuNPs to DTT reduction to reduce the Au–thiol bonds and precipitate the gold particles, and then measuring fluorescence and absorbance values of the DNA-containing supernatant and comparing to standard curves. The oligonucleotide-to-particle ratio for all hAuNPs synthesized in this study was determined to be approximately 68 hairpins and 23 T10 ligands per particle.

UV–visible absorbance spectra of hAuNPs exhibited absorbance peaks at nucleic acid- and AuNP-specific peaks of 260 and 520 nm, respectively. Fluorescence spectrophotometry studies confirmed that the hAuNPs were specific for target sequences, with emission of fluorescence consistent with the fluorophore coupled to the 3′ end of the hDNA (Fig. 2A). hAuNPs exhibited dose-specific increases in fluorescence intensity in response to complement concentration and did not react to an appreciable level when incubated with an equal concentration of mismatched DNA (Fig. 2A). DLS analyses indicated an increase in hydrodynamic diameter of 15–16 nm when hDNA was coupled to citrate-capped AuNP (Fig. 2B). When introduced to a specific target, the hairpin opening resulted in a hydrodynamic diameter increase of 6–7 nm, while the presence of a nonspecific target showed no change. Therefore, DLS can be used in conjunction with fluorescence spectrophotometry to validate hAuNP specificity and sensitivity in the presence of the target oligonucleotide sequence.

3.3 Flow Cytometric Analysis of hAuNP Uptake in Live Breast Cancer Cells

To evaluate gene expression, three different cell lines that express MMPs at varying levels were used: MCF-10A, a nontumorigenic epithelial cell line; MCF-7, a poorly invasive breast ductal carcinoma in situ; and MDA-MB-231, a highly malignant invasive ductal carcinoma. The cells were incubated with 0.5 nM hAuNPs and analyzed by flow cytometry (Fig. 3). MDA-MB-231 cells showed the highest expression of all three MMPs

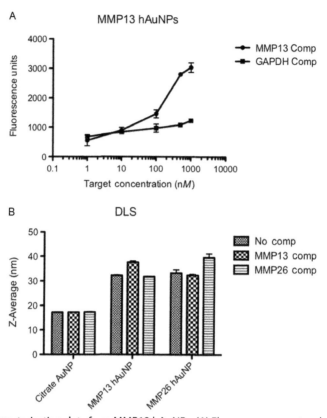

Fig. 2 Characterization data from MMP13 hAuNPs. (A) Fluorescence spectrophotometry studies showed an increase in fluorescent signal due to opening of hairpins when hAuNPs were incubated with their respective complement as opposed to a non-complementary sequence. (B) Dynamic light scattering studies on hAuNPs showed an increase in hydrodynamic diameter of 15–16 nm when hDNA was coupled to citrate-capped AuNP and an increase of 6–7 nm when hAuNPs are introduced to a specific target.

examined, followed by MCF-7 and MCF-10A cells. A scrambled sequence of MMP26, MMP26scram, served as a nonspecific, negative control probe. MMP26scram had relatively consistent intensities between cell lines. This is important to note that it assures that the changes of intensities of the same probe between cell lines are due to differences in mRNA expression and not due to differing rates of hAuNP uptake. These results mirror published expression profiles given for these targets using RT-PCR as shown in Fig. 3.

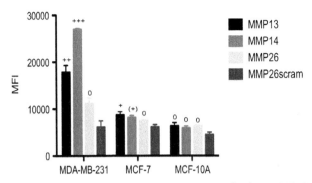

Fig. 3 Flow cytometry analysis of the gene expression of select MMPs in cells using hAuNPs. The *symbols above the columns* refer to relative expression of MMP mRNA by RT-PCR from previously published studies (Hegedus et al., 2008; Kohrmann, Kammerer, Kapp, Dietl, & Anacker, 2009). 0—no expression, (+)—very weak expression, +—weak expression, ++—moderate expression, +++—high expression. (See the color plate.)

3.4 Confocal Microscopy of hAuNP in Live Breast Cancer Cell Lines

Differential expression of MMPs in breast cancer cell lines detected by hAuNP correlated with previously published reports using qRT-PCR analysis. Confocal microscopy studies were performed for each cell line and probe combination (Figs. 4 and 5). Emission from MMP14-specific hAuNPs was greatest in MDA-MB-231 cells, followed by MCF-7 and MCF-10A, consistent with flow cytometric analysis (Fig. 4). The hAuNPs are distributed throughout the cytosol but are not found within the SYTO13-stained nucleus. Differential expression patterns of MMPs in MDA-MB-231 cells also reflected trends observed using flow cytometry and expression levels reported in the literature (Fig. 5). Specifically, MMP14 fluorescence was the highest, followed by MMP13 hAuNP fluorescence intensity, with MMP26 and negative control hAuNP remaining at undetectable levels.

4. CONCLUSIONS

In this report we have demonstrated that hAuNPs can be used to image multiple biomarkers in three different types of living cancer cells. The fluorescence intensities for these probes in cells were measured by flow cytometry and confocal microscopy and exhibited correlation with reported

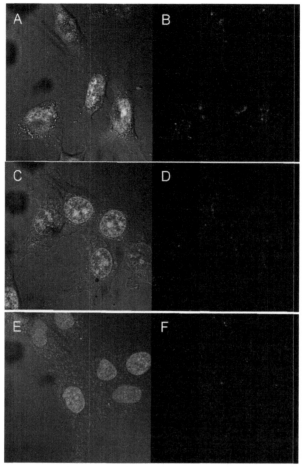

Fig. 4 Intracellular uptake and fluorescence emission of hAuNPs across cell lines. MMP14-specific hAuNP emission (*blue*) decreases from MDA-MB-231 cells (A and B) to MCF-7 (C and D) to MCF-10A (E and F). *Red* fluorescence is GAPDH-specific hAuNP. *Green* is Syto13 nuclear counterstain. 100× magnification. (See the color plate.)

qRT-PCR values for their target mRNA expression, thus confirming the ability to measure relative expression rates of mRNAs in living cells using hAuNPs. This readout can be applied to profile the invasiveness of single cells or tissues from clinical samples and also serve as an indicator for screening anticancer inhibitors. A similar strategy utilizing hAuNPs can be applied to many other systems in which modular detection of multiple mRNAs is advantageous.

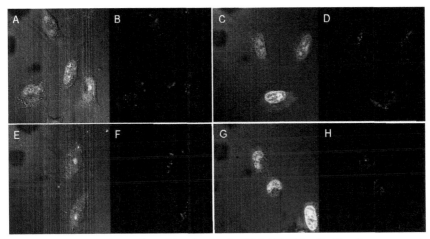

Fig. 5 Comparison of MMP hAuNP activation in MDA-MB-231 cells. *Blue* fluorescence is specific for MMP14 (A and B), MMP13 (C and D), MMP26 (E and F), and the negative control MMP26scram (G and H). *Red* fluorescence is GAPDH-specific hAuNP. *Green* is Syto13 nuclear counterstain. 100× magnification. (See the color plate.)

ACKNOWLEDGMENTS

The authors would like to thank M.F. Richards for critical comments concerning this manuscript. This work was supported in part by the National Institutes of Health (R01EY023397, to A.J.).

REFERENCES

Bratu, D. P. (2003). Molecular beacons light the way: Imaging native mRNAs in living cells. *Discovery Medicine*, *3*(19), 44–47.

Bratu, D. P. (2006). Molecular beacons: Fluorescent probes for detection of endogenous mRNAs in living cells. *Methods in Molecular Biology*, *319*, 1–14. http://dx.doi.org/10.1007/978-1-59259-993-6_1.

Bratu, D. P., Catrina, I. E., & Marras, S. A. (2011). Tiny molecular beacons for in vivo mRNA detection. *Methods in Molecular Biology*, *714*, 141–157. http://dx.doi.org/10.1007/978-1-61779-005-8_9.

Bremer, C., Bredow, S., Mahmood, U., Weissleder, R., & Tung, C. H. (2001). Optical imaging of matrix metalloproteinase-2 activity in tumors: Feasibility study in a mouse model. *Radiology*, *221*(2), 523–529. http://dx.doi.org/10.1148/radiol.2212010368.

Bremer, C., Tung, C. H., & Weissleder, R. (2001). In vivo molecular target assessment of matrix metalloproteinase inhibition. *Nature Medicine*, *7*(6), 743–748. http://dx.doi.org/10.1038/89126.

Brew, K., Dinakarpandian, D., & Nagase, H. (2000). Tissue inhibitors of metalloproteinases: Evolution, structure and function. *Biochimica et Biophysica Acta*, *1477*(1–2), 267–283. http://dx.doi.org/10.1016/S0167-4838(99)00279-4.

Catrina, I. E., Marras, S. A., & Bratu, D. P. (2012). Tiny molecular beacons: LNA/2′-O-methyl RNA chimeric probes for imaging dynamic mRNA processes in living cells. *ACS Chemical Biology*, *7*(9), 1586–1595. http://dx.doi.org/10.1021/cb300178a.

Chen, J., Huang, Y., Zhao, S., Lu, X., & Tian, J. (2012). Gold nanoparticles-based fluorescence resonance energy transfer for competitive immunoassay of biomolecules. *Analyst*, *137*(24), 5885–5890. http://dx.doi.org/10.1039/c2an36108f.

Feng, X., Ma, H., Huang, S., Pan, W., Zhang, X., Tian, F., … Luo, J. (2006). Aqueous-organic phase-transfer of highly stable gold, silver, and platinum nanoparticles and new route for fabrication of gold nanofilms at the oil/water interface and on solid supports. *The Journal of Physical Chemistry. B*, *110*(25), 12311–12317. http://dx.doi.org/10.1021/jp0609885.

Hegedus, L., Cho, H., Xie, X., & Eliceiri, G. L. (2008). Additional MDA-MB-231 breast cancer cell matrix metalloproteinases promote invasiveness. *Journal of Cellular Physiology*, *216*(2), 480–485. http://dx.doi.org/10.1002/jcp.21417.

Jayagopal, A., Halfpenny, K. C., Perez, J. W., & Wright, D. W. (2010). Hairpin DNA-functionalized gold colloids for the imaging of mRNA in live cells. *Journal of the American Chemical Society*, *132*(28), 9789–9796. http://dx.doi.org/10.1021/ja102585v.

Jiang, T., Olson, E. S., Nguyen, Q. T., Roy, M., Jennings, P. A., & Tsien, R. Y. (2004). Tumor imaging by means of proteolytic activation of cell-penetrating peptides. *Proceedings of the National Academy of Sciences of the United States of America*, *101*(51), 17867–17872. http://dx.doi.org/10.1073/pnas.0408191101.

Kohrmann, A., Kammerer, U., Kapp, M., Dietl, J., & Anacker, J. (2009). Expression of matrix metalloproteinases (MMPs) in primary human breast cancer and breast cancer cell lines: New findings and review of the literature. *BMC Cancer*, *9*, 188. http://dx.doi.org/10.1186/1471-2407-9-188.

Lepage, M., Dow, W. C., Melchior, M., You, Y., Fingleton, B., Quarles, C. C., … McIntyre, J. O. (2007). Noninvasive detection of matrix metalloproteinase activity in vivo using a novel magnetic resonance imaging contrast agent with a solubility switch. *Molecular Imaging*, *6*(6), 393–403. http://dx.doi.org/10.2310/7290.2007.00035.

McIntyre, J. O., Fingleton, B., Wells, K. S., Piston, D. W., Lynch, C. C., Gautam, S., & Matrisian, L. M. (2004). Development of a novel fluorogenic proteolytic beacon for in vivo detection and imaging of tumour-associated matrix metalloproteinase-7 activity. *The Biochemical Journal*, *377*(Pt 3), 617–628. http://dx.doi.org/10.1042/BJ20030582.

McIntyre, J. O., & Matrisian, L. M. (2003). Molecular imaging of proteolytic activity in cancer. *Journal of Cellular Biochemistry*, *90*(6), 1087–1097. http://dx.doi.org/10.1002/jcb.10713.

Medina, O. P., Kairemo, K., Valtanen, H., Kangasniemi, A., Kaukinen, S., Ahonen, I., … Koivunen, E. (2005). Radionuclide imaging of tumor xenografts in mice using a gelatinase-targeting peptide. *Anticancer Research*, *25*(1A), 33–42.

Olson, E. S., Aguilera, T. A., Jiang, T., Ellies, L. G., Nguyen, Q. T., Wong, E. H., … Tsien, R. Y. (2009). In vivo characterization of activatable cell penetrating peptides for targeting protease activity in cancer. *Integrative Biology*, *1*(5–6), 382–393. http://dx.doi.org/10.1039/b904890a.

Schellenberger, E., Rudloff, F., Warmuth, C., Taupitz, M., Hamm, B., & Schnorr, J. (2008). Protease-specific nanosensors for magnetic resonance imaging. *Bioconjugate Chemistry*, *19*(12), 2440–2445. http://dx.doi.org/10.1021/bc800330k.

Shi, J., Chan, C., Pang, Y., Ye, W., Tian, F., Lyu, J., … Yang, M. (2015). A fluorescence resonance energy transfer (FRET) biosensor based on graphene quantum dots (GQDs) and gold nanoparticles (AuNPs) for the detection of mecA gene sequence of Staphylococcus aureus. *Biosensors & Bioelectronics*, *67*, 595–600. http://dx.doi.org/10.1016/j.bios.2014.09.059.

Sprague, J. E., Li, W. P., Liang, K., Achilefu, S., & Anderson, C. J. (2006). In vitro and in vivo investigation of matrix metalloproteinase expression in metastatic tumor models. *Nuclear Medicine and Biology*, *33*(2), 227–237. http://dx.doi.org/10.1016/j.nucmedbio.2005.10.011.

Sternlicht, M. D., & Werb, Z. (2001). How matrix metalloproteinases regulate cell behavior. *Annual Review of Cell and Developmental Biology*, *17*, 463–516. http://dx.doi.org/10.1146/annurev.cellbio.17.1.463.

Tsien, R. Y. (2005). Building and breeding molecules to spy on cells and tumors. *FEBS Letters*, *579*(4), 927–932. http://dx.doi.org/10.1016/j.febslet.2004.11.025.

Weissleder, R., & Pittet, M. J. (2008). Imaging in the era of molecular oncology. *Nature*, *452*(7187), 580–589. http://dx.doi.org/10.1038/nature06917.

Weissleder, R., Tung, C. H., Mahmood, U., & Bogdanov, A., Jr. (1999). In vivo imaging of tumors with protease-activated near-infrared fluorescent probes. *Nature Biotechnology*, *17*(4), 375–378. http://dx.doi.org/10.1038/7933.

Xue, M., Wang, X., Duan, L., Gao, W., Ji, L., & Tang, B. (2012). A new nanoprobe based on FRET between functional quantum dots and gold nanoparticles for fluoride anion and its applications for biological imaging. *Biosensors & Bioelectronics*, *36*(1), 168–173. http://dx.doi.org/10.1016/j.bios.2012.04.007.

Yang, Y., Hong, H., Zhang, Y., & Cai, W. (2009). Molecular imaging of proteases in cancer. *Cancer Growth Metastasis*, *2*, 13–27.

Zhu, L., Niu, G., Fang, X., & Chen, X. (2010). Preclinical molecular imaging of tumor angiogenesis. *The Quarterly Journal of Nuclear Medicine and Molecular Imaging*, *54*(3), 291–308.

CHAPTER FIVE

In Vivo RNA Visualization in Plants Using MS2 Tagging

E.J. Peña*, M. Heinlein[†,1]

*Instituto de Biotecnología y Biología Molecular, CCT—La Plata CONICET, Fac. Cs. Exactas, U.N.L.P., La Plata, Argentina
[†]Institut de Biologie Moléculaire des Plantes du CNRS, Université de Strasbourg, Strasbourg, France
[1]Corresponding author: e-mail address: manfred.heinlein@ibmp-cnrs.unistra.fr

Contents

1. Introduction	106
2. Plasmid Constructs: Cloning of SL-Tagged RNA and NLS:MCP:FP into Plant Expression Vectors	111
2.1 Materials	111
2.2 Protocol	112
3. Transient Expression in *N. benthamiana* Leaves	114
3.1 Materials	115
3.2 Protocol	115
4. Imaging and Time-Lapse Image Acquisition	116
4.1 Materials	117
4.2 Protocol	117
5. Summary and Perspectives	119
Acknowledgments	119
References	119

Abstract

Intracellular trafficking and asymmetric localization of RNA molecules within cells are a prevalent process across phyla involved in developmental control and signaling and thus in the determination of cell fate. In addition to intracellular localization, plants support the trafficking of RNA molecules also between cells through plasmodesmata (PD), which has important roles in the cell-to-cell and systemic communication during plant growth and development. Viruses have developed strategies to exploit the underlying plant RNA transport mechanisms for the cell-to-cell and systemic dissemination of infection. In vivo RNA visualization methods have revolutionized the study of RNA dynamics in living cells. However, their application in plants is still in its infancy. To gain insights into the RNA transport mechanisms in plants, we study the localization and transport of *Tobacco mosaic virus* RNA using MS2 tagging. This technique involves the tagging of the RNA of interest with repeats of an RNA stem–loop (SL) that is derived from the origin of assembly of the bacteriophage MS2 and recruits the MS2 coat protein (MCP). Thus, expression of MCP fused to a fluorescent marker allows the specific visualization of the SL-carrying RNA. Here we describe a detailed protocol for *Agrobacterium tumefaciens*-mediated transient expression and in vivo visualization of MS2-tagged mRNAs in *Nicotiana benthamiana* leaves.

1. INTRODUCTION

RNA transport and localization mechanisms determine the subcellular distribution of RNA molecules and thereby establish control over localized protein synthesis and cell polarity (Kloc, Zearfoss, & Etkin, 2002; Palacios & St Johnston, 2001). Surprisingly, more than 70% of 3370 genes analyzed by in situ hybridization during early embryogenesis in *Drosophila* encoded localized mRNAs (Lecuyer et al., 2007). A growing list of localized mRNAs appear to be key factors in development and their encoded proteins can have undesired effects if expressed elsewhere (Martin & Ephrussi, 2009). Similar to cells in animal models and humans, also cells in plants have the capacity to transport and localize RNA molecules (Okita & Choi, 2002). In addition, plants transport RNA molecules between cells (Kim, Canio, Kessler, & Sinha, 2001; Melnyk, Molnar, & Baulcombe, 2011; Ruiz-Medrano, Xoconostle-Cazares, & Lucas, 1999). Intercellular transport occurs through plasmodesmata (PD), membranous pores in the plant cell walls that provide membrane and cytoplasmic continuity between adjacent cells. The system of PD is connected to the phloem and establishes an intercellular and systemic communication network through which plants coordinate developmental processes and responses to the environment at the tissue and organismal level. A wide range of mRNAs and small RNAs have been shown to move cell-to-cell and over long distances through the symplastic pathway (Banerjee et al., 2006; Dunoyer et al., 2010; Hannapel, Sharma, & Lin, 2013; Lucas et al., 1995; Melnyk et al., 2011). A recent work using next generation sequencing of heterografted *Arabidopsis thaliana* ecotypes showed that approximately 7.000 of the 33.602 known genes of Arabidopsis encode mobile RNAs (Thieme et al., 2015). The underlying mechanisms to target RNA complexes to and through PDs and the factors involved in these processes are not well understood and are under intense study, especially by using the spread of RNA virus infections as models (Heinlein, 2015a, 2015b; Niehl & Heinlein, 2011; Peña & Heinlein, 2012).

The use of genetically encoded fluorescent proteins (FPs) together with advances in time-lapse microscopy revolutionized cell biology (Chamberlain & Hahn, 2000; Lippincott-Schwartz, 2001). These technological advances allow the study of dynamic biological processes with subcellular resolution, single-molecule sensitivity, and under unaltered in vivo conditions. The in vivo visualization of proteins is straightforward since proteins can be translationally fused to an FP. However, unlike proteins, RNAs cannot be expressed as fluorescent molecules. The visualization of RNA molecules rather depends on in vitro labeling with a fluorescent dye followed by introduction into the cell or on indirect in vivo labeling methods. Given that plant cells are encased by rigid cell walls, the introduction of in vitro-labeled RNA molecules into cells depends on invasive delivery methods such as microinjection, microprojectile bombardment, or transfection of protoplasts isolated upon digestion of the cell walls with cell wall-degrading enzymes. The RNA localization patterns observed upon application of such invasive methods may not reflect the natural amount, trafficking, and localization behavior of the addressed RNA in question. A method for indirect RNA labeling involves the use of specific hybridization probes (eg, molecular beacons Bao, Rhee, & Tsourkas, 2009). However, this method is again invasive and also requires extensive optimization, especially if applied in vivo (Tilsner, 2015). A less invasive approach for indirect in vivo RNA labeling employs genetically encoded RNA-binding proteins (RBPs) that bind to specific RNA motifs (Urbanek, Galka-Marciniak, Olejniczak, & Krzyzosiak, 2014). An RBP that has been developed for RNA imaging in various organisms is the RNA-binding domain of the human translational repressor Pumilio 1, which binds to its native RNA target sequence with high affinity. Interestingly, this RNA-binding domain can be engineered to act as an FP-tagged RBP with any designed sequence-binding specificity (Cheong & Hall, 2006) and thus can be adapted to determine the localization of any native RNA molecule. However, Pumilio RNA-binding domains show a degree of sequence promiscuity, which poses the risk of unspecific background labeling. A possibility to reduce unspecific background is to employ Pumilio variants that bind to adjacent target sites of an mRNA (Ozawa, Natori, Sato, & Umezawa, 2007). By fusing the two variants with two halves of split FP and using bimolecular fluorescence complementation (BiFC) for detection, visualization is restricted to the desired mRNA molecule, whereas the nondesired RNA molecules that may bind only one of the variants, or two variants at distant sites, remain undetected. This approach (Pumilio-BiFC) has been used for the subcellular detection of *Tobacco mosaic virus* (TMV) and *Potato virus X* RNA molecules within replication sites (Tilsner et al., 2009) and for the localization of *Turnip mosaic virus* (TuMV) RNA at the chloroplast envelope (Wei et al., 2010). Unlike Pumilio-BIFC, two-component RNA labeling systems, such as MS2 (Bertrand et al., 1998), λN (Daigle & Ellenberg, 2007), or BglG (Chen et al., 2009), depend on the incorporation of specific protein-binding RNA sequences into the target RNA. These sequences are specifically recognized and bound by their corresponding coexpressed and FP-tagged RBP, thus allowing the in vivo visualization of the specific RNA under investigation. Another interesting two-component approach, but not requiring the coexpression of RBPs, employs RNA aptamers that exhibit fluorescence upon binding small molecules. Several RNA aptamers that bind and switch on the fluorescence of small fluorophore molecules have been described (Babendure, Adams, & Tsien, 2003; Constantin et al., 2008; Da Costa, Andreiev, & Dieckmann, 2013; Lux, Peña, Bolze, Heinlein, & Nicoud, 2012; Sando, Narita, Hayami, & Aoyama, 2008). An initial problem of nonspecific activation of fluorescence by cellular components led to the development of fluorophores that resemble the fluorophore found in green fluorescent protein (GFP) and which are switched on upon binding to structured RNA aptamers, such as "spinach" (Paige, Wu, & Jaffrey, 2011), "broccoli" (Filonov, Moon, Svensen, & Jaffrey, 2014), or "mango" (Dolgosheina et al., 2014). However, these promising new aptamer techniques have not yet been successfully applied to plants. A major limiting factor may again be the cell wall across which the fluorophore needs to be introduced to enter the cell. The use of RNA labeling technologies in plants has been recently reviewed (Tilsner, 2015).

Here we describe a protocol for the in vivo localization of transiently expressed RNA using MS2 tagging. The method was developed in yeast and has been extensively used in model organisms (Bertrand et al., 1998; Buxbaum, Haimovich, & Singer, 2015), including plants (Hamada et al., 2003; Michaud et al., 2014; Sambade et al., 2008). It relies on the fusion of the target RNA with a tandem repetition of a protein-binding motif consisting of 19 nucleotides derived from the origin of assembly (OAS) of the bacteriophage MS2. Upon transcription, this motif adopts a stem–loop structure (MS2 SL) that recruits the MS2 phage coat protein (MCP). Thus, by expressing MCP in fusion to a fluorescent reporter

(MCP:FP), the tagged RNA can be localized (Bertrand et al., 1998). We optimized the MS2 tagging RNA detection system for the visualization of the transiently expressed mRNA of TMV Movement Protein (MP; Sambade et al., 2008). TMV moves its RNA genome between cells through PD in the form of a nonencapsidated viral ribonucleoprotein complex (vRNP) and, therefore, represents an excellent model to study the cellular mechanism involved in RNP production, localization, and transport in plants (Heinlein, 2015a; Peña & Heinlein, 2012). The MP is required for the cell-to-cell movement of the virus—it targets PD and is part of the vRNP that spreads between cells. TMV derivatives that express functional MP:FP fusion proteins and mutations in the MP affecting intercellular transport of viral RNA were used to characterize the subcellular localization of MP involved in MP function during infection (Boyko, Ferralli, Ashby, Schellenbaum, & Heinlein, 2000; Boyko, Ferralli, & Heinlein, 2000; Boyko, Hu, Seemanpillai, Ashby, & Heinlein, 2007). These studies revealed that MP function in viral movement involves the formation MP particles at junctions of microtubules (MTs) with the cortical endoplasmic reticulum (ER)–actin network (cortical MT-associated ER sites, cMERs). Following formation, the particles exhibit intermittent motions along ER membranes and pause their movements at cMERs (Boyko et al., 2007; Niehl, Peña, Amari, & Heinlein, 2013; Peña & Heinlein, 2013; Sambade et al., 2008). The formation and trafficking of these particles along the ER and to PD can be reproduced in the absence of viral infection, ie, upon transient expression of MP. We used the MS2 tagging system to visualize the MP mRNA in vivo. This led to the demonstration that the MP particles contain MP-encoding mRNA and that the mRNA accumulated together with MP at PD (Sambade et al., 2008).

The in vivo visualization of RNA molecules by the MS2 method requires the coexpression of the RNA under investigation (in our case, MP mRNA) tagged with MS2 SL sequences together with fluorescent reporter-tagged MS2 CP (MCP:FP). The number and location of SL repeats to be added to the mRNA molecule need to be properly designed and tested. Addition of a larger number of SL repeats will increase the signal-to-noise ratio of the experiment. However, while it has been shown that 24 SL repeats are sufficient to achieve single-molecule detection (Fusco et al., 2003), a high number of repetitive sequences can affect mRNA maturation and increase the risk of recombination during cloning procedures. The modification of the RNA of interest should avoid any interference with *cis*-acting sequences that interact with cellular RBPs and are required for normal mRNA maturation and function (eg, zip codes). Because of these reasons, it is advisable to test different SL repeat constructs and verify the expected size of the mature mRNA and, if possible, the activity of the encoded protein (in our case, of MP in virus movement). The MCP:FP is usually tagged with a nuclear localization signal (NLS, NLS:MCP:FP) to restrict its localization to the nucleus and allow export from the nucleus only if recruited by the SL-tagged RNA. This is expected to decrease the cytoplasmic background and thereby to increase the signal-to-noise ratio for the labeled RNA. A high nuclear concentration of NLS:MCP:FP may also enhance the loading of SL-tagged mRNA with NLS:MCP:FP upon transcription and, thus, RNA signal strength in the cytoplasm.

To introduce the MS2 system into plant leaves, we use *Agrobacterium tumefaciens*-mediated transient expression. *A. tumefaciens* transforms plant cells by transferring T-DNA, a region of the tumor-inducing (Ti) plasmid, into host plant cells. This T-DNA usually contains genes that reprogram the plant cells to grow into a tumor and produce a food source for the bacteria. These genes are dispensable for T-DNA transfer and can be replaced by genes of interest, in our case the SL-tagged MP gene and the gene for NLS:MCP:FP. Transient expression takes advantage of the large number of T-DNA copies that reach the nucleus for transcription by plant RNA polymerases, resulting in a high protein expression level after very short periods of time (Fischer et al., 1999; Kapila, De Rycke, Van Montagu, & Angenon, 1997). For transient expression, a suspension of Agrobacteria containing the plasmid of interest is injected into the plant leaves (agroinfiltration).

To achieve a natural expression pattern, the use of a native promoter for the expression of SL-tagged mRNA is recommended. This may prevent overexpression and, thus, the risk of mislocalization due to the saturation of RBPs and trafficking factors. Moreover, weak promoters should be used for expression of NLS:MCP:FP in order to avoid unspecific fluorescence background (Dahm, Zeitelhofer, Gotze, Kiebler, & Macchi, 2008; Hamada et al., 2003). However, we use the strong *Cauliflower mosaic virus* 35S promoter, which is widely used for transient expression upon agroinfiltration in plant cells (Karimi, Depicker, & Hilson, 2007). Desired expression levels can be obtained by carefully selecting the incubation time after agroinfiltration of leaf tissues as well as by adjusting the amount of infiltrated bacteria and thus the number of T-DNA copies transferred per cell.

To obtain faithful information about the localization and dynamics of the RNA under investigation, several experimental controls are essential. The cytoplasmic NLS:MCP:FP signal obtained with SL-tagged RNA should be verified by controlling the absence of the cytoplasmic signal in cells expressing the respective nontagged control RNA. The expression conditions must be rigorously tested and optimized to ensure the absence of any cytoplasmic signal in the presence of nontagged RNA, whereas a clear and significant signal should be observed in the presence of the SL-tagged RNA.

Below, we describe the protocol that was successfully used for the visualization of TMV MP mRNA (Sambade et al., 2008) and which can be generally applied for in vivo mRNA localization and trafficking studies in plants. The protocol includes the use of *Agrobacterium*-mediated transient expression in *Nicotiana benthamiana* leaf epidermal cells and the application of Gateway® cloning (Life Technologies).

2. PLASMID CONSTRUCTS: CLONING OF SL-TAGGED RNA AND NLS:MCP:FP INTO PLANT EXPRESSION VECTORS

RNA visualization using MS2 involves the coexpression of the SL-tagged RNA of interest together with the NLS:MCP:FP for in vivo detection. Based on available plant Gateway® cloning-compatible destination vector (pMDC32; Curtis & Grossniklaus, 2003), we describe here the construction of a destination vector for tagging RNA targets with 12 SL repeats and its use for expression of an SL-tagged mRNA encoding MP fused to red fluorescent protein (MP:mRFP). The construction of the vector for expression of NLS:MCP:eGFP (pB7-NLS:MCP:eGFP) has been published previously (Sambade et al., 2008).

2.1 Materials
- Standard equipment and reagents for high-fidelity PCR amplification (eg, Phusion High-Fidelity PCR Master Mix, Thermo Fisher # F531S), cloning PCR products (eg, pGem®-T Easy, Promega # TM042) and for the isolation of plasmid DNA (eg, NucleoSpin® Plasmid kit, Macherey-Nagel # 740588).
- Incubators set to 37°C, for both solid and liquid cultures (180–200 rpm).

- LB medium: 10 g/L bacto-tryptone (SIGMA # T9410), 5 g/L yeast extract (SIGMA # Y1625), 85.6 mM NaCl (SIGMA # S9888), pH 7.0. Add 15 g/L bacto-agar for solid medium.
- Antibiotic stock solutions (1000×, filtered through a 0.22-μm filter to sterilize):
  - Zeocin: 20 mg/mL in distilled water (Thermo Fisher # R250-01).
  - Ampicillin: 100 mg/mL in distilled water (SIGMA # A0166).
  - Kanamycin: 50 mg/mL in distilled water (SIGMA # K1377).
  - Spectinomycin: 50 mg/mL in distilled water (SIGMA # S4014).
  - Chloramphenicol: 34 mg/mL in 100% ethanol (SIGMA # C0378).
- DNA restriction enzymes *Pac*I (NEB # R0547) and *Sac*I (NEB # R0156) and T4 DNA ligase (NEB # M0202).
- Plasmids containing MCP gene and MS2 SL sequences (tandem of 6, 12, or 24 SLs; original plasmids can be requested from the laboratory of Robert Singer at http://www.addgene.org/Robert_Singer/).
- Primers used to amplify 12xMS2 SL:
    MS2-12SLfw 5′-ttaattaacgggccctatatatggatcc
    MS2-12SLrev 5′-gagctccgctgatatcgatcgcgcgc
- pDONR™/Zeo Vector (Thermo Fisher # 12535-035).
- BP™ and LR™ clonase enzymes (Thermo Fisher # 11789 and # 11791).
- pMDC32 destination vector for *A. tumefaciens-mediated* transient expression. Plant destination vectors compatible with Gateway® cloning can be requested at http://www.arabidopsis.org/abrc/catalog/vector_1.html or http://gateway.psb.ugent.be/.
- *Escherichia coli* strain Stbl2™, with genetic modifications to minimize genetic recombination of unstable sequences (Thermo Fisher # 10268-019).
- *E. coli* strain Survival™ 2, for the amplification of *ccdB* gene-containing vectors (Thermo Fisher # A10460).

2.2 Protocol

2.2.1 Generation of a Destination Vector for the Expression of 3′-12xSLs-Tagged RNAs (pMDC32-3′-MS2-12xSL)

1. Using high-fidelity DNA polymerization enzymes, PCR amplify the 12xSL tandem repeat with forward primer MS2-12SLfw and reverse primer MS2-12SLrev using plasmid pSL-MS2-12 (http://www.addgene.org/27119/; Bertrand et al., 1998) as template. The forward

and reverse primers contain *PacI* and *SacI* restriction sites, respectively (underlined).
2. Gel-purify and clone the amplified 708 bp long DNA fragment into a PCR cloning vector. Transform competent *E. coli* Stbl2™ (low recombination frequency strain) and incubate at 37°C overnight (ON) on LB agar plates with appropriate antibiotic selection (100 µg/mL ampicillin for pGem®-T Easy).
3. Pick four to five colonies and grow bacteria ON in 2–5 mL liquid LB medium at 37°C with agitation (180–200 rpm), under antibiotic selection (100 µg/mL ampicillin for pGem®-T Easy).
4. Harvest bacteria, isolate plasmid DNA, and amplify by PCR the inserted DNA sequence (T7 and SP6 universal primers for pGem®-T Easy). Use four to five samples for DNA sequence and select a plasmid verified for the presence of anticipated DNA sequences with no mutations and the presence of the correct number of SL repeats.
5. Use restriction enzymes and purify the *PacI/SacI* DNA fragment after agarose gel electrophoresis. Ligate the fragment into a previously *PacI/SacI*-digested pMDC32 vector.
6. Transform competent *E. coli* Survival™ 2 cells and grow ON at 37°C on LB agar plates containing 50 µg/mL kanamycin and 34 µg/mL chloramphenicol.
7. Grow bacterial cultures from several individual colonies at 37°C and 180–200 rpm and under antibiotic selection (50 µg/mL kanamycin and 34 µg/mL chloramphenicol), isolate plasmid DNA and verify the correct insertion by DNA sequencing.

2.2.2 Generation of an Entry Vector Containing the Sequence of Your Target RNA of Interest

1. Design primers for the amplification of the DNA sequence encoding your RNA of interest and add *att* B sequences (required for BP™ Gateway® reaction) following the manufacturer's instructions (http://tools.lifetechnologies.com/content/sfs/manuals/gatewayman.pdf). If the RNA of interest encodes a protein fused to an FP, the cells in which the RNA is expressed can be easily identified by microscopy.
2. Use the primers to amplify the DNA encoding your RNA of interest by PCR using the appropriate source DNA as template, purify the product after agarose gel electrophoresis and insert the product into the plasmid pDONR™/Zeo by recombination using BP™ clonase according to the

manufacturer's instructions. Transform competent *E. coli cells* and grow colonies ON at 37°C on LB agar plates containing 20 μg/mL of zeocin.
3. Grow liquid cultures from several bacterial colonies, isolate plasmid DNA, and verify the insert by DNA sequencing.

2.2.3 Generation of the SL-Tagged RNA Expression Vector
1. Perform LR reactions between the entry vector and (a) the destination plasmid pMDC32 and (b) the destination plasmid pMDC32-3'-MS2-12xSL. Recombination with pMDC32 creates pMDC32-RNA, a vector for expression of the RNA of interest without SL repeats (for control experiments), whereas the recombination with pMDC32-3'-MS2-12xSL creates pMDC32-RNA-SL, thus a vector for expression of the RNA of interest fused to 12 SL repeats.
2. Transform competent *E. coli* Stbl2™ cells and select the cells on LB agar plates containing 50 μg/mL of kanamycin.
3. Grow liquid bacterial cultures from several colonies under antibiotic selection (50 μg/mL of kanamycin) and at 180–200 rpm, isolate plasmid DNA, and verify the constructs by DNA sequencing.

3. TRANSIENT EXPRESSION IN *N. BENTHAMIANA* LEAVES

Agrobacterium-mediated transient gene expression in *N. benthamiana* leaves is a fast, flexible, and reproducible approach to obtain high protein expression levels. The mixing of agrobacterium cultures containing different plant expression vectors can be used for the coexpression of several transgenes in almost all cells of the infiltrated leaf region at the same time (Kapila et al., 1997). Alternatively, bacteria can be sequentially transformed with more than one plasmid and the double-transformed bacteria can be maintained if the plasmids carry different antibiotic resistance genes for selection (Buschmann, Green, Sambade, Doonan, & Lloyd, 2011). For inoculation of plant leaves, the suspension of agrobacteria is injected into the intercellular spaces using a syringe without needle (agroinfiltration). The highest expression level is reached at 48–72 h postinfiltration. Expression can be enhanced and sustained in time by coinfiltration of an agrobacterial strain encoding a silencing suppressor protein (eg, *Tomato bushy stunt virus* P19; Voinnet, Rivas, Mestre, & Baulcombe, 2003).

The determination of RNA subcellular localization may require the coexpression of specific cellular markers. This can be achieved either by using stably transformed plant lines such as *N. benthamiana* expressing *tua:*

GFP for visualization of MTs (Gillespie et al., 2002) or GFP targeted to the ER (line 16c; Ruiz, Voinnet, & Baulcombe, 1998), or by the simultaneous transient expression of fluorescent markers. Several markers for the specific labeling of MTs, actin filaments, ER, PD, etc., are available (eg, Martin et al., 2009; Mathur, 2007).

3.1 Materials
- *A. tumefaciens* strain GV3101 (Hellens, Mullineaux, & Klee, 2000).
- Plasmids pMDC32-RNA and pMDC32-RNA-SL obtained in Section 2 and plasmid pB7-NLS:MS2CP:GFP (available upon request).
- Incubators set to 28°C, for both solid and solid cultures (180–200 rpm).
- 50 mL Falcon tubes (Fisher Scientific # 14-432-22).
- LB agar and liquid media (see Section 2.1).
- Antibiotic stock solutions (see Section 2.1); in addition:
    - Rifampicin: 50 mg/mL in DMSO (SIGMA # R3501).
    - Gentamicin: 50 mg/mL in distilled water (SIGMA # G1397).
- Spectrophotometer.
- Centrifuge for 50 mL tubes.
- Sterile distilled water.
- Chamber or greenhouse for the cultivation of *N. benthamiana* plants (16/8 h light/dark, 24–22°C).
- Four- to five-week-old *N. benthamiana* plants.
- 1.0–2.5 mL syringes for tissue infiltration.
- Transgenic plants or transformed *A. tumefaciens* lines for stable or transient expression of fluorescent subcellular markers.

3.2 Protocol
Agroinfiltration of plant tissues for transient coexpression of the SL-tagged RNA and GFP-tagged RBP
1. Transform *A. tumefaciens* strain GV3101 with (i) pMDC32-RNA, (ii) pMDC32-RNA-SL, and (iii) pB7-NLS:MCP:GFP and select each of the, respectively, transformed bacteria on LB agar plates containing 50 μg/mL rifampicin, 50 μg/mL gentamycin, and either 50 μg/mL kanamycin for selection of bacteria containing the pMDC32 vectors or 50 μg/mL spectinomycin for selection of bacteria containing pB7-NLS:MS2CP:GFP. Incubate for 48 h at 28°C without agitation.
2. Select individual colonies and grow liquid cultures under antibiotic selection on a shaker at 28°C and at 180–200 rpm.

3. Harvest the bacteria by centrifugation at room temperature (10 min at $4000 \times g$), resuspend the bacterial pellets in sterile distilled water, and determine the concentration by absorbance at 600 nm (OD_{600}).
4. Set up mixtures of the different bacteria for agroinfiltration. The final concentration of each bacterial clone in the infiltration mixture should correspond to the desired expression level for each plasmid. High levels of expression of the SL-tagged RNA will increase the efficiency of RNA detection and a low expression of NLS:MCP:GFP may be required to avoid the occurrence of nonspecific background signal. For the labeling of the mRNA encoding the MP of TMV, we use a mixture of agrobacteria with pMDC32-RNA-SL (or pMDC32-RNA, as control) at $OD_{600} = 0.5$ and agrobacteria with pNLS:MCP:GFP at $OD_{600} = 0.1$. If required, increase expression levels by resuspending the harvested agrobacteria in infiltration medium (10 mM MES pH 5.6, 10 mM $MgCl_2$) supplemented with 150 µM acetosyringone and incubate at room temperature for at least 2 h to allow the induction of *A. tumefaciens* virulence (*vir*) genes (Rogowsky, Close, Chimera, Shaw, & Kado, 1987).
5. Use a 1.0- to 2.5-mL syringe without needle to infiltrate the fifth to seventh leaves of 4- to 5-week-old *N. benthamiana* plants. Use a needle or scalpel to incise small holes into the abaxial side of the plant leaf. Then, press the syringe tip to the incision and inject the bacterial solution into the leaf airspace with gentle pressure. Inject the bacterial mixture for expression of pNLS:MCP:GFP together with pMDC32-RNA-SL into one half of the leaf and the bacterial mixture for expression of pNLS:MCP:GFP together with pMDC32-RNA (control without SL repeats) into the other half of the leaf.
6. Incubate the infiltrated plants at 22–18 °C in 16 h/8 h light/dark for 1–4 days postinfiltration before microscopical analysis. The time of incubation after agroinfiltration required for achieving the best possible signal-to-noise ratio for fluorescence imaging must be empirically determined.

4. IMAGING AND TIME-LAPSE IMAGE ACQUISITION

N. benthamiana epidermal cells are large, jigsaw puzzle-shaped cells with a central vacuole that restricts the cytoplasm to a thin cortical layer subjacent to the plasma membrane. This cortical layer contains the cortical ER–actin network and the interphase cortical MT array. RNA particles move either by diffusion or by specific transport in association with molecular motors. The tracking of fluorescent RNA particle movements in vivo by fluorescence microscopy requires high sensitivity and fast image acquisition capacity without inducing phototoxic effects. We describe here the sample preparation, microscope setups, and imaging conditions we use for the in vivo visualization of TMV MP mRNA particles in epidermal cells of *N. benthamiana* leaves.

4.1 Materials

- Glass slides (75 × 25 mm) and coverslips (50 mm × 24 mm, 0.14 mm thick).
- Cork borer (0.7–1.0 cm diameter).
- Sterile water.
- Vacuum pump with vacuum desiccator.
- Confocal or epifluorescence microscope equipped for time-lapse image acquisition and with high numerical aperture objectives. We used (1) a Nikon TE2000 inverted fluorescence microscope equipped with a Roper CoolSnap CCD camera, piezo-driven Z-focus, and a 60× 1.45 aperture TIRF objective (excitation/emission wavelengths are 460–500/510–560 nm for GFP and 550–600/615–665 nm for mRFP) and a Dual-View beam splitter (Optical Insights) to acquire images through two fluorescence channels at the same time or (2) a Zeiss LSM780 laser-scanning microscope equipped with a plan-apochromat 63×/1.4 oil-immersion objective, 488 and 561 nm excitation lasers and a detection system optimized for eGFP and mRFP, respectively, and with Zen Imaging Software (http://www.zeiss.com/).
- Computer for data analysis with ImageJ software (http://rsbweb.nih.gov/ij/).

4.2 Protocol

4.2.1 Sample Preparation, Imaging, and Time-Lapse Image Acquisition

1. Use the cork borer to excise leaf discs from infiltrated leaf areas. Prepare in parallel leaf samples expressing SL-tagged RNA and leaf samples expressing the nontagged control RNA. Comparison of GFP fluorescence patterns between these sample types is required to identify the specific signal coming from SL-tagged RNA and to distinguish it from unspecific background signal.

2. Place leaf discs on a microscope slide with the adaxial side facing the glass, cover with a coverslip, and use sticky tape for fixation. Fill the free space between glasses with sterile water.
3. Remove air from intercellular spaces in the leaf discs by placing the prepared samples into a vacuum desiccator and by evacuating the air (−0.8 bar). Bubbles should be released from the plant tissue. Allow air to enter the desiccator slowly. The leaf apoplastic space should now be filled with water as can be seen by a dark green appearance of the leaf discs.
4. Place the samples onto the stage of the fluorescence microscope. Observations should finish within 30 min after leaf discs were taken. After this time a new sample should be prepared to continue.

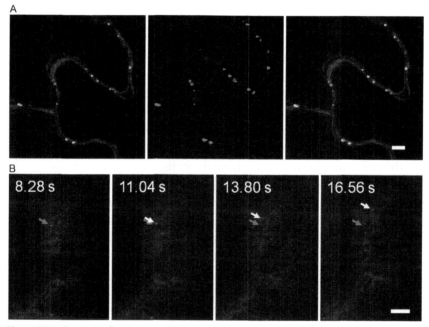

Fig. 1 Visualization of TMV MP:mRFP mRNA by MS2 tagging in *N. benthamiana* epidermal cells. (A) MP:mRFP mRNA (labeled with NLS:MCP:GFP, *green, left*) and MP:mRFP (*red, middle*) coincide to *small dots* (presumably PD) at the cell wall (*yellow, right*). Size bar, 5 μm. (B) Example of video frames showing the colocalization of MP:mRFP and MP:mRFP mRNA (labeled by NLS:MCP:GFP) in mobile particles present in the cortical cytoplasm. *Red* and *green* channels are merged. The particles appear in *yellowish color*. MP:mRFP is shown also in association with microtubules, which are seen as *red-colored* filaments. *Gray arrows* indicate the mRNA particle at the location within the first frame and *white arrows* indicate the particle location in each subsequent time frame. Size bar, 5 μm. (See the color plate.)

5. Use a low magnification objective (eg, 20×) to verify that tissues expressing the SL-tagged RNA and the tissues expressing nontagged RNA express similar levels of NLS:MCP:GFP. At this magnification, it might already be possible to detect differences in the distribution of NLS:MCP:GFP between the two types of tissues.
6. Use a high magnification objective (63–100 ×) and start by imaging the localization of the RNA at PD within the lateral cell walls (see examples shown in Fig. 1A). To visualize and record the cytoplasmic transport of tagged RNA particles, record time-lapse image series at high sampling rates at cortical planes (see examples shown in Fig. 1B and Movie S1). Use ImageJ (http://rsbweb.nih.gov/ij/) or similar software for image analysis. ImageJ is a broadly used open-source software capable of handling multidimensional data in most of the microscopy imaging file formats. A multitude of plug-ins/macros are available at the software website. Approaches for the tracking and analysis of individual MCP-labeled RNA particle movements have been addressed by Park, Buxbaum, and Singer (2010).

5. SUMMARY AND PERSPECTIVES

The MS2 method provides a powerful method for the detection and analysis of RNA molecules in vivo. Two-component systems like MS2 also offer the possibility to biochemically isolate the tagged RNA molecules and to characterize associated proteins that are part of the RNA particle. These proteins likely have important roles in RNA particle formation, structure and regulation, as well as in transport, anchoring, turnover, or translation of RNA.

ACKNOWLEDGMENTS
This work was supported by a grant from the Agence National de la Recherche (ANR, Grant ANR-08-BLAN-244) to M.I.I. and by a binational grant from ANR—Consejo Nacional de Investigaciones Científicas y Técnicas (CONICET, Grant PICS-2014) to E.J.P. and M.H.

REFERENCES
Babendure, J. R., Adams, S. R., & Tsien, R. Y. (2003). Aptamers switch on fluorescence of triphenylmethane dyes. *Journal of the American Chemical Society*, *125*, 14716–14717.
Banerjee, A. K., Chatterjee, M., Yu, Y., Suh, S. G., Miller, W. A., & Hannapel, D. J. (2006). Dynamics of a mobile RNA of potato involved in a long-distance signaling pathway. *Plant Cell*, *18*, 3443–3457.

Bao, G., Rhee, W. J., & Tsourkas, A. (2009). Fluorescent probes for live-cell RNA detection. *Annual Review of Biomedical Engineering, 11*, 25–47.

Bertrand, E., Chartrand, P., Schaefer, M., Shenoy, S. M., Singer, R. H., & Long, R. M. (1998). Localization of ASH1 mRNA particles in living yeast. *Molecular Cell, 2*, 437–445.

Boyko, V., Ferralli, J., Ashby, J., Schellenbaum, P., & Heinlein, M. (2000a). Function of microtubules in intercellular transport of plant virus RNA. *Nature Cell Biology, 2*, 826–832.

Boyko, V., Ferralli, J., & Heinlein, M. (2000b). Cell-to-cell movement of TMV RNA is temperature-dependent and corresponds to the association of movement protein with microtubules. *The Plant Journal, 22*, 315–325.

Boyko, V., Hu, Q., Seemanpillai, M., Ashby, J., & Heinlein, M. (2007). Validation of microtubule-associated *Tobacco mosaic virus* RNA movement and involvement of microtubule-aligned particle trafficking. *The Plant Journal, 51*, 589–603.

Buschmann, H., Green, P., Sambade, A., Doonan, J. H., & Lloyd, C. W. (2011). Cytoskeletal dynamics in interphase, mitosis and cytokinesis analysed through Agrobacterium-mediated transient transformation of tobacco BY-2 cells. *The New Phytologist, 190*, 258–267.

Buxbaum, A. R., Haimovich, G., & Singer, R. H. (2015). In the right place at the right time: Visualizing and understanding mRNA localization. *Nature Reviews. Molecular Cell Biology, 16*, 95–109.

Chamberlain, C., & Hahn, K. M. (2000). Watching proteins in the wild: Fluorescence methods to study protein dynamics in living cells. *Traffic, 1*, 755–762.

Chen, J., Nikolaitchik, O., Singh, J., Wright, A., Bencsics, C. E., Coffin, J. M., et al. (2009). High efficiency of HIV-1 genomic RNA packaging and heterozygote formation revealed by single virion analysis. *Proceedings of the National Academy of Sciences of the United States of America, 106*, 13535–13540.

Cheong, C. G., & Hall, T. M. (2006). Engineering RNA sequence specificity of Pumilio repeats. *Proceedings of the National Academy of Sciences of the United States of America, 103*, 13635–13639.

Constantin, T. P., Silva, G. L., Robertson, K. L., Hamilton, T. P., Fague, K., Waggoner, A. S., et al. (2008). Synthesis of new fluorogenic cyanine dyes and incorporation into RNA fluoromodules. *Organic Letters, 10*, 1561–1564.

Curtis, M. D., & Grossniklaus, U. (2003). A gateway cloning vector set for high-throughput functional analysis of genes in planta. *Plant Physiology, 133*, 462–469.

Da Costa, J. B., Andreiev, A. I., & Dieckmann, T. (2013). Thermodynamics and kinetics of adaptive binding in the malachite green RNA aptamer. *Biochemistry, 52*, 6575–6583.

Dahm, R., Zeitelhofer, M., Gotze, B., Kiebler, M. A., & Macchi, P. (2008). Visualizing mRNA localization and local protein translation in neurons. *Methods in Cell Biology, 85*, 293–327.

Daigle, N., & Ellenberg, J. (2007). LambdaN-GFP: An RNA reporter system for live-cell imaging. *Nature Methods, 4*, 633–636.

Dolgosheina, E. V., Jeng, S. C., Panchapakesan, S. S., Cojocaru, R., Chen, P. S., Wilson, P. D., et al. (2014). RNA mango aptamer-fluorophore: A bright, high-affinity complex for RNA labeling and tracking. *ACS Chemical Biology, 9*, 2412–2420.

Dunoyer, P., Schott, G., Himber, C., Meyer, D., Takeda, A., Carrington, J. C., et al. (2010). Small RNA duplexes function as mobile silencing signals between plant cells. *Science, 328*, 912–916.

Filonov, G. S., Moon, J. D., Svensen, N., & Jaffrey, S. R. (2014). Broccoli: Rapid selection of an RNA mimic of green fluorescent protein by fluorescence-based selection and directed evolution. *Journal of the American Chemical Society, 136*, 16299–16308.

Fischer, R., Vaquero-Martin, C., Sack, M., Drossard, J., Emans, N., & Commandeur, U. (1999). Towards molecular farming in the future: Transient protein expression in plants. *Biotechnology and Applied Biochemistry, 30*(Pt. 2), 113–116.
Fusco, D., Accornero, N., Lavoie, B., Shenoy, S. M., Blanchard, J. M., Singer, R. H., et al. (2003). Single mRNA molecules demonstrate probabilistic movement in living mammalian cells. *Current Biology, 13*, 161–167.
Gillespie, T., Boevink, P., Haupt, S., Roberts, A. G., Toth, R., Valentine, T., et al. (2002). Functional analysis of a DNA-shuffled movement protein reveals that microtubules are dispensable for the cell-to-cell movement of *Tobacco mosaic virus*. *Plant Cell, 14*, 1207–1222.
Hamada, S., Ishiyama, K., Choi, S. B., Wang, C., Singh, S., Kawai, N., et al. (2003). The transport of prolamine RNAs to prolamine protein bodies in living rice endosperm cells. *Plant Cell, 15*, 2253–2264.
Hannapel, D. J., Sharma, P., & Lin, T. (2013). Phloem-mobile messenger RNAs and root development. *Frontiers in Plant Science, 4*, 257.
Heinlein, M. (2015a). Plant virus replication and movement. *Virology, 479–480*, 657–671.
Heinlein, M. (2015b). Plasmodesmata: Channels for viruses on the move. *Methods in Molecular Biology, 1217*, 25–52.
Hellens, R., Mullineaux, P., & Klee, H. (2000). Technical focus: A guide to Agrobacterium binary Ti vectors. *Trends in Plant Science, 5*, 446–451.
Kapila, J., De Rycke, R., Van Montagu, M., & Angenon, G. (1997). An Agrobacterium-mediated transient gene expression system for intact leaves. *Plant Science, 122*, 101–108.
Karimi, M., Depicker, A., & Hilson, P. (2007). Recombinational cloning with plant gateway vectors. *Plant Physiology, 145*, 1144–1154.
Kim, M., Canio, W., Kessler, S., & Sinha, N. (2001). Developmental changes due to long-distance movement of a homeobox fusion transcript in tomato. *Science, 293*, 287–289.
Kloc, M., Zearfoss, N. R., & Etkin, L. D. (2002). Mechanisms of subcellular mRNA localization. *Cell, 108*, 533–544.
Lecuyer, E., Yoshida, H., Parthasarathy, N., Alm, C., Babak, T., Cerovina, T., et al. (2007). Global analysis of mRNA localization reveals a prominent role in organizing cellular architecture and function. *Cell, 131*, 174–187.
Lippincott-Schwartz, J. (2001). Studying protein dynamics in living cells. *Nature Reviews. Molecular Cell Biology, 2*, 444–456.
Lucas, W. J., Bouche-Pillon, S., Jackson, D. P., Nguyen, L., Baker, L., Ding, B., et al. (1995). Selective trafficking of KNOTTED1 homeodomain protein and its mRNA through plasmodesmata. *Science, 270*, 1980–1983.
Lux, J., Peña, E. J., Bolze, F., Heinlein, M., & Nicoud, J. F. (2012). Malachite green derivatives for two-photon RNA detection. *Chembiochem, 13*, 1206–1213.
Martin, K. C., & Ephrussi, A. (2009). mRNA localization: Gene expression in the spatial dimension. *Cell, 136*, 719–730.
Martin, K., Kopperud, K., Chakrabarty, R., Banerjee, R., Brooks, R., & Goodin, M. M. (2009). Transient expression in *Nicotiana benthamiana* fluorescent marker lines provides enhanced definition of protein localization, movement and interactions *in planta*. *The Plant Journal, 59*, 150–162.
Mathur, J. (2007). The illuminated plant cell. *Trends in Plant Science, 12*, 506–513.
Melnyk, C. W., Molnar, A., & Baulcombe, D. C. (2011). Intercellular and systemic movement of RNA silencing signals. *The EMBO Journal, 30*, 3553–3563.
Michaud, M., Ubrig, E., Filleur, S., Erhardt, M., Ephritikhine, G., Marechal-Drouard, L., et al. (2014). Differential targeting of VDAC3 mRNA isoforms influences mitochondria morphology. *Proceedings of the National Academy of Sciences of the United States of America, 111*, 8991–8996.

Niehl, A., & Heinlein, M. (2011). Cellular pathways for viral transport through plasmodesmata. *Protoplasma, 248,* 75–99.
Niehl, A., Peña, E. J., Amari, K., & Heinlein, M. (2013). Microtubules in viral replication and transport. *The Plant Journal, 75,* 290–308.
Okita, T. W., & Choi, S. B. (2002). mRNA localization in plants: Targeting to the cell's cortical region and beyond. *Current Opinion in Plant Biology, 5,* 553–559.
Ozawa, T., Natori, Y., Sato, M., & Umezawa, Y. (2007). Imaging dynamics of endogenous mitochondrial RNA in single living cells. *Nature Methods, 4,* 413–419.
Paige, J. S., Wu, K. Y., & Jaffrey, S. R. (2011). RNA mimics of green fluorescent protein. *Science, 333,* 642–646.
Palacios, I. M., & St Johnston, D. (2001). Getting the message across: The intracellular localization of mRNAs in higher eukaryotes. *Annual Review of Cell and Developmental Biology, 17,* 569–614.
Park, H. Y., Buxbaum, A. R., & Singer, R. H. (2010). Single mRNA tracking in live cells. *Methods in Enzymology, 472,* 387–406.
Peña, E. J., & Heinlein, M. (2012). RNA transport during TMV cell-to-cell movement. *Frontiers in Plant Science, 3,* 193.
Peña, E. J., & Heinlein, M. (2013). Cortical microtubule-associated ER sites: Organization centers of cell polarity and communication. *Current Opinion in Plant Biology, 16,* 764–773.
Rogowsky, P. M., Close, T. J., Chimera, J. A., Shaw, J. J., & Kado, C. I. (1987). Regulation of the vir genes of *Agrobacterium tumefaciens* plasmid pTiC58. *Journal of Bacteriology, 169,* 5101–5112.
Ruiz, M. T., Voinnet, O., & Baulcombe, D. C. (1998). Initiation and maintenance of virus-induced gene silencing. *Plant Cell, 10,* 937–946.
Ruiz-Medrano, R., Xoconostle-Cazares, B., & Lucas, W. J. (1999). Phloem long-distance transport of CmNACP mRNA: Implications for supracellular regulation in plants. *Development, 126,* 4405–4419.
Sambade, A., Brandner, K., Hofmann, C., Seemanpillai, M., Mutterer, J., & Heinlein, M. (2008). Transport of TMV movement protein particles associated with the targeting of RNA to plasmodesmata. *Traffic, 9,* 2073–2088.
Sando, S., Narita, A., Hayami, M., & Aoyama, Y. (2008). Transcription monitoring using fused RNA with a dye-binding light-up aptamer as a tag: A blue fluorescent RNA. *Chemical Communications, 33,* 3858–3860.
Thieme, C. J., Rojas-Triana, M., Stecyk, E., Schudoma, C., Zhang, W., Yang, L., et al. (2015). Endogenous Arabidopsis messenger RNAs transported to distant tissues. *Nature Plants, 1,* 15025.
Tilsner, J. (2015). Techniques for RNA *in vivo* imaging in plants. *Journal of Microscopy, 258,* 1–5.
Tilsner, J., Linnik, O., Christensen, N. M., Bell, K., Roberts, I. M., Lacomme, C., et al. (2009). Live-cell imaging of viral RNA genomes using a Pumilio-based reporter. *The Plant Journal, 57,* 758–770.
Urbanek, M. O., Galka-Marciniak, P., Olejniczak, M., & Krzyzosiak, W. J. (2014). RNA imaging in living cells—Methods and applications. *RNA Biology, 11,* 1083–1095.
Voinnet, O., Rivas, S., Mestre, P., & Baulcombe, D. (2003). An enhanced transient expression system in plants based on suppression of gene silencing by the p19 protein of *Tomato bushy stunt virus*. *The Plant Journal, 33,* 949–956.
Wei, T., Huang, T. S., McNeil, J., Laliberte, J. F., Hong, J., Nelson, R. S., et al. (2010). Sequential recruitment of the endoplasmic reticulum and chloroplasts for plant potyvirus replication. *Journal of Virology, 84,* 799–809.

CHAPTER SIX

TRICK: A Single-Molecule Method for Imaging the First Round of Translation in Living Cells and Animals

J.M. Halstead[*,1], **J.H. Wilbertz**[*,†,1], **F. Wippich**[‡], **T. Lionnet**[§], **A. Ephrussi**[‡], **J.A. Chao**[*,2]

[*]Friedrich Miescher Institute for Biomedical Research, Basel, Switzerland
[†]University of Basel, Basel, Switzerland
[‡]European Molecular Biology Laboratory, Heidelberg, Germany
[§]Transcription Imaging Consortium, HHMI Janelia Research Campus, Ashburn, VA, United States
[2]Corresponding author: e-mail address: jeffrey.chao@fmi.ch

Contents

1. Introduction	124
2. Design of TRICK Reporter mRNAs	125
3. TRICK Experiment in Mammalian Cells	128
3.1 Expression of TRICK Reporter Transcripts	128
3.2 Expression of Coat Proteins Fused to Fluorescent Proteins	129
3.3 Considerations and Challenges of TRICK in Primary Cells	131
3.4 Controls	132
4. Microscopy	132
4.1 Imaging Modality	132
4.2 Light Source	135
4.3 Signal Detection	136
4.4 Temperature and CO_2 Control	137
5. Data Collection	138
5.1 Considerations for Single-Molecule Detection and Tracking	138
5.2 Considerations for Long Time-Lapse Experiments	139
6. Analysis	140
6.1 Single-Molecule Detection and Tracking	141
6.2 Determining Colocalization of Tracked Two-Colored mRNA Particles	143
6.3 Controls	145
7. TRICK Experiment in HeLa Cells to Determine Fraction of Untranslated mRNAs	145
7.1 Preparation of Cells for Live-Cell Imaging	145
7.2 Image Acquisition	146
7.3 Image Analysis	147

[1] These authors contributed equally to this work.

8. TRICK Experiment in *Drosophila*	148
8.1 Imaging and Analysis	149
8.2 Controls	151
9. Outlook	152
Acknowledgments	153
References	153

Abstract

The life of an mRNA is dynamic within a cell. The development of quantitative fluorescent microscopy techniques to image single molecules of RNA has allowed many aspects of the mRNA lifecycle to be directly observed in living cells. Recent advances in live-cell multicolor RNA imaging, however, have now made it possible to investigate RNA metabolism in greater detail. In this chapter, we present an overview of the design and implementation of the translating RNA imaging by coat protein knockoff RNA biosensor, which allows untranslated mRNAs to be distinguished from ones that have undergone a round of translation. The methods required for establishing this system in mammalian cell lines and *Drosophila melanogaster* oocytes are described here, but the principles may be applied to any experimental system.

1. INTRODUCTION

Messenger RNA (mRNA) translation is tightly regulated to produce protein at the correct time and place with appropriate abundance. While many of the principles of translation regulation have been elucidated from ensemble biochemical measurements, understanding the mechanisms controlling where and when an mRNA is translated within a single cell is an emerging research goal. Indeed, considerable evidence now suggests that regulation of translation is devolved to specific cytoplasmic compartments, and that well-timed mRNA translation at a defined location is critical to several physiological processes, ranging from synaptic plasticity (Holt & Schuman, 2013), axis specification (Kumano, 2012), and cell motility (Liao, Mingle, Van De Water, & Liu, 2015). Understanding the mechanistic basis of localized translation regulation therefore requires spatial and temporal information to be extracted from single translation events in single living cells.

While technical advances have expanded the toolbox available to measure translation, spatial and temporal information are rarely simultaneously acquired. Ribosome profiling maps transcriptome-wide ribosome occupancy to subcodon resolution, providing a powerful readout of translation on the timescale of minutes (Ingolia, Ghaemmaghami, Newman, & Weissman, 2009). Spatial information, however, is either restricted to particular organelles (Jan, Williams, & Weissman, 2014) or lost altogether. Imaging assays for translation provide broad spatial information by labeling components of the translational machinery (Pan, Kirillov, & Cooperman, 2007), or the nascent polypeptide itself (David et al., 2012; Dieterich et al., 2010; Terskikh et al., 2000). While these approaches have elegantly demonstrated that protein production is locally regulated within subcellular domains, they do not measure single translation events in real time. Recently, the fluorescence of single protein molecules has been used to detect local translation events; however, maturation of fluorescent proteins (FPs) occurs several minutes after translation (Ifrim, Williams, & Bassell, 2015; Tatavarty et al., 2012; Yu, Xiao, Ren, Lao, & Xie, 2006).

To detect single translation events with high temporal and spatial resolution in living cells, we developed an RNA biosensor that enables identification of untranslated mRNAs from ones that have undergone at least one round of translation (Halstead et al., 2015). This technique relies on the ribosome removing a unique fluorescent signal from the coding sequence of a transcript during the first round of translation. We refer to this methodology as translating RNA imaging by coat protein knockoff (TRICK).

In this chapter, we describe the steps to design, express, acquire, and analyze data from the TRICK system. Particular attention is given to expressing TRICK transgenes at levels appropriate for single-molecule RNA imaging and to acquiring and analyzing two-color imaging data. Though emphasis is given to establishing the TRICK system in mammalian cell lines and *Drosophila melanogaster*, many of the principles described here are applicable to other experimental systems.

2. DESIGN OF TRICK REPORTER mRNAs

Considering translation from the perspective of a transcript has the advantage that robust methods have been developed that allow detection of single molecules of mRNA in living cells using fluorescent microscopy. The highly specific interaction between the MS2 bacteriophage coat protein (MCP) and its cognate RNA stem-loop has been extensively used to image RNAs in many experimental systems ranging from bacteria to mouse neurons (Urbanek, Galka-Marciniak, Olejniczak, & Krzyzosiak, 2014). This strategy relies on insertion of multiple copies of the MS2 stem-loop (usually 24) within the 3′-untranslated region (UTR) of a transcript of interest and the binding of a chimeric fusion of MCP with a FP to these sequences. Since the transcripts are bound by many fluorescent MCP-FPs, the mRNAs appear as bright diffraction-limited spots that can be detected above the background of the unbound MCP-FP. The inclusion of a nuclear localization sequence (NLS) to MCP-FP increases imaging sensitivity, because the unbound NLS-MCP-FP is concentrated in the nucleus, which enables rapid labeling of MS2 transcripts during transcription and reduces background in the cytoplasm (Fusco et al., 2003).

In order to take advantage of the spectrum of FPs that have been created, a number of other RNA–protein complexes have been engineered to visualize mRNAs (Chen et al., 2009; Daigle & Ellenberg, 2007; Takizawa & Vale, 2000). Given the success of the MS2 system, we developed the PP7 bacteriophage coat protein (PCP) that binds to a unique RNA stem-loop as an orthogonal RNA-labeling system (Chao, Patskovsky, Almo, & Singer, 2008; Larson, Zenklusen, Wu, Chao, & Singer, 2011). Using the MS2 and PP7 systems together allows simultaneous single-molecule RNA imaging of two distinct species of transcripts within a living cell and also enables the possibility of labeling a single mRNA within two different regions of the transcript (Coulon et al., 2014; Hocine, Raymond, Zenklusen, Chao, & Singer, 2013; Martin, Rino, Carvalho, Kirchhausen, & Carmo-Fonseca, 2013).

The act of translation requires that a ribosome traverses the coding sequence of an mRNA thereby decoding the information contained within a transcript in order to synthesize polypeptides. Consequently, any RNA-binding proteins that are bound to the transcript within the coding sequence must be removed by the ribosome. In particular, the exon junction complex, a multiprotein complex deposited upstream of exon–exon boundaries during splicing, is displaced by the ribosome during the first round of translation and therefore identifies transcripts that have never been translated (Ishigaki, Li, Serin, & Maquat, 2001). We reasoned that it should be possible to engineer a fluorescent RNA biosensor based upon this principle that would enable untranslated mRNAs to be distinguished from ones that had undergone at least one round of translation.

In order to construct an RNA biosensor to image the first round of translation, we designed PP7 stem-loops that could be translated by the ribosome that permitted the labeling of a transcript within the coding sequence (PP7) and the 3′-UTR (MS2) (Fig. 1A). When this transcript is untranslated, it will be bound by the two coat proteins fused to distinct FPs (eg, NLS-PCP-GFP and NLS-MCP-RFP) and will appear yellow when imaged because it will be labeled by both green and red FPs (Fig. 1B). During the first round of

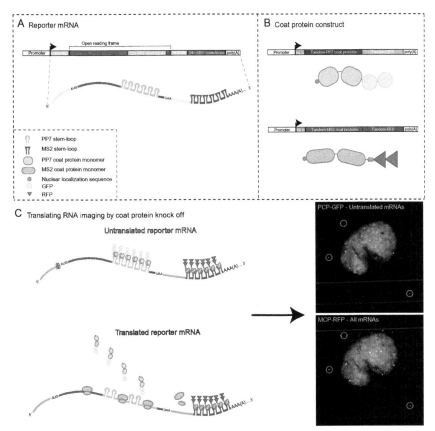

Fig. 1 Translating RNA imaging by coat protein knockoff. (A) TRICK reporter transcript contains translatable PP7 stem-loops in the open reading frame and MS2 stem-loops in the 3′-UTR. (B) PP7 and MS2 bacteriophage coat proteins are fused to spectrally distinct fluorescent proteins (eg, NLS-PCP-GFP and NLS-MCP-RFP). The addition of nuclear localization sequences (NLS) results in accumulation of unbound fluorescent proteins in the nucleus. (C) Untranslated mRNAs are bound by both fluorescent fusion proteins while translated mRNAs are labeled with only NLS-MCP-RFP. Dual-color RNA imaging can distinguish single molecules of untranslated mRNAs that are fluorescent in both *green* and *red channels* (*yellow circles*) from those that have been translated and are detected in the *red channel* alone. (See the color plate.)

translation, however, the ribosome will strip the NLS-PCP-GFP signal from the transcript resulting in the appearance of a red mRNA that is labeled with only NLS-MCP-RFP. Since the concentration of NLS-PCP-GFP is low in the cytoplasm, rebinding of NLS-PCP-GFP is not favorable and the displaced FP returns to the nucleus (Fig. 1C).

Since it is possible to design translatable RNA sequences for any of the RNA–protein complexes used to image RNAs, we have outlined the principles we used to design the PP7 stem-loops. The first MS2 stem-loops used to image RNAs were constructed by PCR by making two copies of the stem-loops that could then be multimerized by ligation with DNA fragments digested using restriction enzymes that generate compatible cohesive ends (eg, *Bam*HI and *Bgl*II) (Bertrand et al., 1998). Consequently, adjacent MS2 stem-loops were spaced by 20 nucleotides due to limitations, at the time, on the synthesis of large DNA oligonucleotides. A translating ribosome, however, produces a footprint of ~30 nts on a transcript, and we found that tight spacing of stem-loops results in a significant block of translation (Ingolia et al., 2009; Steitz, 1969). By spacing PP7 stem-loops farther apart (40 nts), we found that the ribosome could efficiently translate through them. Since the PP7 stem-loops will encode a polypeptide, we removed stop codons from the open reading frame and optimized codon usage for expression in mammalian cells by considering both codon frequency and coding potential. RNA folding was assessed using the *mfold* software to confirm that all PP7 stem-loops were predicted to fold correctly (Zuker, 2003).

Initially, a cassette containing six copies of the PP7 stem-loops was synthesized and tested for its translatability when fused to the C-terminus of a reporter gene. While western blot analysis revealed that the PP7 stem-loops were efficiently translated, imaging of this construct was difficult due to the low signal-to-noise ratio resulting from the small number of stem-loops. A 12×PP7 cassette was then generated which improved imaging of single mRNA molecules without inducing adverse effects on translation. In instances when addition of the polypeptide encoded by the PP7 stem-loops is not desirable (eg, labeling of endogenous genes), the addition of self-cleaving 2A sequences between the C-terminus of the protein of interest and the stem-loops may be advantageous (Kim et al., 2011). We have also found that it is possible to place the PP7 stem-loops within the N-terminus of a reporter gene, which may facilitate experiments designed to measure translation initiation rates by reducing the effect of ribosome elongation.

3. TRICK EXPERIMENT IN MAMMALIAN CELLS

3.1 Expression of TRICK Reporter Transcripts

The TRICK system provides a fluorescent readout from three transgenes, a reporter mRNA and two coat proteins, each of which must be expressed at appropriate levels in the same cell. Selecting the appropriate promoter to drive expression of the reporter mRNA is therefore key to designing a

TRICK experiment. As the TRICK system reports on the first round of translation only, there are advantages to controlling precisely when the reporter mRNA is expressed. While constitutive expression produces a mixed population of reporter transcripts that have been transcribed at different time points, including those transcribed and translated before imaging began, an inducible promoter can provide a population of nascent mRNAs whose pioneer round of translation is restricted to the experimental window. As a result, inducible expression of reporter mRNA is better suited to many experiments, particularly those examining the temporal regulation of translation. We found that both the tetracycline- and ponasterone A-inducible systems are well suited for performing TRICK experiments (Gossen et al., 1995; No, Yao, & Evans, 1996).

TRICK reporter mRNAs can be expressed from plasmids that are transiently transfected or stably integrated into the genome. Transient transfection is relatively fast and simple, with constitutive reporter mRNA expression peaking 24–48 h after transfection by lipid-based reagents, and can allow different TRICK reporters to be rapidly tested. Delivery of DNA plasmids by transient methods, however, can result in significant cell-to-cell variation in expression levels that can complicate image analysis. As it is possible to determine the translational status of every single TRICK reporter mRNA within a cell, the most meaningful quantitative comparisons are between cells expressing similar numbers of transcripts and at levels comparable to endogenous transcripts. It is possible to transiently transfect and identify cells within a heterogeneous population that meet this criteria, however, this limits data acquisition to a relatively small number of cells per experiment. Instead, reproducible and more uniform expression is best achieved by stably integrating the TRICK reporter into the cell genome by either random or site-specific integration. We have found that site-specific integration of an inducible TRICK reporter by recombinase-mediated cassette exchange to be an efficient method for generating TRICK cell lines. In this system, reporter mRNA expression is consistent across cells thereby limiting experimental variability. We use a HeLa cell line that contains both the rtTA2-M2 tetracycline reverse transactivator required for doxycycline-inducible expression and a single RCME site that allows stable integration of TRICK reporter mRNAs (Weidenfeld et al., 2009).

3.2 Expression of Coat Proteins Fused to Fluorescent Proteins

The crux of the TRICK experiment is to spectrally distinguish translated and untranslated mRNAs. Consequently, the appropriate selection and expression of the coat protein (CP; PCP and MCP)–FP pairs is critical to the success of the experiment. The CP-FPs used in the TRICK experiment must be fused to spectrally distinct fluorophores that are suitable for single-molecule RNA imaging. Both CP-FPs must also be expressed at cellular levels that allow robust labeling of all reporter mRNAs and, importantly for the TRICK experiment, do not favor rebinding of PCP-FP to the transcript once it has been displaced by the ribosome in the cytoplasm.

Single-molecule two-color imaging of mRNAs requires that the coat proteins are fused to fluorophores that satisfy a number of criteria. Foremost, as TRICK requires unambiguous separation of two fluorescent signals, spectrally distinct fluorophores must be selected to label each coat protein. Overlap of emission spectra and cross-excitation of overlapping absorption bands can be avoided by judicial choice of fluorophores. As a result, optimal imaging requires that distinct fluorophores are also matched to the light source and filters that will be used to acquire data.

Within these criteria, brightness, the product of quantum yield and extinction coefficient, and photostability, the decrease in emission over sequential excitation events, are the leading properties for fluorophore selection. This is particularly important for single-particle tracking (SPT) experiments because single mRNAs must be imaged multiple times and photobleaching of one channel can bias analysis. Genetically encoded FPs, such as GFP, exist in a range of spectra (Shaner, Steinbach, & Tsien, 2005) and have been successfully used in a number of single-molecule RNA imaging studies. We have found that NLS-PCP-EGFP and NLS-MCP-TagRFP-T are compatible with two-color RNA imaging with short (≤ 50 ms) exposure times. Other FPs can be used to suit other imaging modalities, however, the properties must fit the experiment parameters. For example, while tdTomato (extinction coefficient $= 138{,}000\ M^{-1}\ cm^{-1}$, quantum yield $= 0.69$) is brighter than TagRFP-T (extinction coefficient $= 81{,}000\ M^{-1}\ cm^{-1}$, quantum yield $= 0.41$), tdTomato is significantly less photostable and less suitable for time-lapse imaging (Shaner et al., 2008). The recent development of alternative intracellular fluorescent labeling technologies, including the Halo-Tag, Snap-Tag, and Clip-Tag, can also be used to conjugate chimeric coat proteins to inorganic dyes (Gautier et al., 2008; Juillerat et al., 2003; Los et al., 2008). This approach allows the coat proteins to be specifically covalently bonded to inorganic dyes such as tetramethylrhodamine-based dyes, which are bright (extinction coefficient $= 101{,}000\ M^{-1}\ cm^{-1}$, quantum yield $= 0.88$) and significantly more photostable than FPs (Grimm et al., 2015).

While inducible promoters are useful for controlling acute and robust reporter mRNA transcription, CP-FPs are best expressed constitutively at low levels. A number of constitutive promoters are commonly used, including SV40, CMV, UbiC, EF1a, PGK, and CAGG variants, of which PGK and UbiC have been found to reproducibly give lower expression levels (Qin et al., 2010). We have found that lentiviral transduction is an efficient means to stably express CP-FPs (Lionnet et al., 2011). Cells can be coinfected with two viruses each encoding a different CP-FP and cells that are double positive can be isolated by fluorescence-activated cell sorting (FACS). Successive rounds of FACS followed by fluorescent microscopy can be used to generate cell lines that express both CP-FPs at levels suitable for two-color single-molecule RNA imaging. The use of tandem single-chain dimers of the MS2 and PP7 coat proteins enables cells to be sorted for very low levels of the FPs without disrupting the RNA-binding function of the coat proteins (Wu, Chao, & Singer, 2012; Wu et al., 2015).

3.3 Considerations and Challenges of TRICK in Primary Cells

Balancing CP-FPs and TRICK reporter mRNA expression is particularly challenging in primary cells. Unless the cells are derived from an animal stably expressing the TRICK transgenes, both the reporter mRNA and CP-FPs must be introduced. While stable transgene integration followed by FACS of positive cells is optimal in cell lines, many primary cells can only be passaged a limited number of times and growing a sufficient number cells from a sorted population may not be possible.

In particular, primary neurons present a number of challenges for establishing single-molecule two-color RNA imaging. First, neurons are postmitotic and transient transfection rates are low. While other means of delivering transgenes (eg, infection using lentiviruses or electroporation) are more efficacious, a small percentage of cells will express the transgenes at appropriate levels that must be individually identified by fluorescence microscopy during each experiment. To increase the number of positive cells in a primary culture, the both CP-FP constructs can be expressed in a single bicistronic plasmid separated by an internal ribosome entry site (IRES), or a 2A peptide sequence. It is necessary to ensure that the CP-FPs are expressed properly because IRES-driven expression is typically lower than cap-dependent translation (Mizuguchi, Xu, Ishii-Watabe, Uchida, & Hayakawa, 2000) and 2A peptide sequences can produce chimeric fusion proteins that will undermine TRICK analysis (Kim et al., 2011).

3.4 Controls

To confirm that any TRICK construct gives a precise readout of translation, it is necessary to test if the insertion of either of the stem-loops or the binding of the CP-FPs perturbs reporter mRNA metabolism. The stability of the TRICK reporter mRNA can be assessed by real-time PCR following inhibition of transcription, either by small-molecules (eg, actinomycin D) or washing away activators of inducible promoters. The translation of the TRICK reporter construct should also be confirmed by western blotting to demonstrate that a protein of appropriate molecular weight is produced. It is also advantageous for the TRICK reporter to encode a protein with a functional readout (eg, fluorescence, bioluminescence, or enzymatic activity). We have found it useful for TRICK reporters to encode Renilla or Firefly luciferase because translation can be measured using a common luciferase assay that can be correlated with imaging data. Control experiments should be performed both with and without stem-loops and with and without coexpression of CP-FPs (Table 1). This approach will identify any element of the TRICK construct that affects reporter stability or translation and provides a starting point for troubleshooting.

4. MICROSCOPY

The success of a TRICK experiment largely depends on its accurate imaging readout. The field of live-cell single-molecule imaging has utilized different light microscopy variants, broadly characterized into epifluorescence wide-field, confocal, and total internal reflection fluorescence (TIRF) microscopy. Each of these microscopy setups offers certain advantages and drawbacks when dual-color single mRNAs, as in TRICK, are to be imaged in live cells.

4.1 Imaging Modality

A straightforward and cost-effective way of imaging single RNA molecules in live cells is conventional wide-field or epifluorescence microscopy. Here, a collimated/parallelized beam of light illuminates the entire thickness of the sample, resulting in a large field-of-view and a high number of molecules that can potentially be tracked. This advantage however comes at the price of a large fraction of out-of-focus light, resulting in high background levels and reduced single-molecule detection. Wide-field microscopy is therefore especially suited for specimens with a low number of fluorescent molecules

Table 1
Controls for TRICK Reporter mRNA Expression

Control Experiment	Reporter mRNA Expression	CP-FP Expression	Method
Is reporter mRNA stability affected by stem-loops?	1. TRICK reporter lacking stem-loops 2. TRICK reporter with stem-loops in the coding region and 3′-UTR	None	Measurement of mRNA half-life by real-time PCR. Northern blotting of mRNA
Is reporter mRNA stability affected by CP-FP binding?	TRICK reporter with stem-loops in the coding region and 3′-UTR	1. CP-FP bound to the coding region 2. CP-FP bound to the 3′-UTR 3. CP-FPs bound to both the coding region and 3′-UTR	Measurement of mRNA half-life by real-time PCR. Northern blotting of mRNA
Is reporter mRNA translation affected by stem-loops?	TRICK reporter with stem-loops in the coding region and 3′-UTR	None	Measurement of protein MW and levels by Western blot
Is reporter mRNA translation affected by CP-FP binding	TRICK reporter with stem-loops in the coding region and 3′-UTR	1. CP-FP bound only to the coding region 2. CP-FP bound only to the 3′-UTR 3. CP-FPs bound to both the coding region and 3′-UTR	Measurement of protein MW and levels by Western blot
Is the translated reporter protein functional?	TRICK reporter with stem-loops in the coding region and 3′-UTR	CP-FPs bound to both the coding region and 3′-UTR	Functional readout of protein activity (eg, fluorescence, bioluminescence, enzymatic activity)

and a thin imaging volume (a few microns or less), causing low background fluorescence. Consequently, cells with a small thickness such as yeast, bacterial, and epithelial cells or certain cellular extensions such as axons are well suited to be imaged by wide-field microscopy.

As opposed to wide-field microscopy, a confocal illumination scheme effectively reduces out-of-focus light and thereby enables the recording of high-contrast images in all spatial dimensions. Focusing the laser beam to the size of a diffraction-limited spot and using a pinhole aperture that rejects all light emitted outside of the focal volume are the two key elements to the spatial filtering of out-of-focus light. However, the confocal illumination scheme suffers from two major drawbacks. Since only a single diffraction-limited spot is illuminated at a time, the focal plane needs to be scanned point-by-point by either moving the sample or the laser beam, as in traditional laser scanning confocal microscopy (LSCM). Secondly, the pinhole aperture filters out a large fraction of the total light used for sample excitation. In combination, loss of emitted light and point-by-point scanning lead to long pixel dwell times to collect enough photons and long scanning times because relevant field-of-views typically require 10^5–10^6 pixels, resulting in a low frame rate that can become limiting when imaging rapidly moving mRNA molecules. Spinning-disk confocal microscopy has been developed as a faster alternative, and we have successfully used it for TRICK experiments. In spinning-disk microscopy, a first disk with a large number of spirally oriented microlenses is rotating at high speed and essentially focuses thousands of split laser beams on the sample. A second spinning disk, with a spiral array of pinholes, is aligned to the first and blocks all out-of-focus light. As a result, the fast, synchronized disk rotation and spiral orientation of microlenses and pinholes enables the near-simultaneous scanning of a single imaging plane. While allowing the high temporal resolution (subsecond frame intervals) required for TRICK experiments, a spinning-disk confocal microscope suffers from reduced fluorescence intensity detection due to beam splitting and the use of pinholes. Rapid photobleaching due to out-of-focus fluorescence excitation is another major limitation of all confocal microscopy approaches.

Reducing photobleaching is imperative during TRICK experiments. The pair of fluorophores that fluorescently label reporter mRNAs usually have different sensitivities to photobleaching. If one fluorophore bleaches faster than the other, the detection efficiency of the corresponding fluorophore will vary over the time of one experiment and might result in a bias toward incorrect identification of translational state. One way to reduce photobleaching (as well background fluorescence due to out-of-focus excitation) is to use an optical sectioning method called TIRF microscopy. TIRF is based on the observation that a collimated laser beam propagating through one medium when reaching a second medium is reflected at the interface, if a large enough angle and appropriately different refractive indices of the two media are chosen. The reflection creates an exponentially decaying evanescent wave in the z-direction within the sample, allowing only the excitation of molecules within a few hundred nanometers within the sample and leading to an excellent reduction of bleaching and high contrast. While TIRF can be a good option for live-cell imaging, it allows only the imaging of molecules close to the sample-glass interface, making it impossible to image and detect TRICK transcripts deep inside of the cell. Furthermore, the rapidly decreasing intensity of the evanescent wave with increasing distance from the coverslip leads to a wide distribution of fluorescent particle brightnesses—a substantial challenge for quantitative analysis. Another optical sectioning method is more widely applicable to imaging TRICK reporter mRNAs and allows for the detection of translation in all cellular compartments. Here, an inclined laser beam just below the critical angle for TIRF is used. Instead of creating an evanescent wave, the laser beam now passes directly into the sample, but in the form of a highly inclined laminated optical sheet (HILO) with a thickness in the low micrometer range (Tokunaga, Imamoto, & Sakata-Sogawa, 2008). A rotating mirror ensures that the sample is illuminated from all angles. Because any form of point scanning is not necessary, high frame rates can be achieved. We have successfully applied HILO to perform TRICK experiments in live cells. HILO offers low amounts of out-of-focus light, low bleaching, and fast image acquisition rates.

4.2 Light Source

Since a large number of fluorophores with excitation–emission spectra ranging from near-ultraviolet, visible, to near-infrared regions are currently available (Kremers, Gilbert, Cranfill, Davidson, & Piston, 2011; Xia, Li, & Fang, 2013), it is often most appropriate to match the required fluorophores to the already existing illumination setup.

Two spectrally distinct lasers constitute the illumination source of choice (especially for confocal imaging) since they provide a narrowly defined excitation wavelength range, low divergence when passing through the optical setup, and high excitation power. Gas lasers such argon or krypton ion lasers or solid-state lasers such as diode lasers are most commonly used. For dual-color imaging, the most important criterion when choosing a light source is to maximize the fluorophores excitation efficiency while minimizing excitation cross talk. For dual-color mRNA imaging with the described MS2 and PP7 system, 488 and 561-nm emitting lasers have been found to be appropriate to excite the commonly used FPs such as GFP or RFP-variants while providing sufficiently separated emission spectra. More information on how to optimize the laser excitation of various fluorophores for live-cell imaging has been described elsewhere (Xia et al., 2013).

For wide-field microscopy, cost-effective alternatives to laser excitation can be used, eg, arc-discharge lamps such as mercury or xenon arc lamps, or more recently light-emitting diodes (LEDs) (Cho et al., 2013; Gerhardt, Mai, Lamas-Linares, & Kurtsiefer, 2011; Higashida et al., 2008). The commonly used mercury-based arc lamps typically display intensity peaks at certain wavelengths and therefore do not provide an even intensity across the full light spectrum. These considerations are important when deciding on a light source for simultaneous dual-color imaging. Xenon-based arc lamps display a more even intensity profile, but lack the shorter wavelength range typically employed for fluorescent microscopy. A third form of arc-discharge-based light sources is metal-halide lamps. This type of lamp combines the properties of xenon and mercury, resulting in an even, high intensity emission across the entire light spectrum from ultraviolet to infrared. The use of LEDs for dual-color RNA imaging is currently limited since the detection of single molecules has so far only been demonstrated in the close blue and green spectra (Gerhardt et al., 2011; Kuo, Kuyper, Allen, Fiorini, & Chiu, 2004). Recently, white LEDs have been used for super-resolution microscopy in live cells (Cho et al., 2013) and might represent an attractive alternative to laser light sources for dual-color single-molecule imaging in the future.

4.3 Signal Detection

For TRICK experiments, it is key to detect both fluorescence channels unambiguously to avoid a systematic analysis bias toward one of the channels and incorrect conclusions about the translation state. Signal bleed-through due to overlapping fluorophore emission spectra can be effectively minimized by appropriate design of the bandpass excitation and emission barrier filters and should be controlled for. mRNA particles can travel relatively fast with diffusion coefficients up to 3.42 $\mu m^2 \ s^{-1}$ (Ma et al., 2013) and transport rates of $\approx 1.3 \ \mu m \ s^{-1}$ (Park et al., 2014). The mRNA's image on the camera chip can therefore move by one pixel or more between frames, even when imaging at subsecond intervals. Consequently, it is crucial to image the two-color channels simultaneously in order to unambiguously identify dual-labeled particles. As a consequence, filter cube switching is not an option for TRICK imaging. Instead, the best setup for the microscope emission path is to first separate the emission light from the excitation source with a multiline dichroic mirror, followed by a second dichroic mirror that will split the collected fluorescence in two beams (eg, GFP vs RFP fluorescence). The two beams can then be imaged by two separate cameras or recombined on the two halves of the same camera chip.

The two most used camera types include scientific complementary metal oxide semiconductor (sCMOS) and electron multiplying charge coupled device (EMCCD) cameras. Due to the coupling of photon detection and voltage conversion at the physical pixel level on the chip, sCMOS cameras allow for high-speed imaging of more than 100 frames s^{-1}. The consequence is a relatively high internally generated background (dark noise) and readout noise, limiting the photon sensitivity under low light conditions, as is the case for single-molecule RNA imaging. In contrast, EMCCD cameras achieve single-photon sensitivity. Despite having a lower chip readout speed, which is generally still permissive for single mRNA tracking, EMCCD cameras currently offer significant advantages over other types of camera sensors, especially when applied under low emitted light conditions.

TRICK experiments require the acquisition of both fluorescent channels simultaneously in order to guarantee signal colocalization. The solution of choice to image relatively large biological structures (eg, an entire mammalian cell, typically tens of microns across) is to separately collect the fluorescence from the two channels on two well-aligned cameras. At a magnification that permits the detection of single molecules, the use of two cameras provides a large enough field-of-view in each fluorescent channel to fully capture the sample. A practical alternative to the costly dual-camera approach is the use of a split chip on a single camera. Here, the image is divided into two spatially equivalent, but separate halves on the camera chip. Postimaging registration of both halves of the chip into a single image can be performed.

4.4 Temperature and CO_2 Control

Translation is very sensitive to changes in the extracellular environment. TRICK live-cell experiments therefore require strictly physiological conditions during imaging. Mammalian cells generally require an incubation temperature of 37°C and a CO_2 atmosphere of 5–7% to maintain an appropriate pH, depending on the growth medium. The humidity level should also be controlled in order to prevent excessive evaporation and drying. Several commercial systems are available to keep the biological sample under these physiological conditions during imaging. A simple and inexpensive method is to use heating elements directly adjacent to the objective and sample dish. Repeated cycles of heating and cooling, however, lead to thermal movements of the objective and the sample itself leading to a loss of focus over time. While less economical, a Plexiglas incubation box, which fully encloses the microscope body, is more suitable for live-cell imaging. A second cover that seals with the stage can then be placed on top of the specimen to regulate CO_2 and humidity within this restricted volume.

5. DATA COLLECTION

The quantification of data obtained from TRICK experiments involves two major steps. First, all mRNA particles per imaging frame need to be detected in each of the two respective channels. Then the detected particle positions in consecutive frames are combined to give individual mRNA trajectories in a computational process termed SPT. To avoid detection biases toward one channel based on the different fluorophore properties, the acquired imaging data needs to be of high quality by fulfilling the two following major criteria: first, the signal-to-noise ratio (SNR) needs to be as high as possible, while, second, the imaging frame rate needs to be sufficiently high to connect the positions of individual mRNAs over time (Park, Buxbaum, & Singer, 2010).

5.1 Considerations for Single-Molecule Detection and Tracking

A major limitation of live-cell imaging experiments is that a high SNR and frame rate will lead to rapid photobleaching and phototoxicity caused by an excessive amount of photons hitting the specimen. Finding a balance between frame rate, exposure time, and excitation power is therefore key to being able to image and subsequently track single mRNAs in two colors (Magidson & Khodjakov, 2013).

Because of the diffraction of light, individual fluorescent particles appear as spots of a few 100 nm in size when imaged through a microscope. The shape of the spot depends on the microscope and the fluorophore and is called the point spread function (PSF). The spatial profile of the PSF on the camera is nearly Gaussian, which makes it possible through fitting to measure the position of each particle center with high accuracy (typically ≈ 40 nm). If the PSF is spread onto a large number of camera pixels, each spot becomes hard to separate from the background and readout noise. Conversely, if the spot on the image is concentrated onto a single pixel, it is hard to achieve a high localization precision through fitting. A high SNR is therefore typically obtained by combining a high numerical aperture objective (which means a high light collection power), and a magnification optimized for single-molecule imaging. In practice, magnifications that yield 100–200 nm per pixel tend to give optimal localization precision for single-molecule tracking (Thompson, Larson, & Webb, 2002). As images of mRNA labeled in separate colors need to be registered, chromatic aberrations pose a significant challenge. They are minimized by using apochromatic objectives and can be corrected postacquisition using adequate calibration techniques, eg, using multicolor beads or fiducial markers within the sample (Grunwald & Singer, 2010).

Diffusion coefficients of mRNAs have been shown to range from 0.009 to 3.42 $\mu m^2 \, s^{-1}$ depending on the type of transcript and subcellular localization (Fusco et al., 2003; Ma et al., 2013; Mor et al., 2010; Shav-Tal et al., 2004). Therefore, exposure times need to be long enough for a good SNR while being short enough to prevent blurring of fast moving mRNA particles. In addition, the frame rate needs to be short enough to reliably track an mRNA's position from frame-to-frame. Exposure times and frame rates should therefore be optimized based on each system's dynamic range, ensuring unbiased detection while minimizing oversampling and photobleaching. In cases where very fast frame rates and short exposure times are crucial, but the chip readout speed is limiting, it is possible to readout only a subarray of the chip, thereby decreasing the total required readout time per frame at the expense of a reduced field-of-view.

Besides optimizing the frame rate, other means to increase the SNR can be used, such as pixel binning. During pixel binning a group of pixels on the camera chip (eg, 2×2 pixels) are binned together and assigned to a single pixel value during the readout of the chip. A 2×2 binning for example increases the signal fourfold (four times more photons per pixel) at the cost of a twofold lower resolution (pixel information is lost). Finally, optimizing the camera gain and the chip readout speed can also improve the SNR and dynamic range. Slower chip readout speeds result in lower readout noise and can be compensated for by utilizing only a subarray of the chip, as described earlier.

5.2 Considerations for Long Time-Lapse Experiments

The TRICK technique can be used to study translation regulation with subcellular resolution over a wide range of time scales: from fast, single-molecule dynamics movies (subsecond frame rates covering a few seconds to minutes) to longer, time-lapsed acquisitions matching the dynamics of cellular responses (minutes to hours). Minimizing photobleaching by the previously described imaging optimizations is key for successful longer time-lapse experiments. Although the detailed mechanism of photobleaching is not entirely known, photo-oxidation of the fluorophores is thought to be caused by reactive oxygen species (ROS). The addition of chemical compounds such as ascorbic acid (vitamin C) (Vigers, Coue, & McIntosh, 1988), trolox (a vitamin E derivative) (Rasnik, McKinney, & Ha, 2006), mercaptoethylamine (Widengren, Chmyrov, Eggeling, Lofdahl, & Seidel, 2007), or enzymatic deoxygenation systems (Aitken, Marshall, & Puglisi, 2008) has been shown to delay photobleaching. Since oxygen plays an important role in cell physiology, the use of ROS scavengers can have unwanted effects that need to be taken into account.

Stage stability over long time periods in all three spatial dimensions is crucial for long-term live-cell experiments. Especially when heated incubation chambers are used, it is important to thermally equilibrate the whole microscope body, including the stage, prior to the experiment in order to prevent thermal drift and to avoid permanent refocusing. Motorized Piezo stages that are equipped with reflection-based rather than image-based autofocus systems minimize manual interventions during an experiment.

Some adherent cell types growing on coverslips can move extensively in the x,y dimension even on the order of minutes and might require the use of cell motion tracking (Rabut & Ellenberg, 2004). Several software packages that can be coupled to the appropriate microscope hardware are currently available by commercial suppliers.

6. ANALYSIS

Dual-color single-molecule mRNA imaging during a TRICK experiment allows direct observation of two distinct translational states depending on the presence of one or both fluorophores. In order to reconstruct the trajectories of individual mRNA particles and determine their translation state, multiple computational steps must be performed (Fig. 2). First one needs to identify and localize discrete particles on each acquired image (Fig. 2A); this step outputs a list of particle positions for each time point and color channel (Fig. 2B). The second step consists of tracking the particles, which means connecting together the spot positions that correspond to the same particle at different times, yielding a list of trajectories for each color channel (Fig. 2C). Finally, one sorts the trajectories present in both channels

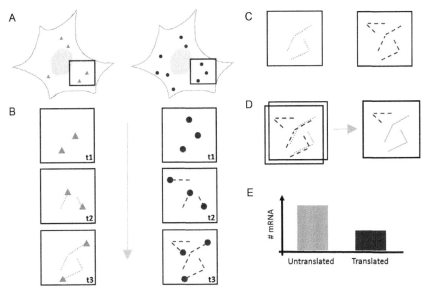

Fig. 2 Schematic depiction of the analysis workflow for a TRICK experiment in living cells. (A) Each cell is imaged simultaneously in two colors, resulting in two fluorescent channels. Here, the mRNAs labeled with NLS-PCP-GFP and NLS-MCP-RFP are depicted by *triangles* and *circles*, respectively. (B) After image acquisition all mRNA diffraction-limited spots in each imaging frame are detected. Spot detection is best performed individually for each fluorescent channel. (C) Next, spot tracking is performed. Here, the spot positions in each frame are taken into account and tracks for each mRNA molecule are calculated. (D) The resulting tracks from both fluorescent channels originating from the same cell are then assessed for colocalization. *Red* (*dark gray* in the print version) and *green* (*light gray* in the print version) mRNA tracks represent the same dual-colored molecule when a significant overlap exists. (E) mRNAs that have been determined to be dual-colored are considered to be untranslated, while single *red* (*dark gray* in the print version) colored mRNAs have been translated at least once.

(corresponding to dual-labeled untranslated mRNA) from those present in a single channel only (single-labeled translated mRNAs) (Fig. 2D and E). Measuring the spatiotemporal evolution of single- vs dual-color-labeled molecules then gives an indication about the localization and translational status of the mRNAs.

6.1 Single-Molecule Detection and Tracking

Detection and tracking of single particles should be performed for each fluorescent channel individually (Fig. 2A and B). A number of commercially and freely available software packages exist that combine detection and tracking of all single particles in each frame within a given image sequence in two consecutive steps in a semiautomated manner. Different particle tracking approaches have recently been extensively tested and reviewed by Chenouard et al. (2014). In general particle detection and tracking should be performed on unprocessed data that have been recorded according to the earlier described principles in order to maximize the SNR and to fulfill the Nyquist criterion on temporal sampling so that individual particles can be tracked over time (Park et al., 2010). The precise particle detection and tracking methodology needs to be chosen based on the data quality, particle density, and intended tracking time frame. For the detection of single particles several approaches exist. Maxima-based detection relies on the identification of the highest local pixel values, which are then defined as spots. Thresholding utilizes the principle of a particle's higher intensity over the surrounding background based on an appropriate intensity threshold. More accurate (and computationally intensive) approaches involve PSF fitting and centroid estimation. Fitting often relies on the PSF-based fitting of a Gaussian intensity distribution to each spot candidate or uses other linear or nonlinear models. Centroid estimation detects diffraction-limited spots by determination of the radial spot center, which often does not coincide with the local maximum and is a reliable method to distinguish neighboring spots (Parthasarathy, 2012).

Once spots have been detected and their positions evaluated, one needs to connect the spots in order to generate trajectories (Fig. 2C). The simplest method to achieve this connects each spot with its nearest neighbor in the next frame, allowing for only a limited displacement range (based on knowledge of the typical transport properties of the biological species), and a given number of gaps—false negatives are common in single-molecule tracking because fluorescence of a particle might be intermittently obscured by noise or background, resulting in a missed detection. Multiframe or multitrack and motion model-based tracking approaches are more sophisticated techniques that go beyond frame-to-frame nearest-neighbor linking and are suitable for live-cell mRNA imaging. They are robust against partial detection failures and crossing trajectories. Multiframe or multitrack approaches take the history of a tracked particle into account in order to match it to a future estimated trajectory. Motion-based models fit the trajectories to typical single-particle movement patterns such as Brownian motion, corralled diffusion, or directed motion (Park et al., 2010). The most robust particle tracking methods rely on a combination of several of the aforementioned detection and tracking approaches (Chenouard et al., 2014). Because it is technically difficult using wide-field epifluorescence or confocal microscopy to acquire 3D volumes at frame rates compatible with single-molecule tracking with sufficient SNR, single mRNA tracking has mostly been performed in 2D planes where particle movement in and out-of-focus limits the observation time to a few seconds in ideal cases. However, innovative microscopy approaches have recently been developed to overcome this issue and collect 3D trajectories of mRNA particles (Smith et al., 2015; Spille et al., 2015). These imaging modalities circumvent the problem of particles moving out-of-focus and are able to generate longer trajectories; the tracking analysis techniques are conceptually identical to those used for 2D tracking.

Although long trajectories are ideal to investigate the fate of individual mRNA molecules, short observations times are not necessarily limiting to determine the translation state of a two-color-labeled mRNAs (Fig. 2). Short trajectories of a few 100 ms are often sufficient to reliably determine the degree of colocalization of an mRNA population within a cell.

6.2 Determining Colocalization of Tracked Two-Colored mRNA Particles

The determination of colocalization between both fluorescent channels is key to determine translation, as in the case of TRICK or other dynamic properties probed by a two-color mRNA imaging experiment. Since every field-of-view typically contains a large number of tracked mRNAs, colocalization between the two trajectory data sets is best performed in an automated fashion (Deschout et al., 2013; Dupont, Stirnnagel, Lindemann, & Lamb, 2013; Koyama-Honda et al., 2005).

One important first step in colocalization is to ensure one can accurately register the two-color channels. This calibration is usually performed by imaging small ($\approx$ 100 nm) fluorescent beads or gold nanorods that emit a broad spectrum of light spanning the two channels used in the experiment. Each bead produces one diffraction-limited spot in each channel. By using detection algorithms to measure the position of each bead image in the separate channels, one can calculate the spatial transformation needed to precisely map the position of the red beads onto the green ones. This process corrects for systematic chromatic aberrations (specific to one microscope because of the properties of its lenses, but invariant over time) and misalignments. As microscopes can substantially drift over time, it is important to perform these calibration routines frequently, ideally on a daily basis if one desires to achieve high registration accuracies. Typical registration errors can be as small as 5–10 nm. The spatial transformation generated during the calibration is then applied to the measured spots after the experiment, before matching them to the second channel.

Algorithms carrying out this kind of colocalization have been described in more detail before (Espenel et al., 2008; Halstead et al., 2015). Although in principle one could assess colocalization at each time point by matching positions of green and red spots, this strategy is very sensitive to missed or spurious detections, which are not uncommon when tracking individual molecules in the low SNR regime. We found that a more effective approach consists of matching trajectories rather than individual spot positions (Fig. 2D). The reason is that tracking algorithms are designed to accommodate both false negatives (short gaps are usually allowed in trajectories) and false positives (only trajectories longer than a few frames are considered, which cleans up spurious detections). As a result, matching trajectories rather than spots is a robust way to assess colocalization at the single-molecule level.

The algorithm to match trajectories consists of measuring the spatiotemporal overlap between all green and red trajectories. If two trajectories are found to be within a certain distance threshold of one another (typical value in our experiments is 100 nm) for a certain number of frames (typical value in our experiments is three frames), then they are scored as colocalized. Even though the number of frames used as our colocalization criterion might seem small, longer colocalized trajectories are usually visibly moving together for their entire duration. Perfect trajectory overlap is often not achievable, because of the uncertainty in measuring the position of each spot (in our conditions, around 40 nm in x,y), and the error in registering the two channels together (≈ 15 nm). This algorithm works best when the particles are bright and well separated in space, but is robust in a wide range of SNR and concentrations typical of single-molecule tracking. One advantage of TRICK is that out of three potential trajectory combinations (red only, green only, and colocalized red + green), one only expects to observe two: red only (3'-UTR label only for translated mRNAs) and colocalized green + red (both ORF and 3'-UTR label for untranslated mRNAs) because the 3'-UTR red label always remains bound to its target. Therefore, the measurement carries an internal control: the number of colocalized trajectories over the total number of green trajectories (green only and colocalized) is a direct metric of the experimental sensitivity (typical values in our experiments 80–90%). The results can be expressed as colocalization percentages per cell (Fig. 2E), indicating for example the amount of translated mRNA at different time points after reporter induction. Trajectories can then be analyzed separately to investigate the relative importance of location, transport, and dynamic properties of various mRNA translation states.

6.3 Controls

It is important to bracket imaging experiments with positive and negative controls. Imaging and performing particle detection on cells lacking CF-FPs should not reveal fluorescent signal. Similarly, imaging cells expressing CF-FPs, but not the reporter mRNA, should show diffuse CP-FP signal in the nucleus only, and no bright single particles. To determine if every mRNA is detectable via live-cell imaging, TRICK can be combined with single-molecule fluorescence in situ hybridization (mRNA FISH). In fixed cells, multiple singly labeled fluorescent FISH probes robustly detect single reporter mRNAs, which should correspond to the same number of mRNAs detected in live cells by CP-FPs. Imaging the entire cell volume in a 3D stack can confirm that every reporter mRNA is fluorescently labeled by coat proteins.

Under steady state conditions in mammalian cells ≈6% of our standard reporter mRNAs are untranslated and so fluoresce in both red and green channels, while ≈94% of mRNAs are translated and appear in the red channel only. Imaging cells expressing only one CP-FP in both channels should yield no colocalization and controls for fluorophore cross talk. As a positive control for detection of colocalization, both stem-loop cassettes can be inserted into the reporter 3′-UTR, which should result in 100% of mRNAs that fluoresce in both channels independent of translation. This is a particularly useful control for optimizing the colocalization of two-color trajectories from SPT data.

Inhibitors of translation can demonstrate that the TRICK signal (loss of fluorescence from the coding sequence) is translation-dependent and serve as a powerful control. Small-molecule inhibitors affecting different steps of translation can be used as complementary controls (eg, puromycin causes premature termination and cycloheximide halts elongation).

7. TRICK EXPERIMENT IN HeLa CELLS TO DETERMINE FRACTION OF UNTRANSLATED mRNAs

7.1 Preparation of Cells for Live-Cell Imaging

Materials
- Tetracycline-inducible HeLa cells stably expressing NLS-PCP-GFP, NLS-MCP-Halo, and a TRICK reporter mRNA containing PP7 (coding region) and MS2 (noncoding region) stem-loops

- Dulbecco's Modified Eagle Medium (DMEM; Life Technologies, 10566-016) supplemented with 10% (v/v) Tet-free FBS (Clontech, 631106) and 1% (v/v) penicillin and streptomycin (pen/strep)
- 35 mm μ-Dish (Ibidi, 81158)
- Automated cell counter and counting slides (Biorad, D9891-1G)
- Doxycycline (Sigma, D9891-1G)
- JF_{549} (HHMI Janelia Research Campus)

Day 1

1. HeLa cells are grown using standard cell culture techniques as adherent monolayers in DMEM + 10% FBS + 1% Pen/Strep.
2. Prepare a cell suspension of HeLa cells at a density of 20,000 cells mL^{-1} and ensure dissociation into single cells.
3. Seed 2 mL of cell solution per 35 mm imaging dish. Care should be taken in order to obtain a homogenous distribution of cells within the dish.
4. Incubate 2 days at 37°C and 5% CO_2. Shorter incubation periods are also possible depending on the time it takes for cells to attach and spread on the surface of the imaging dish.

Day 3

1. Prewarm PBS and DMEM + 10% FBS to 37°C.
2. Halo-label cells by addition of 1 mL 100 nM JF_{549} in DMEM + 10% FBS.
3. Return cells to incubator (37°C, 5% CO_2) for 15 min.
4. Remove medium and wash cells 3 × with PBS.
5. Replace medium with 37°C warm DMEM + 10% FBS containing 1 μg mL^{-1} doxycycline to induce TRICK reporter expression.

7.2 Image Acquisition

Materials

- Olympus IX81 inverted microscope (Olympus) equipped with a Yokogawa CSU-X1 scanhead (Yokogawa) and Borealis modification (Andor)
- Dichroic beam-splitter in scanhead (Semrock Di01-T488/568-13x15x0.5)
- 100 × 1.45NA PlanApo TIRFM oil immersion objective (Olympus)
- Two back-illuminated EvolveDelta EMCCD cameras (Photometrics)
- Emission filters for GFP (Semrock, FF01-617/73-25) and JF_{549} (Semrock, FF02-525/40-25) fluorescence

- Beam-splitter between cameras (Chroma, 565DCXR)
- Solid-state lasers (100 mW 491 nm and 100 mW 561 nm; Cobolt)
- Motorized X,Y,Z-Piezo controlled stage (ASI)
- Incubation box around microscope providing heating and CO_2 regulation (Life Imaging Services)

Day 3
1. Equilibrate microscope imaging chamber to 37°C and 5% CO_2.
2. Select cells for imaging using MCP-Halo channel by identifying cells that contain well-resolved diffraction-limited particles (spot width ~2 pixels). Image using low laser power to limit photobleaching before acquisition of TRICK data.
3. Simultaneously image cells in both channels using laser powers, camera gain, and exposure times compatible for SPT. Exposure times should be 40–50 ms or less, in order to ensure that fast moving mRNPs can still be unambiguously tracked between subsequent frames. Laser power optimization is a tradeoff: high laser powers result in bright, well-resolved particles that are easier to track, but induce rapid photobleaching. Once adequate exposure and laser power settings have been set, the camera gain should be optimized to provide the largest possible intensity dynamic range without saturating the detector.

7.3 Image Analysis

Materials
- Broad-emitting beads, Tetraspeck microspheres mounted on a slide (ThermoFisher Scientific T-14792)
- ImageJ with TrackMate plugin
- Matlab (Mathworks) software and scripts

Day 4 (Tracking)
1. Ensure that both channels are precisely registered. Cameras should be aligned prior to image acquisition using multicolor beads mounted on a standard slide. Any residual systematic offset between channels can be corrected using the translate function within ImageJ.
2. Particle tracking can be performed on a small number of frames (typically 3–5) to prevent biasing the analysis toward immobile particles.
3. Define a region of interest for analysis (eg, a single cell, nucleus, or cytoplasm).
4. Filter out randomly distributed noise using the FFT bandpass filter within ImageJ. Filter small objects below 3 pixels to reduce noise.

5. Detect spots using the Laplacian of Gaussian (LoG) detector in TrackMate (ImageJ). Spot size and thresholds should be optimized for detection of single mRNA particles.
6. Detected spots can be joined into trajectories in TrackMate using the linear assignment problem (LAP) tracker. The parameters linking max distance, gap-closing distance, and gap-closing max frame gap should be optimized as necessary to increase or reduce tracking stringency.
7. Use the visual inspector to ensure that particles are appropriately tracked.
8. Export tracking data as a spreadsheet.

Day 4 (Colocalization)
9. Colocalization analysis of trajectories is performed in Matlab (Mathworks) with custom written scripts.
10. Two tracks are considered to be colocalized if at least two spots of the green trajectory are within a pixel in x,y of a red trajectory.
11. Colocalization is then evaluated for accuracy by assessing individual colocalized trajectories.
12. Orphan red channel trajectories are identified as the translated mRNA fraction while the colocalized trajectories represent the mRNA fraction that has remained untranslated.

8. TRICK EXPERIMENT IN *DROSOPHILA*

Maternally deposited mRNAs of *D. melanogaster* encoding embryonic axis determinants such as *oskar*, *bicoid*, *gurken*, and *nanos* are frequently used as model systems to study mRNA transport and translational regulation. In the past, transport of these mRNAs has been successfully studied using transgenic animals expressing MS2-tagged reporter mRNAs (Forrest & Gavis, 2003; Jaramillo, Weil, Goodhouse, Gavis, & Schupbach, 2008; Weil, Forrest, & Gavis, 2006; Zimyanin et al., 2008). However, the insertion of MS2-binding sites has to be planned carefully in order to not destroy important *cis*-regulatory elements that are often located in the 3′-UTR and essential for proper transport and translational control (eg, oocyte entry signal (Jambor, Mueller, Bullock, & Ephrussi, 2014); translational control element (Gavis, Lunsford, Bergsten, & Lehmann, 1996)). Notably, some mRNAs such as *oskar* require splicing for transport and translational control (Ghosh, Marchand, Gaspar, & Ephrussi, 2012; Hachet & Ephrussi, 2004). We therefore always modify genomic DNA fragments.

To show the feasibility of imaging the first round of translation in *Drosophila*, we chose the *oskar* mRNA, which is produced in the nurse cells and transported over a long distance in order to localize to the posterior pole of the developing oocyte, where it is finally translated. We used a genomic rescuing construct of 6.45 kb (Ephrussi & Lehmann, 1992) in which 6×MS2 loops were inserted into the 3′-UTR (Fig. 3A). This insertion has been previously used to study transport and has been shown to give rise to functional Oskar protein (Lin et al., 2008). Using the endogenous *oskar* promoter ensures that expression of the TRICK reporter is comparable to wild-type levels. In order to generate the functional *osk-TRICK* reporter, we inserted 12×PP7-binding sites in frame into the coding region (Fig. 3A). To generate transgenic animals, the full genomic region was subjected to P element-mediated germline transformation.

In order to provide the coat proteins necessary for labeling of the TRICK mRNA in the nurse cell nuclei, we express NLS-MCP-RFP and NLS-PCP-GFP fusion proteins under the control of a weak maternally active promoter, such as the *hsp83* promoter (Forrest & Gavis, 2003). Importantly, only this moderate expression of the coat proteins ensures no labeling artifacts, as seen by UAS-Gal4 driven constructs that produce nonspecific motile particles even in the absence of MS2-labeled mRNA (Xu, Brechbiel, & Gavis, 2013).

8.1 Imaging and Analysis
Materials
- 1 × PBS
- 16% paraformaldehyde solution (Electron Microscopy Sciences, #15710)
- Tween20 (Sigma, T9284)
- Triton X-100 (Sigma, P1379)
- BSA (bovine serum albumin, Sigma, A2153)
- Glass slides and coverslips
- Mounting solution (eg, Shandon Immu-Mount, Fisher Scientific, 9990402)

Protocol for *Drosophila* Oocytes
1. Dissect ovaries from well-fed female flies expressing *TRICK* mRNA, NLS-MCP-RFP, and NLS-PCP-GFP in 1 × PBS.
2. Replace PBS with fixative (1 × PBS supplemented with 4% paraformaldehyde) and incubate for 20 min.
3. Wash twice with PBST (1 × PBS, 0.1% Tween20).
4. Permeabilize ovaries for 1 h in 1 × PBS with 1% Triton X-100.
5. Wash twice with PBST (1 × PBS, 0.1% Tween20).
6. Block with blocking buffer (1 × PBST with 0.5% BSA) for 30 min.
7. Remove blocking buffer and add primary antibody (eg, anti-Oskar) in blocking buffer for 2 h.

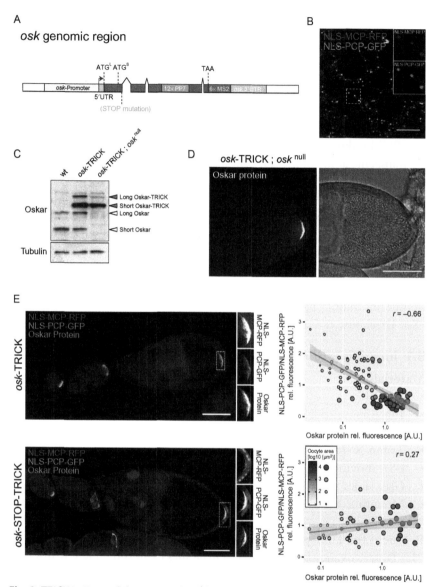

Fig. 3 TRICK in *Drosophila* oocytes. (A) Schematic of a genomic *osk*-TRICK reporter construct. The alternative translational start sites producing the long and short Oskar isoforms (ATG^L and ATG^S), the insertion site of 12×PP7 in the coding region, 6×MS2-binding sites right after the stop codon (TAA), and the position of the stop codon mutation used for control experiments are indicated. (B) Imaging of individual mRNPs in the ooplasm of an egg-chamber expressing *osk*-TRICK mRNA, NLS-MCP-RFP and NLS-PCP-GFP using FP-booster, scale bar 5 μm. (C) Insertion of 12xPP7-binding sites does not disturb translation of *oskar* mRNA. Western blot analysis of ovarian samples

8. Wash three times with PBST.
9. Incubate ovaries with secondary antibody conjugated to fluorophores spectrally distinct from EGFP and TagRFP-T (eg, Cy5, Alexa 647) in blocking buffer for 1 h.
10. Wash three times with PBST.
11. Separate individual egg-chambers, mount them on a glass slide using mounting solution and cover with a coverslip.
12. Acquire images on a standard wide-field or confocal microscope.
13. Images are further processed and the fluorescent signals and oocyte size of individual egg-chambers are measured using ImageJ (http://rsb.info.nih.gov/ij/). The ratio of NLS-PCP-GFP per NLS-MCP-RFP is calculated and plotted against the protein signal intensity and oocyte size.

Optional: *Drosophila* egg-chambers are relatively thick (from ~50 to >100 µm), which can present challenges for imaging. In order to obtain a good SNR, immunostaining with direct-coupled antibodies (eg, Anti-GFP-CF488A (Sigma, SAB4600051), RFP-booster (ChromoTek, rba594-100)) against FPs of the coat proteins can help to overcome this issue in fixed egg-chambers. This allows higher resolution imaging of individual RNA–protein particles containing *osk*-TRICK mRNA by confocal microscopy followed by deconvolution (Fig. 3B). Single-particle analysis can be carried out as described earlier.

8.2 Controls

As a first test for any defects in translation of the TRICK reporter transgenes, we recommend to use western blot analysis to identify the fusion protein derived from the *osk*-TRICK reporter, which appears with an increase in molecular weight of approximately 30 kDa caused by the extra polypeptide sequence derived from the 12×PP7-binding sites insertion (see Fig. 3C, middle lane). The use of mutant alleles to deplete any wild-type protein

---

from wild-type flies, flies expressing *osk*-TRICK and *osk*-TRICK in an *osk*^{null} background. (D) Immunostaining against Oskar protein in an egg-chamber expressing *osk*-TRICK in an *osk*^{null} background shows exclusive synthesis of Oskar protein at the posterior pole of the oocyte. (E) Quantification of fluorescent signals from NLS-PCP-GFP/NLS-MCP-RFP, Oskar protein immunostaining and oocyte area (color- and size-coded) of individual oocytes. The correlation of the TRICK reporter readout with Oskar protein and oocyte area observed in *osk*-TRICK expressing egg-chambers (*upper* plot) is abolished by the introduction of the STOP mutation prior the PP7-binding sites (*lower* plot). Pearson correlation coefficient (*r*), scale bars 50 µm. (See the color plate.)

allows standard immuno-fluorescence techniques and the use of well-established antibodies to detect protein derived exclusively from the TRICK reporter (Fig. 3C, right lane). This is important when considering the use of constructs in which the coding region and the PP7 loops are separated by a self-cleaving 2A peptide, making reporter and wild-type protein difficult to distinguish by mass.

Only at the posterior pole of the oocyte is translational repression of *oskar* mRNA relieved and Oskar protein produced. This process requires a precise orchestration of the transport machinery and translational regulators. Therefore, correct localization of Oskar protein exclusively at the posterior pole is a significant indicator of undisturbed regulation of the localized translation of the TRICK reporter construct. The use of mutant alleles (eg, osk^{null}) and standard immuno-fluorescence allows detection of Oskar protein solely derived from the *oskar*-TRICK reporter mRNA and confirms the correct localization of the protein independent of wild-type transcript and protein (Fig. 3D). This demonstrates that the introduction of PP7- and MS2-binding sites has no impact on the transport, translational repression, and translational activation of *osk*-TRICK mRNA.

Oskar protein first appears during mid-oogenesis (Kim-Ha, Kerr, & Macdonald, 1995), allowing a precise readout of the TRICK reporter performance to report on the translational status of *osk*-TRICK mRNA of individual egg-chambers during different developmental stages. A comparison of oocyte area, fluorescent intensities of Oskar protein immunostaining, and the NLS-PCP-GFP to NLS-MCP-RFP ratio shows the correlation of loss of NLS-PCP-GFP signal with oocyte size and Oskar protein appearance (Fig. 3E). In order to demonstrate that the observed loss of the NLS-PCP-GFP signal in later stage oocytes depends on active translation, introduction of a STOP codon by site-directed mutagenesis upstream of the PP7-binding sites of the *osk-TRICK* mRNA (*osk*-STOP-TRICK) should be used (Fig. 3E). Similar to TRICK experiments in cultured mammalian cells, small-molecule inhibitors of translation can also be used.

TRICK reporter mRNAs can be used to monitor the first round of translation with single mRNP resolution in the animal model system *Drosophila*, a powerful resource with established genetic toolboxes and well-studied examples of localized translation.

9. OUTLOOK

The development of multiple orthogonal fluorescent labeling methodologies for imaging single molecules of RNA in living cells has made it possible to perform more detailed analyses of RNA metabolism. RNA biosensors, which go beyond simply being able to observe mRNAs, enable direct measurement of specific events in an mRNA's life. We have engineered the TRICK system that reports on the first round of translation, but we envision that conceptually similar approaches may also be applied to mRNA turnover and other aspects of RNA metabolism. These advances coupled to the revolution in genome engineering tools will allow the complete lives of endogenous mRNAs to be imaged in living cells.

ACKNOWLEDGMENTS

This work was supported by the Novartis Research Foundation and the Swiss National Science Foundation (J.A.C); the European Molecular Biology Lab (A.E.); and postdoctoral fellowships from the Peter and Traudl Engelhorn Foundation (J.M.H) and the EMBL Interdisciplinary Postdoc Program (EIPOD) under Marie Curie COFUND actions (F.W.).

REFERENCES

Aitken, C. E., Marshall, R. A., & Puglisi, J. D. (2008). An oxygen scavenging system for improvement of dye stability in single-molecule fluorescence experiments. *Biophysical Journal*, 94(5), 1826–1835.

Bertrand, E., Chartrand, P., Schaefer, M., Shenoy, S. M., Singer, R. H., & Long, R. M. (1998). Localization of ASH1 mRNA particles in living yeast. *Molecular Cell*, 2(4), 437–445.

Chao, J. A., Patskovsky, Y., Almo, S. C., & Singer, R. H. (2008). Structural basis for the coevolution of a viral RNA-protein complex. *Nature Structural & Molecular Biology*, 15(1), 103–105.

Chen, J., Nikolaitchik, O., Singh, J., Wright, A., Bencsics, C. E., Coffin, J. M., ... Hu, W. S. (2009). High efficiency of HIV-1 genomic RNA packaging and heterozygote formation revealed by single virion analysis. *Proceedings of the National Academy of Sciences of the United States of America*, 106(32), 13535–13540.

Chenouard, N., Smal, I., de Chaumont, F., Maska, M., Sbalzarini, I. F., Gong, Y., ... Meijering, E. (2014). Objective comparison of particle tracking methods. *Nature Methods*, 11(3), 281–289.

Cho, S., Jang, J., Song, C., Lee, H., Ganesan, P., Yoon, T. Y., ... Park, Y. (2013). Simple super-resolution live-cell imaging based on diffusion-assisted Forster resonance energy transfer. *Scientific Reports*, 3, 1208.

Coulon, A., Ferguson, M. L., de Turris, V., Palangat, M., Chow, C. C., & Larson, D. R. (2014). Kinetic competition during the transcription cycle results in stochastic RNA processing. *Elife*, 3, e03939.

Daigle, N., & Ellenberg, J. (2007). LambdaN-GFP: An RNA reporter system for live-cell imaging. *Nature Methods*, 4(8), 633–636.

David, A., Dolan, B. P., Hickman, H. D., Knowlton, J. J., Clavarino, G., Pierre, P., ... Yewdell, J. W. (2012). Nuclear translation visualized by ribosome-bound nascent chain puromycylation. *The Journal of Cell Biology*, 197(1), 45–57.

Deschout, H., Martens, T., Vercauteren, D., Remaut, K., Demeester, J., De Smedt, S. C., ... Braeckmans, K. (2013). Correlation of dual colour single particle trajectories for improved detection and analysis of interactions in living cells. *International Journal of Molecular Sciences*, 14(8), 16485–16514.

Dieterich, D. C., Hodas, J. J., Gouzer, G., Shadrin, I. Y., Ngo, J. T., Triller, A., ... Schuman, E. M. (2010). In situ visualization and dynamics of newly synthesized proteins in rat hippocampal neurons. *Nature Neuroscience, 13*(7), 897–905.

Dupont, A., Stirnnagel, K., Lindemann, D., & Lamb, D. C. (2013). Tracking image correlation: Combining single-particle tracking and image correlation. *Biophysical Journal, 104*(11), 2373–2382.

Ephrussi, A., & Lehmann, R. (1992). Induction of germ cell formation by oskar. *Nature, 358*(6385), 387–392.

Espenel, C., Margeat, E., Dosset, P., Arduise, C., Le Grimellec, C., Royer, C. A., ... Milhiet, P. E. (2008). Single-molecule analysis of CD9 dynamics and partitioning reveals multiple modes of interaction in the tetraspanin web. *The Journal of Cell Biology, 182*(4), 765–776.

Forrest, K. M., & Gavis, E. R. (2003). Live imaging of endogenous RNA reveals a diffusion and entrapment mechanism for nanos mRNA localization in *Drosophila*. *Current Biology, 13*(14), 1159–1168.

Fusco, D., Accornero, N., Lavoie, B., Shenoy, S. M., Blanchard, J. M., Singer, R. H., & Bertrand, E. (2003). Single mRNA molecules demonstrate probabilistic movement in living mammalian cells. *Current Biology, 13*(2), 161–167.

Gautier, A., Juillerat, A., Heinis, C., Correa, I. R., Kindermann, M., Beaufils, F., & Johnsson, K. (2008). An engineered protein tag for multiprotein labeling in living cells. *Chemistry & Biology, 15*(2), 128–136.

Gavis, E. R., Lunsford, L., Bergsten, S. E., & Lehmann, R. (1996). A conserved 90 nucleotide element mediates translational repression of nanos RNA. *Development, 122*(9), 2791–2800.

Gerhardt, I., Mai, L., Lamas-Linares, A., & Kurtsiefer, C. (2011). Detection of single molecules illuminated by a light-emitting diode. *Sensors (Basel), 11*(1), 905–916.

Ghosh, S., Marchand, V., Gaspar, I., & Ephrussi, A. (2012). Control of RNP motility and localization by a splicing-dependent structure in oskar mRNA. *Nature Structural & Molecular Biology, 19*(4), 441–449.

Gossen, M., Freundlieb, S., Bender, G., Muller, G., Hillen, W., & Bujard, H. (1995). Transcriptional activation by tetracyclines in mammalian cells. *Science, 268*(5218), 1766–1769.

Grimm, J. B., English, B. P., Chen, J., Slaughter, J. P., Zhang, Z., Revyakin, A., & Lavis, L. D. (2015). A general method to improve fluorophores for live-cell and single-molecule microscopy. *Nature Methods, 12*(3), 244–250. 243 pp. following 250 pp.

Grunwald, D., & Singer, R. H. (2010). In vivo imaging of labelled endogenous beta-actin mRNA during nucleocytoplasmic transport. *Nature, 467*(7315), 604–607.

Hachet, O., & Ephrussi, A. (2004). Splicing of oskar RNA in the nucleus is coupled to its cytoplasmic localization. *Nature, 428*(6986), 959–963.

Halstead, J. M., Lionnet, T., Wilbertz, J. H., Wippich, F., Ephrussi, A., Singer, R. H., & Chao, J. A. (2015). Translation. An RNA biosensor for imaging the first round of translation from single cells to living animals. *Science, 347*(6228), 1367–1671.

Higashida, C., Suetsugu, S., Tsuji, T., Monypenny, J., Narumiya, S., & Watanabe, N. (2008). G-actin regulates rapid induction of actin nucleation by mDia1 to restore cellular actin polymers. *Journal of Cell Science, 121*(Pt. 20), 3403–3412.

Hocine, S., Raymond, P., Zenklusen, D., Chao, J. A., & Singer, R. H. (2013). Single-molecule analysis of gene expression using two-color RNA labeling in live yeast. *Nature Methods, 10*(2), 119–121.

Holt, C. E., & Schuman, E. M. (2013). The central dogma decentralized: New perspectives on RNA function and local translation in neurons. *Neuron, 80*(3), 648–657.

Ifrim, M. F., Williams, K. R., & Bassell, G. J. (2015). Single-molecule imaging of PSD-95 mRNA translation in dendrites and its dysregulation in a mouse model of fragile

X syndrome. *The Journal of Neuroscience, 35*(18), 7116–7130. http://dx.doi.org/10.1523/JNEUROSCI.2802-14.2015.

Ingolia, N. T., Ghaemmaghami, S., Newman, J. R., & Weissman, J. S. (2009). Genome-wide analysis in vivo of translation with nucleotide resolution using ribosome profiling. *Science, 324*(5924), 218–223.

Ishigaki, Y., Li, X., Serin, G., & Maquat, L. E. (2001). Evidence for a pioneer round of mRNA translation: mRNAs subject to nonsense-mediated decay in mammalian cells are bound by CBP80 and CBP20. *Cell, 106*(5), 607–617.

Jambor, H., Mueller, S., Bullock, S. L., & Ephrussi, A. (2014). A stem-loop structure directs oskar mRNA to microtubule minus ends. *RNA, 20*(4), 429–439.

Jan, C. H., Williams, C. C., & Weissman, J. S. (2014). Principles of ER cotranslational translocation revealed by proximity-specific ribosome profiling. *Science, 346*(6210), 1257521.

Jaramillo, A. M., Weil, T. T., Goodhouse, J., Gavis, E. R., & Schupbach, T. (2008). The dynamics of fluorescently labeled endogenous gurken mRNA in Drosophila. *Journal of Cell Science, 121*(Pt. 6), 887–894.

Juillerat, A., Gronemeyer, T., Keppler, A., Gendreizig, S., Pick, H., Vogel, H., & Johnsson, K. (2003). Directed evolution of O-6-alkylguanine-DNA alkyltransferase for efficient labeling of fusion proteins with small molecules in vivo. *Chemistry & Biology, 10*(4), 313–317.

Kim, J. H., Lee, S. R., Li, L. H., Park, H. J., Park, J. H., Lee, K. Y., ... Choi, S. Y. (2011). High cleavage efficiency of a 2A peptide derived from porcine teschovirus-1 in human cell lines, zebrafish and mice. *PloS One, 6*(4), e18556.

Kim-Ha, J., Kerr, K., & Macdonald, P. M. (1995). Translational regulation of oskar mRNA by bruno, an ovarian RNA-binding protein, is essential. *Cell, 81*(3), 403–412.

Koyama-Honda, I., Ritchie, K., Fujiwara, T., Iino, R., Murakoshi, H., Kasai, R. S., & Kusumi, A. (2005). Fluorescence imaging for monitoring the colocalization of two single molecules in living cells. *Biophysical Journal, 88*(3), 2126–2136.

Kremers, G. J., Gilbert, S. G., Cranfill, P. J., Davidson, M. W., & Piston, D. W. (2011). Fluorescent proteins at a glance. *Journal of Cell Science, 124*(Pt. 2), 157–160.

Kumano, G. (2012). Polarizing animal cells via mRNA localization in oogenesis and early development. *Development, Growth & Differentiation, 54*(1), 1–18.

Kuo, J. S., Kuyper, C. L., Allen, P. B., Fiorini, G. S., & Chiu, D. T. (2004). High-power blue/UV light-emitting diodes as excitation sources for sensitive detection. *Electrophoresis, 25*(21–22), 3796–3804.

Larson, D. R., Zenklusen, D., Wu, B., Chao, J. A., & Singer, R. H. (2011). Real-time observation of transcription initiation and elongation on an endogenous yeast gene. *Science, 332*(6028), 475–478.

Liao, G., Mingle, L., Van De Water, L., & Liu, G. (2015). Control of cell migration through mRNA localization and local translation. *Wiley Interdisciplinary Reviews. RNA, 6*(1), 1–15.

Lin, M. D., Jiao, X., Grima, D., Newbury, S. F., Kiledjian, M., & Chou, T. B. (2008). Drosophila processing bodies in oogenesis. *Developmental Biology, 322*(2), 276–288.

Lionnet, T., Czaplinski, K., Darzacq, X., Shav-Tal, Y., Wells, A. L., Chao, J. A., ... Singer, R. H. (2011). A transgenic mouse for in vivo detection of endogenous labeled mRNA. *Nature Methods, 8*(2), 165–170.

Los, G. V., Encell, L. P., McDougall, M. G., Hartzell, D. D., Karassina, N., Zimprich, C., ... Wood, K. V. (2008). HaloTag: A novel protein labeling technology for cell imaging and protein analysis. *ACS Chemical Biology, 3*(6), 373–382.

Ma, J., Liu, Z., Michelotti, N., Pitchiaya, S., Veerapaneni, R., Androsavich, J. R., ... Yang, W. (2013). High-resolution three-dimensional mapping of mRNA export through the nuclear pore. *Nature Communications, 4*, 2414.

Magidson, V., & Khodjakov, A. (2013). Circumventing photodamage in live-cell microscopy. *Methods in Cell Biology*, *114*, 545–560.

Martin, R. M., Rino, J., Carvalho, C., Kirchhausen, T., & Carmo-Fonseca, M. (2013). Live-cell visualization of pre-mRNA splicing with single-molecule sensitivity. *Cell Reports*, *4*(6), 1144–1155.

Mizuguchi, H., Xu, Z., Ishii-Watabe, A., Uchida, E., & Hayakawa, T. (2000). IRES-dependent second gene expression is significantly lower than cap-dependent first gene expression in a bicistronic vector. *Molecular Therapy*, *1*(4), 376–382.

Mor, A., Suliman, S., Ben-Yishay, R., Yunger, S., Brody, Y., & Shav-Tal, Y. (2010). Dynamics of single mRNP nucleocytoplasmic transport and export through the nuclear pore in living cells. *Nature Cell Biology*, *12*(6), 543–552.

No, D., Yao, T. P., & Evans, R. M. (1996). Ecdysone-inducible gene expression in mammalian cells and transgenic mice. *Proceedings of the National Academy of Sciences of the United States of America*, *93*(8), 3346–3351.

Pan, D., Kirillov, S. V., & Cooperman, B. S. (2007). Kinetically competent intermediates in the translocation step of protein synthesis. *Molecular Cell*, *25*(4), 519–529.

Park, H. Y., Buxbaum, A. R., & Singer, R. H. (2010). Single mRNA tracking in live cells. *Methods in Enzymology*, *472*, 387–406.

Park, H. Y., Lim, H., Yoon, Y. J., Follenzi, A., Nwokafor, C., Lopez-Jones, M., ... Singer, R. H. (2014). Visualization of dynamics of single endogenous mRNA labeled in live mouse. *Science*, *343*(6169), 422–424.

Parthasarathy, R. (2012). Rapid, accurate particle tracking by calculation of radial symmetry centers. *Nature Methods*, *9*(7), 724–726.

Qin, J. Y., Zhang, L., Clift, K. L., Hulur, I., Xiang, A. P., Ren, B. Z., & Lahn, B. T. (2010). Systematic comparison of constitutive promoters and the doxycycline-inducible promoter. *PLoS One*, *5*(5), e10611.

Rabut, G., & Ellenberg, J. (2004). Automatic real-time three-dimensional cell tracking by fluorescence microscopy. *Journal of Microscopy*, *216*(Pt. 2), 131–137.

Rasnik, I., McKinney, S. A., & Ha, T. (2006). Nonblinking and long-lasting single-molecule fluorescence imaging. *Nature Methods*, *3*(11), 891–893.

Shaner, N. C., Lin, M. Z., McKeown, M. R., Steinbach, P. A., Hazelwood, K. L., Davidson, M. W., & Tsien, R. Y. (2008). Improving the photostability of bright monomeric orange and red fluorescent proteins. *Nature Methods*, *5*(6), 545–551.

Shaner, N. C., Steinbach, P. A., & Tsien, R. Y. (2005). A guide to choosing fluorescent proteins. *Nature Methods*, *2*(12), 905–909.

Shav-Tal, Y., Darzacq, X., Shenoy, S. M., Fusco, D., Janicki, S. M., Spector, D. L., & Singer, R. H. (2004). Dynamics of single mRNPs in nuclei of living cells. *Science*, *304*(5678), 1797–1800.

Smith, C. S., Preibisch, S., Joseph, A., Abrahamsson, S., Rieger, B., Myers, E., ... Grunwald, D. (2015). Nuclear accessibility of beta-actin mRNA is measured by 3D single-molecule real-time tracking. *The Journal of Cell Biology*, *209*(4), 609–619.

Spille, J. H., Kaminski, T. P., Scherer, K., Rinne, J. S., Heckel, A., & Kubitscheck, U. (2015). Direct observation of mobility state transitions in RNA trajectories by sensitive single molecule feedback tracking. *Nucleic Acids Research*, *43*(2), e14.

Steitz, J. A. (1969). Nucleotide sequences of the ribosomal binding sites of bacteriophage R17 RNA. *Cold Spring Harbor Symposia on Quantitative Biology*, *34*, 621–630.

Takizawa, P. A., & Vale, R. D. (2000). The myosin motor, Myo4p, binds Ash1 mRNA via the adapter protein, She3p. *Proceedings of the National Academy of Sciences of the United States of America*, *97*(10), 5273–5278.

Tatavarty, V., Ifrim, M. F., Levin, M., Korza, G., Barbarese, E., Yu, J., & Carson, J. H. (2012). Single-molecule imaging of translational output from individual RNA granules in neurons. *Molecular Biology of the Cell*, *23*(5), 918–929.

Terskikh, A., Fradkov, A., Ermakova, G., Zaraisky, A., Tan, P., Kajava, A. V., ... Siebert, P. (2000). "Fluorescent timer": Protein that changes color with time. *Science*, *290*(5496), 1585–1588.

Thompson, R. E., Larson, D. R., & Webb, W. W. (2002). Precise nanometer localization analysis for individual fluorescent probes. *Biophysical Journal*, *82*(5), 2775–2783.

Tokunaga, M., Imamoto, N., & Sakata-Sogawa, K. (2008). Highly inclined thin illumination enables clear single-molecule imaging in cells. *Nature Methods*, *5*(2), 159–161.

Urbanek, M. O., Galka-Marciniak, P., Olejniczak, M., & Krzyzosiak, W. J. (2014). RNA imaging in living cells—Methods and applications. *RNA Biology*, *11*(8), 1083–1095.

Vigers, G. P., Coue, M., & McIntosh, J. R. (1988). Fluorescent microtubules break up under illumination. *The Journal of Cell Biology*, *107*(3), 1011–1024.

Weidenfeld, I., Gossen, M., Low, R., Kentner, D., Berger, S., Gorlich, D., ... Schonig, K. (2009). Inducible expression of coding and inhibitory RNAs from retargetable genomic loci. *Nucleic Acids Research*, *37*(7), e50.

Weil, T. T., Forrest, K. M., & Gavis, E. R. (2006). Localization of bicoid mRNA in late oocytes is maintained by continual active transport. *Developmental Cell*, *11*(2), 251–262.

Widengren, J., Chmyrov, A., Eggeling, C., Lofdahl, P. A., & Seidel, C. A. (2007). Strategies to improve photostabilities in ultrasensitive fluorescence spectroscopy. *The Journal of Physical Chemistry. A*, *111*(3), 429–440.

Wu, B., Chao, J. A., & Singer, R. H. (2012). Fluorescence fluctuation spectroscopy enables quantitative imaging of single mRNAs in living cells. *Biophysical Journal*, *102*(12), 2936–2944.

Wu, B., Miskolci, V., Sato, H., Tutucci, E., Kenworthy, C. A., Donnelly, S. K., ... Hodgson, L. (2015). Synonymous modification results in high-fidelity gene expression of repetitive protein and nucleotide sequences. *Genes & Development*, *29*(8), 876–886.

Xia, T., Li, N., & Fang, X. (2013). Single-molecule fluorescence imaging in living cells. *Annual Review of Physical Chemistry*, *64*, 459–480.

Xu, X., Brechbiel, J. L., & Gavis, E. R. (2013). Dynein-dependent transport of nanos RNA in *Drosophila* sensory neurons requires Rumpelstiltskin and the germ plasm organizer Oskar. *The Journal of Neuroscience*, *33*(37), 14791–14800.

Yu, J., Xiao, J., Ren, X., Lao, K., & Xie, X. S. (2006). Probing gene expression in live cells, one protein molecule at a time. *Science*, *311*(5767), 1600–1603.

Zimyanin, V. L., Belaya, K., Pecreaux, J., Gilchrist, M. J., Clark, A., Davis, I., & St Johnston, D. (2008). In vivo imaging of oskar mRNA transport reveals the mechanism of posterior localization. *Cell*, *134*(5), 843–853.

Zuker, M. (2003). Mfold web server for nucleic acid folding and hybridization prediction. *Nucleic Acids Research*, *31*(13), 3406–3415.

CHAPTER SEVEN

Fluctuation Analysis: Dissecting Transcriptional Kinetics with Signal Theory

A. Coulon[*,†,1], D.R. Larson[‡]

[*]Institut Curie, PSL Research University, Laboratoire Physico-Chimie, CNRS UMR 168, Paris, France
[†]Sorbonne Universités, UPMC Univ Paris 06, Paris, France
[‡]Laboratory of Receptor Biology and Gene Expression, National Cancer Institute, NIH, Bethesda, MD, United States
[1]Corresponding author: e-mail address: antoine.coulon@curie.fr

Contents

1. Introduction 160
   1.1 Definitions and Terminology 163
   1.2 What Correlation Functions Can—and Cannot—Do 164
2. Computing and Averaging Correlation Functions 165
   2.1 Single Correlation Functions 165
   2.2 Mean Subtraction of Fluorescence Traces 167
   2.3 Averaging Methods 170
   2.4 Correct Weighting of Time-Delay Points 172
   2.5 Baseline Correction and Renormalization 172
   2.6 Uncertainty, Error Bars, and Bootstrapping 174
3. Interpretation of Correlation Functions 176
   3.1 A Primer for Correlation Function Modeling 176
   3.2 Data Fitting and Model Discrimination 180
4. Common Issues and Pitfalls 182
   4.1 Location of the MS2 and PP7 Cassettes 182
   4.2 Interpreting Single (or Too Few) Traces 184
   4.3 Technical Sources of Fluctuations 185
   4.4 Biased Selection of Data (Cells, TS, Part of Traces) 188
   4.5 Validation by Complementary Measurements 188
5. Conclusion 189
References 190

Abstract

Recent live-cell microscopy techniques now allow the visualization in multiple colors of RNAs as they are transcribed on genes of interest. Following the number of nascent RNAs over time at a single locus reveals complex fluctuations originating from the underlying transcriptional kinetics. We present here a technique based on concepts from signal theory—called fluctuation analysis—to analyze and interpret multicolor transcriptional time traces and extract the temporal signatures of the underlying mechanisms. The principle is to generate, from the time traces, a set of functions called correlation functions. We explain how to compute these functions practically from a set of experimental traces and how to interpret them through different theoretical and computational means. We also present the major difficulties and pitfalls one might encounter with this technique. This approach is capable of extracting mechanistic information hidden in transcriptional fluctuations at multiple timescales and has broad applications for understanding transcriptional kinetics.

1. INTRODUCTION

Enzymatic reactions involved in the making of a mature messenger RNA (mRNA) are numerous. These include reactions to initiate transcription at the promoter, to synthesize the pre-mRNA from the DNA template, to cleave and add a poly(A) tail to the transcript once the 3' end of the gene is reached, and to splice the pre-mRNA into a fully mature mRNA (Craig et al., 2014). In addition to the RNA polymerase II (Pol II) and the spliceosome—the two enzymes that carry out RNA synthesis and splicing, respectively—many others act indirectly on this process, eg, by affecting the topology of DNA or depositing posttranslational marks on proteins such as histones or Pol II itself, which in turn influence the recruitment and function of other enzymes (Craig et al., 2014).

As often in enzymology, the energy-dependent nature of the reactions involved requires an out-of-equilibrium description: some of the synthesis and processing reactions are non- (or very weakly) reversible, hence maintaining a constant flux in the system turning substrates into products (Segel, 1993). In this context, certain reaction pathways can be favored simply because some reactions occur faster than others and not because the product is more energetically favorable as in an equilibrium scheme. This situation is also seen in the preinitiation steps of gene regulation and has implications for accuracy of transcriptional control (Coulon, Chow, Singer, & Larson, 2013). As a consequence, the final product of RNA synthesis and processing critically depends on the temporal coordination between the different events involved. For instance, splicing decisions are expected to be influenced by whether the pairing of splice sites occurs in a first-come-first-served basis as the transcript emerges from the elongating polymerase or happens slower than elongation, hence allowing more flexibility to pair nonadjacent splice sites (Bentley, 2014). A recurrent question in the RNA processing field is whether splicing decisions are governed by this principle of *kinetic competition* or if the cell has developed additional *checkpoint mechanisms* to ensure a predefined order between events (Bentley, 2014). Clearly, answering such questions, given the stochastic and nonequilibrium nature of these processes, requires being able to observe and dissect their dynamics at the single-molecule level.

To address this, we and others have developed tools and methods to visualize transcription and splicing in real time as it occurs in living cells (Coulon et al., 2014; Martin, Rino, Carvalho, Kirchhausen, & Carmo-Fonseca, 2013). The principle is to decorate the RNAs from a gene of interest with fluorescent proteins that are fused to an MS2 bacteriophage coat protein (MCP) that binds to MS2 RNA stem-loops present in the transcripts due to the insertion of a DNA cassette in the gene (Fig. 1A; Bertrand et al., 1998). This method allows detecting both single RNAs diffusing in the nucleoplasm as well as nascent RNAs being synthesized at the transcription site (TS) (Fig. 1B). In the latter case, one can track the TS and follow over time the fluctuations in the amount of nascent transcripts on the gene, originating from the stochastic and discrete nature of the transcription process (Fig. 1C). Combining the MS2 technique with a recent equivalent from the PP7 bacteriophage (Chao, Patskovsky, Almo, & Singer, 2007; Larson, Zenklusen, Wu, Chao, & Singer, 2011), we were able to decorate different RNAs or different regions of the same RNA with distinct fluorophores (Coulon et al., 2014; Lenstra, Coulon, Chow, & Larson, 2015).

Interpreting the two-color time traces resulting from the MS2/PP7 RNA-labeling technique can be nontrivial—so much so that data analysis might only focus on the rare instances in time traces where a single nascent RNA can be distinguished at the TS. An alternative approach, which we favor, consists in extracting information about the synthesis and processing kinetics of single RNAs by analyzing entire time traces using a method based on signal theory, called *fluctuation analysis*. This method allows an unbiased selection of all the observed transcription events, hence resulting in a high statistical power and a detailed description of the underlying kinetics

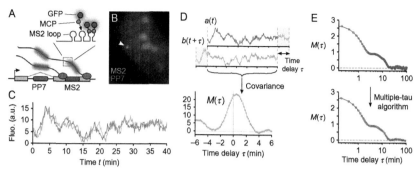

Fig. 1 Transcriptional time traces and correlation function. (A) The MS2 and PP7 RNA-labeling technique consists in inserting, in one or two gene(s) of interest, two DNA cassettes (MS2 and PP7; here at two different locations in the same gene). They produce stem-loop structures in the nascent RNAs, which are bound by an MS2 or PP7 coat protein (MCP and PCP) fused to a fluorescent protein (eg, GFP and mCherry, respectively). (B) The transcription site (*arrow*) appears as a bright diffraction-limited spot in the nucleus in both fluorescence channels. (C) Recording its intensity fluctuations then yields a signal that is proportional to the number of nascent RNAs on the gene over time. (D) Using this signal as an example, the computation of a correlation function (here the covariance function) consists in shifting one signal relatively to the other and calculating the covariance between the values of the overlapping portions of the two signals as a function of the time-delay shift (Eqs. 1 and 2). (E) To analyze fluctuations at multiple timescales in a signal, computing the correlation function with the multiple-tau algorithm yields a somewhat uniform spacing of the time-delay points on a logarithmic scale (simulated data as in Fig. 2A). *Panels (B) to (D): Data from Coulon, A., Ferguson, M. L., de Turris, V., Palangat, M., Chow, C. C., & Larson, D. R. (2014). Kinetic competition during the transcription cycle results in stochastic RNA processing. eLife, 3. http://doi.org/10.7554/eLife.03939.* (See the color plate.)

(Coulon et al., 2014). In addition, it is a very general framework and can be used to interpret transcriptional fluctuations in many contexts, such as gene bursting and the kinetics of sense and antisense transcription of a single gene (Lenstra et al., 2015). In principle, it may also be applied to—or combined with—time traces from single-molecule imaging of protein recruitment at the TS (such as TFs and Pol II). By revealing the temporal relationship between specific molecular events one can now answer a wide range of questions about the mechanisms and regulation of RNA synthesis and processing.

Here, we discuss this fluctuation analysis methods, how to implement it, how to interpret its results, and the main difficulties that may arise. For reagent preparation, single-molecule 4D imaging, and time trace generation, we refer the readers to our earlier description of these methods (Ferguson & Larson, 2013).

1.1 Definitions and Terminology

Fluctuation analysis consists in computing and interpreting functions called *correlation functions* from a set of time traces. The term *correlation function* can have slightly different meanings depending on the context and the field of research. It always consist in some measure of the joint second moment between the values of a signal $a(t)$ and the corresponding values at each time point of a signal $b(t+\tau)$ (ie, $b(t)$ shifted by a delay τ), hence measuring the statistical correlation between fluctuations in the two signals as a function of time separation τ (Fig. 1D). Differences in the precise formulation essentially rely on how the second moment is calculated and normalized (eg, central vs raw moments, and covariance vs coefficient of variation vs Pearson correlation). In the biophysics field, the correlation between two signals is often written

$$G(\tau) = \frac{\langle \delta a(t) \delta b(t+\tau) \rangle}{\langle a(t) \rangle \langle b(t) \rangle} \quad (1)$$

where $\delta a(t) = a(t) - \langle a(t) \rangle$ and $\langle \cdot \rangle$ denotes the temporal mean. Let us note $R(\tau) = \langle a(t) b(t+\tau) \rangle$ and $M(\tau) = \langle \delta a(t) \delta b(t+\tau) \rangle$, respectively, the *raw* moment and the *central* moment (or *covariance*), so that we have

$$G(\tau) = \frac{M(\tau)}{\langle a(t) \rangle \langle b(t) \rangle} = \frac{R(\tau)}{\langle a(t) \rangle \langle b(t) \rangle} - 1 \quad (2)$$

When the two signals $a(t)$ and $b(t)$ are the same, $G(\tau)$ and $M(\tau)$ are, respectively, called an *autocorrelation* and an *autocovariance*, and both are necessarily symmetrical by construction. When the two signals are different, $G(\tau)$ and $M(\tau)$ are, respectively, called a *cross-correlation* and a *cross-covariance* and may be asymmetrical. Note that, when one measures multicolor time traces (eg, a PP7 signal $a(t)$ in red and an MS2 signal $b(t)$ in green), all the pairwise correlations should be calculated (the two red and green autocorrelations and the red–green cross-correlation) since they carry complementary information about the underlying processes.

The formulation of $G(\tau)$ in Eq. (1)—akin to a squared coefficient of variation—yields a dimensionless measure that has the advantage of being insensitive to any arbitrary rescaling of either signal by an unknown multiplicative factor. This situation indeed arises frequently in microscopy data since the correspondence between fluorescence units and actual number of molecules—the *fluorescence-to-RNA conversion factor*—is often unknown and may change from one experiment to the next: eg, depending on the optical setup, the imaging conditions and, in the case of MS2/PP7 time traces, the expression level of coat proteins which can vary substantially between cells. Note that the Pearson correlation (ie, normalization by the product of the standard deviations rather than the means), although also dimensionless, is less informative since it is insensitive to both rescaling and offsetting the values of the signals. When the fluorescence-to-RNA conversion factor is known, the time traces can be expressed in terms of number of RNAs instead of arbitrary fluorescence units. In this case, using the covariance function $M(\tau)$ instead of $G(\tau)$ is preferable since it maximizes the information that this function reflects.

1.2 What Correlation Functions Can—and Cannot—Do

An important point to make first is that, even though a correlation function is made of many transcription events from one or several time traces (typically ~2000 transcripts in Coulon et al., 2014), the result is *not* an average view of the transcriptional kinetics, on the contrary. As an analogy, if X_i are N random variables following a probability distribution with density $P(x)$, then averaging together the Dirac functions $\delta(x - X_i)$ converges toward the full distribution $P(x) = \lim_{N \to \infty} \sum_i \delta(x - X_i)/N$, not its average, with an accuracy that increases with N. From a theoretical point of view, this is exactly what a correlation functions reflects about the stochastic kinetics of the underlying processes (eg, X_i could be the stochastic elongation and release time of single transcripts), with the difference that the elementary functions averaged together are not Dirac functions (cf Section 3.1). In practice, full distributions are difficult to estimate accurately, but one can typically discriminate between distribution shapes (eg, Dirac, exponential, gamma), as well as the order and dependency between stochastic events (Coulon et al., 2014). These aspects are described in more details in Section 3.1.

A caveat of correlation functions to mention upfront is that it reveals the stochastic kinetics of RNAs without any distinction of whether different statistics occur in different portions of the time traces. We are currently extending the fluctuation analysis technique to circumvent this limitation and analyze transcriptional kinetics in a time-dependent manner.

Another difficulty with this approach is that correlation functions reflect *all* types of fluctuations in a given set of signals, including technical ones (bleaching, tracking errors, etc.) and biological ones that are not necessarily the object of the study (eg, cell cycle kinetics). Some of these aspects are specifically discussed in Sections 2.5 and 4.3.

2. COMPUTING AND AVERAGING CORRELATION FUNCTIONS

2.1 Single Correlation Functions

Calculating the numerator of Eq. (1) can be done in several ways. Let us first put aside the mean subtraction of the signals and simply discuss the calculation of $R(\tau) = \langle a(t)b(t+\tau)\rangle$.

2.1.1 Iterative Method

The simplest method is to compute iteratively all the time-delay points. Specifically, if $a_0 \ldots a_{N-1}$ and $b_0 \ldots b_{N-1}$ are the values of the signals $a(t)$ and $b(t)$ at the N measured time points $t \in \{0, \Delta t, \ldots, (N-1)\Delta t\}$, then

$$R(i\Delta t) = \frac{1}{N-i} \sum_{q=0}^{N-i-1} a_q b_{q+i} \tag{3}$$

An important point here is that, as signal $b(t)$ is shifted relatively to signal $a(t)$, one should only use the $N-i$ pairs of time points that overlap between the two signals and discard the overhanging ends (Fig. 1D). We refer to this as *overhang trimming*.

2.1.2 Multiple-Tau Algorithm

If an experiment is meant to probe a broad range of timescales, one does not need the same absolute temporal resolution for fast and slow processes. For instance, if the correlation function at 20-s delay is described with 1-s resolution, it could be described at 500-s delay with 25-s resolution instead of 1-s resolution to still maintain the same *relative* resolution. This concept is behind the *multiple-tau* (or *multi-tau*) algorithm (Wohland, Rigler, & Vogel, 2001). It consists in down-sampling the signals (ie, reducing their temporal resolution) progressively as the correlation function is computed from small to large time delays, yielding a somewhat uniform spacing of the time-delay points of the correlation function on a logarithmic scale, ie, a somewhat constant relative resolution (Fig. 1E). In addition, reducing the resolution at long delays has the advantage of reducing the sampling noise (cf Section 4.2), which is naturally stronger for slower processes. Interestingly, this algorithm comes originally from the hardware correlators built in the 1980s and used to calculate autocorrelations in real time, while a signal is being acquired and without having to store it entirely (Schatzel, 1990).

Although we implement it differently here, this algorithm is useful in cases where a broad range of timescales need to be observed.

The principle of the multiple-tau algorithm is to choose a *resampling frequency* parameter m and to do the following:

(i) Compute $R(i\Delta t)$ as in Eq. (3) for $i = 0, 1, 2, ..., 2m-1$

(ii) When $i = 2m$, down-sample the signals by a factor of 2 as follows:

- $N \leftarrow \left\lfloor \dfrac{N}{2} \right\rfloor$, where $\lfloor \cdot \rfloor$ denotes the integer part
- $a_q \leftarrow \dfrac{a_{2q} + a_{2q+1}}{2}$ and $b_q \leftarrow \dfrac{b_{2q} + b_{2q+1}}{2}$ for $q \in \{0, 1, ..., N-1\}$
- then $\Delta t \leftarrow 2\Delta t$ and $i \leftarrow m$

(iii) Compute $R(i\Delta t)$ as in Eq. (3) for $i = m, ..., 2m-1$ and go to step (ii). Note that, even if the length of the original signal is not a power of 2, *all* the time points will be used initially for the first $2m$ time-delay points. Only at long delays, when down-sampling occurs, one time point at the end is occasionally lost. For instance, if $N = 37$, it will assume the values $37 \xrightarrow{1 \text{ lost}} 18 \longrightarrow 9 \xrightarrow{1 \text{ lost}} 4 \longrightarrow 2 \longrightarrow 1$. This loss of a few time points is generally not a problem since significant effects only occur at the very end of the correlation function.

2.1.3 Fourier Transforms

A very fast and convenient way of computing a correlation function is to use Fourier transforms. Thanks to the Wiener–Khinchin theorem (Van Etten, 2006)

$$R(\tau) = \mathcal{F}^{-1}\left[\overline{\mathcal{F}[a(t)]} \, \mathcal{F}[b(t)]\right] \tag{4}$$

where $\mathcal{F}[\cdot]$ and $\mathcal{F}^{-1}[\cdot]$ are the forward and inverse Fourier transforms and the bar denotes the complex conjugate. In practice, for discrete and finite signals, using the fast Fourier transform (FFT) algorithm (and its inverse form FFT^{-1}) translates Eq. (4) into $\dfrac{1}{N} \text{FFT}^{-1}\left(\overline{\text{FFT}(a_0...a_{N-1})} \, \text{FFT}(b_0...b_{N-1})\right)$. The FFT algorithm is very efficient and makes the computation of the correlation function orders of magnitude faster (execution time grows as $N \log(N)$, as opposed to N^2 for Eq. 3).

However, using Fourier transforms as such is not ideal since it wraps the nonoverlapping ends of the signals when shifting them—referred to as *overhang wrapping*. Indeed, FFT implicitly treat finite signals as infinite periodic signals, hence correlating not only $a_0 ... a_{N-1-i}$ with $b_i ... b_{N-1}$

but also $a_{N-i}\ldots a_{N-1}$ with $b_0\ldots b_{i-1}$. A way around is to extend both signals by padding N zeros at their ends and to normalize the result of the FFT^{-1} by $|N-i|$ instead of N. Another advantage of this method is that, while the two halves of a cross-correlation function (at positive and negative delays) have to be computed independently if using Eq. (3), the Fourier approach provides both halves directly. More explicitly, the computation can be done as follows:

$$R\left(\underbrace{0, \Delta t, \ldots, (N-1)\Delta t}_{\text{positive delays}},\ \underbrace{-N\Delta t, \ldots, -\Delta t}_{\text{negative delays}}\right) = \frac{\text{FFT}^{-1}\left(\overline{\text{FFT}(a_0,\ldots,a_{N-1},\underbrace{0\ldots 0}_{N})}\ \text{FFT}(b_0,\ldots,b_{N-1},\underbrace{0\ldots 0}_{N})\right)}{[N, N-1, N-2, \ldots, 1,\ 0, 1, \ldots, N-2, N-1]} \quad (5)$$

where products and divisions are taken term by term. Eq. (5) gives the exact same result as the iterative method of Eq. (3), but runs for instance >80 times faster on a 1000 time point signal. Down-sampling may then be performed a posteriori at different delays to mimic the result of the multiple-tau algorithm.

Even though computation time is generally not an issue when calculating a few tens of correlation functions on signals with hundreds of time points, the much better efficiency of this technique based on FFTs is useful when computing measurement errors on correlation functions using the bootstrap technique (Section 2.6).

2.2 Mean Subtraction of Fluorescence Traces

As developed in this section and the following one, when going from theory to practice, two important considerations arise from the application of Eq. (1) to signals of finite duration, both coming from an inaccurate estimation of the mean of the signals. As illustrated in Fig. 2 on simulated time traces, these effects become apparent when comparing the correlation function calculated on a long signal with that obtained by averaging the correlation functions calculated on a partition of the same signal. The former corresponds to an ideal case (ie, close to the theoretical situation of an infinite signal), while the latter emulates what happens in practice when we only have a set of finite—and short—time traces that are all obtained in the same experimental conditions.

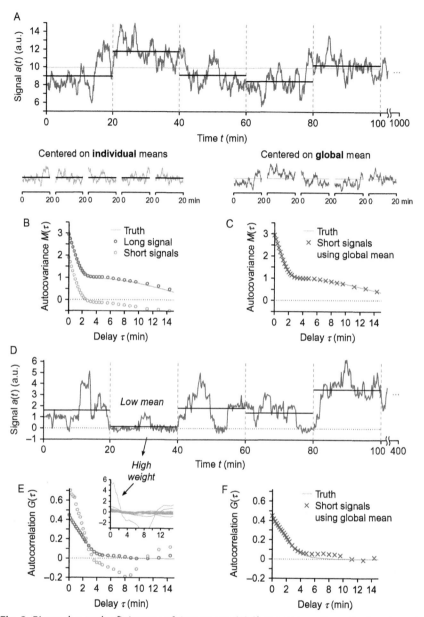

Fig. 2 Biases due to the finiteness of time traces. (A) Shown is a portion of a signal used to illustrate the effect of inaccurate mean estimation. This 1000-min-long signal is partitioned into a set of 20-min-long signals. The true mean of the signal (ie, calculated on the long trace) is shown in *gray* and the inaccurate means of individual short traces are shown in *black*. (B) The autocovariance $M(\tau)$ of the entire signal shown in (A) is close to the expected curve (*red circles* vs *gray curve*). When averaging the autocovariances

Dissecting Transcriptional Kinetics with Fluctuation Analysis

Here, we explain why correlation functions computed on finite signals have, in many cases, an arbitrary and unknown vertical offset. It is first important to realize that, if we subtract two arbitrary constants c_a and c_b from the signals $a(t)$ and $b(t)$, then $\langle (a(t) - c_a)(b(t+\tau) - c_b) \rangle$ equals $\langle a(t)b(t+\tau) \rangle$ up to a constant value that depends on c_a, c_b, $\langle a(t) \rangle$, and $\langle b(t) \rangle$. Hence, when computing the covariance function $M(\tau) = \langle \delta a(t) \delta b(t+\tau) \rangle$, an inaccurate estimation of the means of the signals would simply result in a vertical offset of the curve.

Taking finite-duration time traces of an infinite signal may imply random over- or underestimation of its mean due to sampling error, especially if some of the underlying fluctuations are at frequencies slower or in the same order as the duration of the measured time traces. In the example of Fig. 2A, the signal shows fluctuations as slow as a few tens of minutes, making the estimated mean on each 20-min-long portions (black lines) deviate from the true mean (gray line). As a result, the autocovariances of each portion are shifted toward the x-axis (loosely speaking, traces appear less variable than they should). This leads to an average autocovariance that, although having an accurate shape, is offset vertically by a constant value when compared to the autocovariance of the (virtually) infinite signal (Fig. 2B).

Fluctuations at slow temporal scales are ubiquitous in biological data (especially for in vivo transcription, eg, cell cycle, cell growth and mobility, response to cell culture passages and media changes, etc.). It is hence almost impossible to rule out this phenomenon, making experimental correlation functions always defined up to an unknown offset value. Solutions to this

---

computed on each one of the 20-min-long traces, the resulting autocovariance deviates from the expected curve by a constant offset (*green circles*). (C) Performing the same calculation using a global estimation of the mean of the signals (ie, once over all the short signals) resolves the issue. (D) Another long signal is shown and partitioned into small sections to illustrate another artifact that may arise when averaging correlation functions $G(\tau)$. (E) The average (*green circles*) of the autocorrelation functions obtained on the 20-min-long sections of the signal shown in (D) deviates from the expected curve. This is due to an inaccurate weighting of the individual curves that occurs when averaging autocorrelation functions $G(\tau)$. As illustrated by the inset, the section that has a very low mean in (D) dominates the average. (F) As in (C), estimating the mean globally over all the signals solves the weighting problem. Both examples shown are simulated signals (A: Gaussian noise shaped in the Fourier domain, D: Monte Carlo simulation of transcription with Poisson initiation, distributed transcript dwell time and additive Gaussian noise). The "truth" curves in *gray* in (B), (C), (E), and (F) are the theoretical curves for both simulated situations. (See the color plate.)

issue include (i) a technique to minimizes this phenomenon, presented in the next section, (ii) offsetting back the correlation functions directly, in cases of a good separation between fast and slow timescales leading to a clearly identifiable baseline (Section 2.5), and/or (iii) performing time-lapse imaging at multiple temporal resolutions, including very slow ones, and to paste together the correlation functions from different timescales.

2.3 Averaging Methods

The second consideration resulting from the finiteness of experimental time traces is that biases may arise depending on how the average correlation functions is computed. The intuitive and classical way to calculate $G(\tau)$ from a set of traces (noted $a_j(t)$ and $b_j(t)$ with $j \in [0...n-1]$) is to average together individual correlation functions

$$G(\tau) = \frac{1}{n}\sum_j G_j(\tau) = \frac{1}{n}\sum_j \frac{\langle \delta a_j(t)\delta b_j(t+\tau)\rangle}{\langle a_j(t)\rangle\langle b_j(t)\rangle} \quad (6)$$

However, as illustrated in Fig. 2D and E (in which $a_j(t) = b_j(t)$ for simplicity), the inaccurate estimation of the mean of the signals results in an incorrect weighting of the individual correlation functions when averaging them together. Namely, traces that have, by random chance, a low mean will be artificially given a high weight because of the normalization by $\langle a_j(t)\rangle\langle b_j(t)\rangle$. In extreme cases, these may completely dominate the average (Fig. 2E), and even in nonextreme cases, this bias gives more importance to traces with a lower signal (eg, with fewer and/or shorter transcription events), hence influencing the result of the analysis.

A solution to both this issue and the one described in the previous section can be found if the amount of fluorescence measured per single molecule *can be assumed identical between traces* (ie, same experimental procedure, same imaging conditions, uniform coat protein levels between cells, uniform illumination over the field of view, etc.). In this case, rather than estimating the means of the signals individually on a trace-by-trace basis, the solution is to estimate them once globally: $\bar{a} = \frac{1}{n}\sum_j \langle a_j(t)\rangle$ and $\bar{b} = \frac{1}{n}\sum_j \langle b_j(t)\rangle$, and to use the same values on all the traces:

$$G(\tau) = \frac{1}{n\bar{a}\bar{b}}\sum_j \langle (a_j(t)-\bar{a})(b_j(t+\tau)-\bar{b})\rangle \quad (7)$$

In this case the correlation function computed from a set of short and finite time traces is much closer to what is expected for infinite signals (Fig. 2C and F). This solution, however, only works well if the fluorescence-to-RNA conversion factor is truly identical between traces (although possibly unknown). In practice, even if this factor is only expected to be roughly similar, with small trace-to-trace variations, the use of global means is also preferable (Eq. 7).

To summarize, as depicted on the decision chart of Fig. 3, the experimenter's knowledge on the fluorescence-to-RNA conversion factor is what should guide the choice between using trace-by-trace vs global estimates of the means of the signals (Eqs. 6 and 7), and using correlation functions $G(\tau)$ vs covariance functions $M(\tau)$. In the latter case, global mean estimates should also be used:

$$M(\tau) = \frac{1}{n} \sum_j \left\langle \left(a_j(t) - \overline{a}\right)\left(b_j(t+\tau) - \overline{b}\right)\right\rangle \tag{8}$$

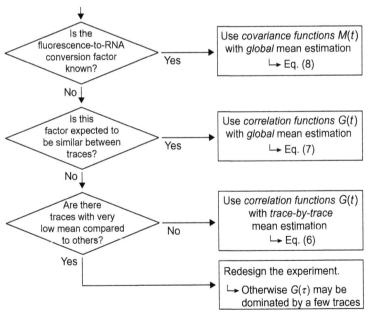

Fig. 3 Decision chart for averaging method. To avoid introducing biases, the most appropriate method for averaging individual correlation functions depends on the experimenter's knowledge of the fluorescence-to-RNA conversion factor, ie, the amount of fluorescence units that corresponds to a single, fully synthesized RNA.

2.4 Correct Weighting of Time-Delay Points

The above description assumes that all the traces $a_j(t)$ and $b_j(t)$ have the same duration. When this is not the case, correlation functions can still be averaged together, but particular care should be taken to assigning the correct weight to the different time-delay points of each correlation functions: For a given delay τ, each correlation function should be given a weight proportional to the number of pairs of time points used in its computation (this number decreases as τ becomes larger due to overhang trimming; see Section 2.1). Let us take an example where traces $a_0(t)$ and $b_0(t)$ have 100 time points each (at $t = 0, \Delta t, 2\Delta t, \ldots, 99\Delta t$) and traces $a_1(t)$ and $b_1(t)$ have 50 time points each, and let us consider for simplicity that the multiple-tau algorithm (Section 2.1) is not used. In this case, $G_0(\tau)$ and $G_1(\tau)$ should have, respectively, a weight of 100 and 50 at $\tau = 0$, a weight of 99 and 49 at $\tau = \Delta t, \ldots$, a weight of 51 and 1 at $\tau = 49\Delta t, \ldots$, a weight of 40 and 0 at $\tau = 60\Delta t, \ldots$, and a weight of 1 and 0 at $\tau = 99\Delta t$. This generalizes into a weight of

$$N_j - i \text{ if positive and } 0 \text{ else, for } G_j(i\Delta t) \tag{9}$$

where N_j denotes the number of time points of traces $a_j(t)$ and $b_j(t)$. This weighting applies to all three formulations of Eqs. (6)–(8) and should also be used in the computation of the global means $\bar{a}$ and $\bar{b}$. When using the multiple-tau algorithm (Section 2.1), the weight in Eq. (9) should be replaced by $\lfloor (N_j - i)/\max\left(2^{\lfloor \log 2(i/m) \rfloor}, 1\right) \rfloor$, where $\lfloor \cdot \rfloor$ denotes the integer part, and the values of N_j and Δt should be used *before resampling*.

2.5 Baseline Correction and Renormalization

Transcriptional time traces often carry many types of biological fluctuations, reflecting distinct phenomena and possibly occurring at multiple timescales (transcription initiation, RNA synthesis, gene bursting, cell cycle, etc.). Experimenters may want to focus on one or a few aspects of this kinetics and ignore or minimize the rest. To achieve this, in addition to choosing an appropriate sampling rate and time trace duration that encompass the timescales of the phenomenon of interest, one may also take advantage of potential timescale separation and scaling properties of the correlation functions.

As an example, in our earlier work (Coulon et al., 2014), we were interested in the kinetics of RNA transcription and splicing from a few seconds to a few hundreds of seconds. Slower kinetics of both biological and technical nature (such as cell cycle dynamics, gene activation/inactivation, or bleaching and imaging artifacts; see Section 4.3) were present, but with a clear timescale separation. Indeed, the correlation functions showed unambiguously a fast dynamics (up to ~4 min and with shapes fully consistent with what was expected for RNA transcription and splicing), followed by a plateau with a very slow decay (Fig. 4C). As explained in

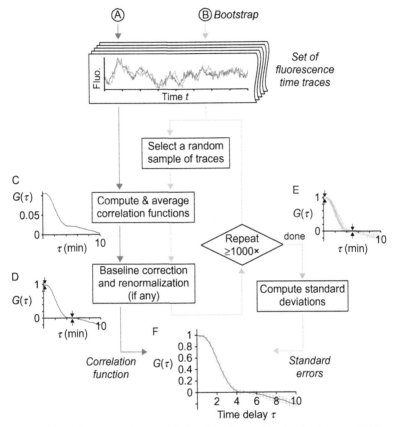

Fig. 4 Flowchart for computing correlation functions with standard errors. (A) From a given set of time traces, one should first compute the average correlation function as appropriate (cf Fig. 3) and then perform the corrections described in Section 2.5 if needed. This yields a "corrected" correlation function. (B) To obtain the standard error by the bootstrap method, one should perform, at least 1000 times, the exact same computation as in (A), using each time a random sample of the time traces (same number of traces as the original set, and randomly drawn with replacement). This yields an estimate of the sample distribution, which standard deviation is the standard error on the correlation function calculated in (A). Intermediate results of the calculations are shown using a set of experimental time traces from Coulon et al. (2014) as an example. Shown are the average correlation function (C) before and (D) after baseline correction and renormalization, (E) multiple average correlation functions (as in D) resulting from the bootstrap loop, and (F) the average correlation function with standard errors.

Section 2.2, the presence of such slow dynamics implies that the calculated correlation functions are defined up to an unknown vertical offset, even if the global mean estimation methods (Section 2.3) is used to minimize this artifact. In the case where a baseline is clearly visible at a certain time delay, if one only wants to focus on phenomena faster than this timescale, then the baseline may be brought to 0 by offsetting the correlation function vertically (eg, by subtracting from $G(\tau)$ its average value observed in the range $|\tau| \in [4 \ldots 6\,\text{min}]$; Fig. 4D). This correction removes an artifactual/unwanted degree of freedom, which will turn out useful for both the computation of the standard error on the correlation functions (next section) and for fitting the data to mathematical models (Section 3.2).

When the slower dynamics decays too fast to make a clear plateau, the safest solution, although not ideal, is to include this phenomenological decay in the fit of the data (Lenstra et al., 2015). Finally, we are currently extending the fluctuation analysis technique to be able to separate source of fluctuations occurring at similar or overlapping timescales, eg, as when transcription initiation is highly nonstationary and undergoes rapid changes such as in a developing organism (Bothma et al., 2014) or during gene induction (Lenstra et al., 2015).

More anecdotal, in certain cases, another unwanted degree of freedom can be eliminated by rescaling the correlation functions: Focusing on postinitiation dynamics in our earlier study (Coulon et al., 2014), data collection/analysis was biased toward cells showing an active TS, hence making irrelevant any measure of transcription initiation rate. In addition, we realized through modeling that (i) whatever the postinitiation dynamics, varying the initiation rate simply rescales vertically all three correlation functions (autocorrelations and cross-correlation) by the same multiplicative factor, and that (ii) an important part of the correlation functions in our case was the precise shape of the cross-correlation $G_{\text{cross}}(\tau)$ around $\tau = 0$. Hence, normalizing all three correlation functions by the same value $G_{\text{cross}}(0)$ eliminates this extra degree of freedom (Fig. 4D). Importantly, the normalization should be performed *after* averaging, so as to avoid introducing inappropriate weights among the different correlation functions (Fig. 4A).

2.6 Uncertainty, Error Bars, and Bootstrapping

Having a measure of uncertainty or confidence interval on a correlation function is crucial. As discussed further in Section 4.2, fallacious features or regularities may appear simply due to low sampling (ie, insufficient amount of data), hence misleading data interpretation. Calculating the uncertainty on a correlation function is however not trivial. Even though computing $\langle a(t)b(t+\tau)\rangle$ consists in taking a temporal average, one should *not* use the standard deviation (or the standard error) of $a(t)b(t+\tau)$ as a measure of uncertainty. Indeed, data points from a time trace are not independent. Only time points separated by a delay longer that the slowest process involved could be considered independent. But since this slowest process is often unknown and likely longer than the measured time traces, the safest solution is to consider independent only data points from distinct traces (ie, distinct cells) as independent. Note, however, that methods have been proposed to estimate the uncertainty on the correlation function from a unique time trace (Guo et al., 2012).

In the very simple case where correlation functions are computed completely independently and then simply averaged together (ie, the method of Eq. 6) without any of the weighting described in Section 2.4 and without baseline correction or renormalization of Section 2.5, then the standard error can be computed directly as the standard deviation of the individual $G_j(\tau)$, divided by $\sqrt{n}$. But in any other case (global mean estimation, traces of different durations, baseline correction, etc.), a *bootstrapping* technique has to be used (Fig. 4B). If one has a pool of n time traces (possibly multicolor traces, eg, $a_j(t)$ and $b_j(t)$), it consists in:

(i) selecting, at random and with replacement, a sample of n time traces within this pool (hence some traces will be selected more than once and some will not be selected),

(ii) performing the whole computation of the correlation function from the beginning (including computation of the mean, if estimated globally) and until the end result (including weighting, baseline correction, etc.),

(iii) reiterating steps (i) and (ii) many times (typically ≥ 1000), and

(iv) computing, at each time-delay τ, the standard deviation (not the standard error) over all the correlation functions obtained in step (ii).

The standard deviation of the sampling distribution (Fig. 4E, estimated at step (iii)) is directly the standard error of the correlation function (Fig. 4F). Confidence intervals can also be computed by taking percentiles instead of standard deviations at step (iv) (eg, for a 90% confidence interval, take the 5th and the 95th percentiles).

On a technical note: In Eqs. (7) and (8), $\langle (a_j(t)-\bar{a})(b_j(t+\tau)-\bar{b})\rangle$ cannot be rewritten as $\langle a_j(t)b_j(t+\tau)\rangle - \bar{a}\bar{b}$ in order to precompute

$\langle a_j(t) b_j(t+\tau) \rangle$ outside of the bootstrap loop; whence the advantage of a fast routine for computing correlation functions, such as the one based on FFT described in Section 2.1.

Finally, since baseline correction and normalization (if any) are included in the bootstrap loop, the correlation function will be clamped at 0 and at 1 at specific time delays, resulting in small error bars at these regions. This can be used at places where one needs to concentrate statistical power on specific features of the correlation functions, as we did in our transcription/splicing study (Coulon et al., 2014) to focus on the shape of the cross-correlation around $\tau = 0$.

3. INTERPRETATION OF CORRELATION FUNCTIONS

A complete discussion of how to model correlation functions is clearly out of the scope of this chapter—at the very least because every experimental system and every biological question is different. Here, we give the reader an introduction to different possible options one can take to extract mechanistic information from correlation functions. We also aim at providing a basic understanding of what affects the shape of a correlation function in rather simple mathematical terms.

3.1 A Primer for Correlation Function Modeling

Transcriptional signals $a(t)$ and $b(t)$ can be viewed as sums of contributions $\hat{a}_p(t)$ and $\hat{b}_p(t)$ from $\hat{n}$ individual RNAs occurring at times t_p

$$a(t) = \sum_p \hat{a}_p(t - t_p) \text{ and } b(t) = \sum_p \hat{b}_p(t - t_p) \tag{10}$$

When transcription initiation t_p occurs at random with a constant rate k over time, it is said to follow a (homogeneous) Poisson process. In this case, the covariance function can be written simply as the mean of all the covariances of individual RNAs, multiplied by k

$$M(\tau) = k \sum_{p=0}^{\hat{n}-1} \frac{\hat{M}_p(\tau)}{\hat{n}} \tag{11}$$

where $\hat{M}_p(\tau)$ is the covariance function between $\hat{a}_p(t)$ and $\hat{b}_p(t)$. (Note that, since $\hat{a}_p(t)$ and $\hat{b}_p(t)$ are square integrable signals, this covariance has a slightly different formulation, ie, it uses a temporal sum $\hat{M}_p(\tau) = \int_{-\infty}^{\infty} \hat{a}_p(t) \hat{b}_p(t+\tau) dt$ instead of a temporal average.)

This equation is central for understanding what correlation functions can reveal and is the basis for further derivations. Here, we present how Eq. (11) can be used in three different manners.

3.1.1 Understanding the Geometry of the Correlation Functions

As an illustration, let us consider the very simple example shown in Fig. 5A. Here, the fluorescence time profile of each RNA is a rectangular function of duration X_p (ie, $\hat{a}_p(t) = \hat{b}_p(t) = 1$ for $t \in [0, X_p]$ and 0 elsewhere). The dwell time X_p of the RNA at the TS (which includes elongation and a potential retention at the 3' end of the gene) is a random variable following a probability distribution with density $P(x)$ and a mean μ. In this case, the autocovariance of each RNA—only described for $\tau \geq 0$ since an autocorrelation is always symmetrical—is the triangle function $\hat{M}_p(\tau) = X_p - \tau$ for $\tau \leq X_p$ and 0 elsewhere (Fig. 5A). From Eq. (11), one can understand simple geometrical properties of the covariance function $M(\tau)$, such as that it starts at $M(0) = k\mu$ (or, if using a correlation function, $G(0) = 1/k\mu$) with a tangent that crosses the τ-axis at $\tau = \mu$ (Fig. 5B). Hence, the first few points of the correlation function already reveal two key parameters of the system: the transcription initiation rate k and the average dwell time μ of the RNA at the TS.

This approach does not impose a simplistic description of the fluorescence time profiles of RNAs as in the example above. For instance, using more realistic time profiles, we took a similar approach in our previous work on transcription and splicing (Coulon et al., 2014) and were able to show that a key measurement for our study (ie, the fraction of RNAs that are spliced before being released) is given by a simple geometrical feature of the correlation functions: the change of slope of the cross-correlation at $\tau = 0$.

To develop further the example of Fig. 5, one can also show that the way $M(\tau)$ deviates from its tangent at the origin reflects the shape of the distribution $P(x)$ of dwell times: if narrowly distributed (Fig. 5B), then $M(\tau)$ follows its tangent closely and makes a marked angle when approaching 0; if broadly distributed (Fig. 5C), this angle is smoother, making $M(\tau)$ deviate more from its tangent. This can even be generalized by realizing that, in theory, the curvature of the correlation function directly yields the full

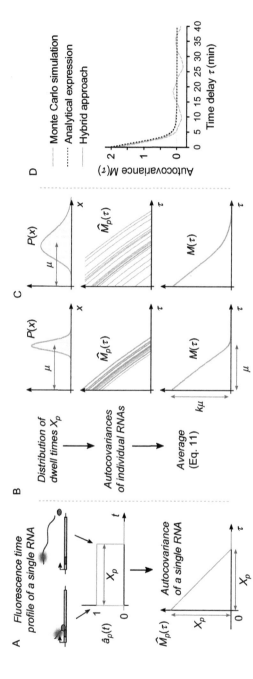

Fig. 5 Principle of correlation function modeling. Although illustrated on a simplified single-color situation, the principle for modeling correlation functions presented here is general and holds for more complex descriptions. (A) Considering that the fluorescence time profile $\hat{a}_p(t)$ recorded at the TS for a single nascent RNA is a rectangular function (rising when the MS2 cassette is transcribed and dropping when the RNA is release), then the covariance function $\hat{M}_p(\tau)$ of this time profile is a triangle function. (B and C) When transcription initiation is considered homogeneous over time, the covariance function $M(\tau)$ can be understood as the average between individual correlation functions of single RNAs (Eq. 11). On the example of (A), if the dwell time X_p of individual RNAs is distributed, then the shape of $M(\tau)$ reveals information about initiation rate and dwell time distribution (mean, variability, etc.). (D) Several methods can be used to predict correlation functions from a given mechanistic scenario. As an alternative to a full Monte Carlo simulation approach, giving rather noisy results, and to analytical expressions, sometime difficult to derive, the hybrid approach described in Section 3.1.3 is both simple and precise. In this example, the hybrid method was performed over 100 single-RNA time profiles, and the simulation was performed over a signal that comprises 500 RNAs. The total computation time was similar in both cases. The analytical expression used is Eq. (12).

distribution $kP(\tau) = \dfrac{d^2 M(\tau)}{d\tau^2}$. However, in practice, differentiating experimental data is problematic since it amplifies the noise, so much that only global features of the distribution can be generally extracted (eg, mean, variance, possibly skewness, etc.).

3.1.2 Analytical Expressions from Mechanistic Models

Even though a lot can be understood from the geometry of the correlation functions without making any strong assumption, it is also useful to take the opposite approach by assuming a given mechanistic model to assess if the predicted shape of the correlation functions can reproduce that of experimental data. In this context, one can derive analytical expressions of correlation functions from a given description of the underlying transcriptional kinetics. For a detailed example, the reader can refer to the supplementary derivations of Coulon et al. (2014). Briefly, we assumed different mechanistic models by describing the timing between specific events (eg, elongation through the MS2 and PP7 cassettes, removal of an intron, release of the RNA from the TS) with interdependent time distributions. We were able to express the analytical form of the correlation functions as convolutions between these distributions. As an illustration, on the simple example of Fig. 5, this approach yields

$$M(\tau) = kH(-\tau) * H(-\tau) * P(\tau) \qquad (12)$$

defined over $\tau \geq 0$, where $H(x)$ is the Heaviside function (ie, 1 if $x \geq 0$ and 0 elsewhere) and where $*$ denotes the convolution product.

This approach has several notable advantages over a simulation approach, especially for data fitting purposes: it is very fast to compute (once one has an analytical expression) and gives an exact result, hence allowing a proper parameter exploration in the fitting procedure. It also reveals which aspects of the kinetics affect the different parts of the correlation curves. However, this approach is rather mathematically cumbersome and does not offer a lot of flexibility: small modifications to the underlying mechanistic assumptions can sometime require one to rederive the equations from the beginning.

3.1.3 Hybrid Monte Carlo Approach

A much simpler alternative to the analytical method described earlier is to calculate Eq. (11) through a Monte Carlo approach. Indeed, no matter how elaborate the description of the fluorescence time profiles and the underlying mechanistic model (eg, with complicated fluorescence time profiles and intricate interdependent random variables), the correlation function $M(\tau)$ is always simply the average of individual correlation functions $\hat{M}_p(\tau)$. It can hence be computed numerically from a set of individual time profiles that were randomly generated from the assumed mechanistic model. To explain this in different terms, one could perform a full Monte Carlo simulation by (i) drawing randomly all the transcription initiation times and (ii) all the fluorescence time profiles of individual RNAs, then (iii) summing them up as in Eq. (10) to obtain a simulated time trace, and finally (iv) computing its correlation function. This approach has the disadvantage of giving a rather noisy result, hence imposing to run many (or very long) simulations to reach a good converge. Instead, the hybrid method described here consist in only performing step (ii) a number of times, to compute individual correlation functions for each time profile generated, and to average them as in Eq. (11). This approach yields a much more precise estimate of the correlation function $M(\tau)$ than the full simulation approach. Fig. 5D shows a comparison between the two methods where both curves were obtained in similar amount of computation time. The precision of the result depends of course on the number of individual time profiles generated, but a few hundreds already give rather precise results (eg, the hybrid curve in Fig. 5D is generated from 100 time profiles). This hybrid approach is both easy to implement (no mathematical derivation) and very flexible (the model can be modified easily), making it an attractive alternative to the analytical approach presented earlier for fitting experimental correlation functions.

All what we just discussed in this section derives from Eq. (11), which assumes that the random transcription initiation events t_p occurs homogeneously over time. This is often not the case in reality since transcriptional activity may be time dependent, for instance if the gene of interest is induced over time, is cell cycle dependent, or switches stochastically *on* and *off* (bursting). Mathematically, taking this into account modifies Eq. (11)—we are currently working on analytical models that include these types of fluctuations. However, as explained in Section 2.5, if these fluctuations are slow compared to the timescale of single-RNA transcription, one can get rid of them and the assumption of homogeneous initiation events is then appropriate.

3.2 Data Fitting and Model Discrimination

Whether generated analytically or from the hybrid method described in the previous section, correlation functions predicted from a given mechanistic model should be compared quantitatively with experimental ones. This allows both discriminating between competing models and obtaining numerical values for the underlying physical parameters.

Since auto- and cross-correlation functions may carry information on different aspects of the transcriptional kinetics, one should predict all of them simultaneously from a given model and set of parameters, and fit them globally at once to the experimental correlation functions. Making sure to use the standard errors calculated as in Section 2.6 and to only include the relevant time-delay points (ie, up to the plateau if using baseline subtraction as in Section 2.5), one may use a regular nonlinear least square fit. It consists in minimizing

$$\chi^2 = \sum_{\tau, G} \frac{\left(G_{\exp}(\tau) - G_{\text{theo}}(\tau)\right)^2}{\sigma\left[G_{\exp}(\tau)\right]^2} \qquad (13)$$

where $G_{\text{theo}}(\tau)$ and $G_{\exp}(\tau)$ are the theoretical and experimental correlation functions, $\sigma[\cdot]$ represents the standard error, and where the sum is taken over all relevant time-delay points and auto-/cross-correlation functions.

One important note is that a fit should always be examined visually before putting trust in the results. Especially when fitting multiple correlation functions at once (auto- and cross-correlations) with only a few parameters, one should ensure that certain features in the curves (whether real or artifactual) do not dominate the fit, hence preventing the most relevant parts from being reproduced accurately. It is hence advisable to know what part of the curves reflect what aspect(s) of the underlying processes, and to always use judgment when considering the result of a fit.

To discriminate between models, one cannot directly compare their χ^2 values. Indeed, if two competing models fit equally well the experimental data, the simplest one is more plausible and should be preferred. A simple way to take this into account is to find the model that minimizes the Bayesian information criterion (Konishi & Kitagawa, 2008)

$$\text{BIC} = \chi^2 + N_{\text{param}} \log N_{\text{pts}} \qquad (14)$$

where N_{param} and N_{pts} are, respectively, the number of parameters in the model and the number of time-delay points used in the fit. Importantly, when deciding between models one should not blindly rely on the BIC. The effect of experimental perturbations should be assessed and complementary measures other than correlation functions should be used to validate or discriminate further between the retained models (see Section 4.5).

4. COMMON ISSUES AND PITFALLS

Fluctuation analysis is a powerful technique, but to apply it successfully, one needs to be aware of a number of potential difficulties and pitfalls. Section 2 focused on how to compute correlation functions properly as to avoid certain biases and artifacts. This section describes other potential issues pertaining to the design of the experimental system, the imaging conditions, and the interpretation of the resulting correlation functions.

4.1 Location of the MS2 and PP7 Cassettes

It is essential to have data analysis considerations in mind from the design stage of a project. Indeed, choices on the position and length of the MS2 and PP7 DNA cassettes to be inserted in the gene(s) of interest will crucially determine what can be concluded from the data. Poor design may make the analysis difficult and/or the interpretation ambiguous. Even though the focus of this chapter is on data analysis, a brief discussion how to design the experimental system is important here.

Design choices essentially concern the position and length of the MS2 and PP7 cassettes. These should be inserted in noncoding regions: $5'$- and $3'$-untranslated regions (UTRs) and introns. A translatable version of the PP7 cassette can also be used to place loops in open reading frames (ORFs) (Halstead et al., 2015). Every application being different, no general advice can be given and the best strategy is necessarily case specific. In this regard, it is advised to make computational predictions and/or simulations (Section 3) to understand how choices in the design will affect the ability of the approach to reflect the phenomenon of interest and/or to discriminate between competing hypotheses. From an analysis point of view, one should consider:
- *signal intensity*: the possibility to have a bright and easy to track TS,
- *measurement sensitivity*: the ability to detect subtle fluctuations coming from single RNAs, and
- *temporal resolution*: setting up the limit on the timescale that can be probed on the underlying processes.

The length of the cassettes, for instance, impacts the number of fluorophores per RNA molecule, hence enhancing both sensitivity and signal intensity for given imaging conditions. Another factor, the number of labeled RNAs simultaneously present at the TS may improve its brightness but will impair in return the possibility of detecting single-RNA contributions. In practice, single-RNA sensitivity is not necessary to apply fluctuation analysis, but the more one sees fluctuations due to the finite and low number of RNAs, the better the method will work. Finally, the time it takes for a single cassette to be transcribed and the time each labeled RNA dwells at the TS—both related to intensity and sensitivity—will also impact on the temporal resolution of what the data reflect.

To illustrate these points, let us give two simple examples. First, to observe the fluctuations of transcription initiation rates over time (eg, bursting, regulatory coupling between genes), placing a cassette in the 5' UTR will result in a strong signal (most nascent RNAs are labeled), but will mask fluctuations that are faster than the dwell time of transcripts at the TS (which comprise elongation and transcript release times). Placing the cassette in the 3' UTR will have the opposite effect: the signal will be weaker because only the polymerases passed the cassette in the 3' UTR will have a labeled RNA, but the resulting shorter dwell time will allow resolving faster fluctuations of initiation rate. However, in this latter case, any process occurring during elongation (eg, pausing, variable elongation rates) will also affect the measurement, making it more difficult to interpret. Placing the cassette in an intron toward the 5' end of the gene minimizes this problem while potentially keeping the advantage of a short dwell time (if the intron is spliced rapidly). In all cases, the dwell time includes unknown factors (eg, release or splicing times) that need to be taken into account for the analysis.

Another instructive example is the measure of elongation kinetics. In this case, placing the two MS2 and PP7 cassettes in both UTRs will reflect the time to elongate throughout the gene body, but without resolving the many potential pauses and variations in elongation rate along the gene. On the contrary, placing both cassettes directly around a given sequence of interest will reveal the instantaneous elongation rate and polymerase pausing kinetics over this particular sequence. Although much more revealing, this latter case is more difficult to implement. Indeed, in the former case, the long distance between the two cassettes will make the time-delay measurement rather precise, while, in the latter case, because the delay between the two signals is comparable with the time to elongate through a single cassette (ie, when the signal ramps up as the loop are being transcribed), the precision of the measurement will crucially depend the length of the cassette (ie, the steepness of the ramps). Specifically, if the cassettes are shorter, the results are more precise, hence imposing a compromise with measurement sensitivity. Also, the less the RNA dwells at the TS after passing both cassettes, the better the stochastic delay between the two cassettes can be resolved, imposing here another compromise with signal intensity.

Finally, in addition to data analysis considerations, another factor to take obviously into account is the risk of affecting endogenous processes. When choosing the location of the MS2 and PP7 cassettes, one should ensure not to disrupt functional sequences in the UTRs, introns, and ORFs (Bentley, 2014; Porrua & Libri, 2015). Length of the cassettes has also been observed to affect whether the modified RNAs behave like the endogenous ones (decay, nuclear export, aggregation, etc.). We have not yet reached a consensus for good practice since it seems context dependent. It is hence advised to perform single-molecule RNA FISH (see Section 4.5) to compare the statistics of endogenous and modified RNAs (levels, localization, number at TS, etc.) and ensure that both behave similarly.

4.2 Interpreting Single (or Too Few) Traces

When gathering experimental data to be analyzed through fluctuation analysis, it is often tempting to interpret the data and draw conclusions based on an insufficient amount of data. Typically, one should not interpret the correlation function of a single time trace. Indeed, *sampling noise*—ie, the finite-size effects coming from the low number of stochastic events in a time trace (even if in the hundreds)—leads to features in the correlation function that only reflect the randomness of these events (Fig. 6). Because sampling noise produces smooth shapes in a correlation function, it does not look like what we are used to call "noise" and often produce oscillations or "bumps" in the correlation curves that only result from a lack of data. For instance, a single

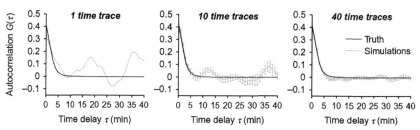

Fig. 6 Convergence when averaging correlation function. Single correlation functions are often misleading since their shape can show features that may look like regularities but are only due to the lack of data. Averaging multiple correlation functions together reduces this noise and allows one to calculate error bars. The more correlation functions are averaged together the more these spurious features disappear, leaving only the true regularities. The examples shown are simulated traces that are 100 data points each.

correlation function may show a bump at a certain time delay, not because there is a regularity at this timescale in the underlying biological process, but simply because, by random chance, there happens to be one *in this particular trace* (Fig. 6). Redoing the analysis on a different time trace obtained on the same experimental system may or may not show this feature. Only if this feature is present in multiple traces and is still present when averaging a number of correlation functions can it be considered a regularity (Fig. 6). To ensure that an averaged correlation function is sufficiently converged and that a given feature is statistically significant, the best way is to compute error bars (standard errors or confidence intervals; Section 2.6). Remember that nonoverlapping standard errors do not necessarily imply high significance, and that a standard error is only expected to overlap with the true, fully converged value in <80% of the cases.

4.3 Technical Sources of Fluctuations

Correlation functions reflect all types of fluctuations in the observed signals. On the positive side, this has the advantage of revealing multiple biological processes at once. On the other hand, any technical source of fluctuation will also show up in a correlation function. We present here the most common types for MS2/PP7 transcriptional time traces, how to identify them and, when possible, how to correct for them:

- *Bleaching*: Inherent to any fluorescence microscopy experiment is the bleaching of fluorophores over time. The result is that the fluorescence intensity of the whole nucleus—and, with it, of the TS—will decay throughout an experiment. If not corrected, bleaching results essentially in an artifactual slow decay (typically exponential) in the correlation functions. It is important here to understand that, in microscopy setups where the whole nuclear volume is illuminated (eg, in widefield microscopy and to some extant in confocal microscopy when acquiring a z-stack), all the fluorophores will bleach equally fast in a nucleus regardless of being bound or not to an RNA. Hence, the relative fluorescence lost by every single RNA is statistically the same as for the whole nucleus. From this observation, it is possible to correct time traces for bleaching prior to any computation. In practice, this requires the total integrated fluorescence of the nucleus to be much higher than that of the TS—typically doable in mammalian cells (Coulon et al., 2014), but not in yeast cells (Lenstra et al., 2015). To perform this correction, sometime referred to as *detrending*, one needs to: (i) isolate the image of the whole nucleus (cropping the time-lapse movie around the nucleus may be sufficient as long as no other nuclei are in the resulting picture, otherwise a masking procedure of the outside of the nucleus can be used); (ii) compute the standard deviation of the pixel fluorescence intensities for each image separately, throughout the course of the time-lapse movie; (iii) ensure that the resulting time course is smooth as to avoid adding extra unwanted fluctuations; and (iv) divide the transcriptional time trace by this time course. The rational behind using the standard deviation is that the mean of a microscopy imaging often includes factors such as the autofluorescence of the media (which has its own bleaching kinetics), the digital offset of the camera/detector, etc., all of which have a negligible spatial standard deviation. Hence the mean of the pixel fluorescence intensities will be affected by these extra factors, while the standard deviation will not.

Acceptable levels of bleaching to be corrected using this technique can be empirically up to a 50–60% attenuation between the first and last frames of a time-lapse movie. More would lead to a strong difference in signal-to-noise ratio between the beginning and the end of the time traces, which may become problematic.

- *Nonhomogeneous illumination of the field*: Often in microscopy setups, illumination is not uniform and tends to be stronger in the center of the field of view and dimmer toward the edges. Hence, not only bleaching rate will be different for different cells, but more importantly, if a cell moves within the field and gets closer and further from the center, this will result in global fluctuations of the observed nuclear intensity and TS intensity that are not due to the transcriptional activity. As long as the spatial inhomogeneities in illumination are larger than a cell nucleus, the detrending procedure described earlier solves this problem by correcting for fluctuations in total nuclear brightness.
- *Measurement noise*: Many factors contribute to inexact measurements of TS fluorescence intensity. This includes Poisson noise of photon collection, current noise in the detectors, numerical inaccuracies in the fitting of the TS, etc. All of these have essentially *white noise* statistics, ie, the error made at a given time point is statistically independent form the error made at all the other time points. The advantage of white noise is that it will show up only in autocorrelation functions (not cross-correlations) and only at the time-delay point $\tau = 0$. All the rest of the correlation functions are unaffected. This noise cannot be corrected without destroying additional information in the signal. Hence the

simplest way to deal with it is to discard the first point ($\tau=0$) of any autocorrelation function, knowing that it is inaccurate.
- *Tracking errors*: Transcriptional time traces are generated from the time-lapse images by detecting the TSs, localizing and fitting them with Gaussians, and generating trajectories by connecting detections across different frames. Whichever software is used (eg, ours is available at http://larsonlab.net/), this process is not error free. Incorrect spot detection and/or the presence of another object in the vicinity of the tracked TS can result in inaccurate tracking. One can distinguish two situations: (i) a brief jump (eg, one frames) away from and then back to the TS, or (ii) a single jump away from the TS that does not come back (or does so after a certain number of frames, eg, ≥ 5 frames). In the former case, one-frame tracking errors have a similar effect as the measurement noise (if they are truly one frame long). That is, they make inaccurate the first point (at $\tau=0$) of correlation functions (both auto- and cross-correlations in case the two channels are tracked simultaneously). Practically, a few of these errors are acceptable per trace. But if too numerous or if one needs a precise measure of the cross-correlation around $\tau=0$, they should be corrected in the tracking procedure. In the cases where the tracking errors are long (case (ii)), the corresponding portions of the traces can be kept as such *only if* the TS is actually inactive (ie, showing no fluorescence signal) during this time period, *and* the inaccurate detections yield a signal close to background. In any other case (eg, the tracking jumps to a nuclear structure yielding a nonnull measured signal, or the tracking stays on the nuclear background while the TS is actually active), it is critical to avoid including such portions of a time trace. Two simple options are to adjust the tracking procedure as to avoid these inaccurate portions (possibly requiring manual user intervention), or to trim the traces to only keep the part where tracking is accurate (see Section 2.4 for dealing with traces of various lengths, and Section 4.4 for the biases this may cause). Note that if this type of inaccurate tracking occurs in the middle of long and otherwise good time trace, one can split the trace in two parts and treat them as two different traces.
- *Volumetric imaging*: Nuclei often being ≥ 10 μm thick in the z-axis direction, the whole nuclear volume is not always covered by the z-stack acquisition. Hence, a TS may diffuse in and out of the imaged volume through the course of the experiments, resulting in fluctuations in the measured intensity that are not due to transcriptional events. It is critical to ensure on all time traces that the TS does not reach the edge of the

imaging z-range. Otherwise, these portions of the traces should be excluded (as described earlier for the long tracking errors). If not taken into account, this type of fluctuations can completely mask the transcriptional kinetics by adding a strong and short-scale decay to all the correlation functions. Along the same lines, choosing an inappropriately large Δz step between z-planes can have a similar effect. Appropriate values depend on the optical setup but are typically ~0.5 μm for a widefield microscope.

A general note on technical fluctuations is that they are not always simple to identify and may be mistaken for biological ones. A good way to confirm if a certain feature in correlation functions is of technical nature is to change the imaging conditions (eg, illumination, step and range of the z-stack, time step, coat protein level, etc.) and see if the feature is affected.

4.4 Biased Selection of Data (Cells, TS, Part of Traces)

When acquiring and analyzing data, one tends to bias the selection of fields of view and/or cells based on how the TS appears. Also, the procedure described in the previous section to exclude portions of traces where tracking is inaccurate (likely when the TS is not or weakly active) introduces a bias in the result—ie, transcription appears more frequent than it is in reality. This is acceptable if the only conclusions drawn from the data are about the postinitiation kinetics of single-RNA synthesis (Coulon et al., 2014). But in this case, it is not possible to make any statement on the frequency of transcription (eg, initiation rate, bursting kinetics, etc.) based on this data alone. To do so, one has to image and analyze cells regardless of their activity, and to generate traces where periods of transcriptional inactivity are indeed measured as such and hence included in the analysis (Lenstra et al., 2015). This is more demanding in terms of image analysis and may require extensive manual user intervention during the tracking procedure.

4.5 Validation by Complementary Measurements

Finally, an important point is that—as holds for any technique—fluctuation analysis should not be used alone. In certain cases, two distinct mechanistic scenarios about the underlying biological process may produce rather similar correlation functions. Not only perturbation experiments should be performed to ensure that the correlation functions are affected as expected, but additional techniques should also be used to corroborate the findings.

A relevant technique to use in this context is single-molecule RNA FISH (Femino, Fay, Fogarty, & Singer, 1998). It consists in hybridizing fluorescently tagged DNA oligonucleotides onto an RNA of interest in fixed cells, leading to the visualization of single RNAs in the cell and allowing the absolute quantification of nascent RNAs at each TS. This technique is also much more amenable to high-throughput acquisition and analysis, giving an unbiased view of the processes under study. Although it does not give access to dynamics, it provides a very complementary picture to fluctuation analysis of MS2/PP7 time traces. It can be performed on a gene already tagged with MS2 or PP7 (by designing oligos against the MS2/PP7 repeats) to validate/complement the conclusions obtained through the live cell approach (Lenstra et al., 2015), or it can be used to observe other genes easily (only requiring to design new oligos) to show how the conclusions of the live-cell measurements can be generalized (Coulon et al., 2014).

5. CONCLUSION

Fluctuation analysis is a powerful method for extracting mechanistic information from complex transcriptional time traces, obtained by MS2 and PP7 RNA labeling, where many RNAs are synthesized simultaneously, each one having its own stochastic transcriptional kinetics, and with potentially multiple biological processes occurring at different timescales. The added value of acquiring multicolor data on a given experimental system is often significant since one can calculate both auto- and cross-correlation functions, hence revealing much more information than an autocorrelation alone.

Implementing this technique is however not always trivial. As we have seen, calculating and interpreting correlation functions correctly requires a good knowledge and understanding of the technique and there are a number of mistakes one can make. A general advice is that, when using fluctuation analysis to interpret experimental data, one should always run numerical simulations. Whether the purpose is

- to troubleshoot the computation of correlation functions (Section 2),
- to estimate the effect of potential technical artifacts (Section 4.3),
- to have an intuition on what the correlation function may reveal for a given experimental system (Section 4.1),
- to test rapidly if a hypothetical mechanistic model is consistent with experimental observations,

- to verify the result of mathematical predictions (Section 3.1),
- or to assess whether the fitting procedure is able to discriminate properly between mechanistic models and to recover the underlying parameters (Section 3.2),

it is always a simple and easy tool to use, with many benefits. It provides a set of time traces to experiment with, where the underlying mechanisms at play are fully known and can be changed freely. To generate simulated signals, one can either use the Gillespie algorithm (Gillespie, 1976) or a more general Monte Carlo approach by drawing random events with the desired statistics and combining them as in Eq. (10).

Fluctuation analysis is a general technique. Its use is not restricted to transcriptional time traces. We can anticipate its future application to other related types of data, upstream and downstream of transcription. This includes, for instance, the simultaneous measurement of transcription (by MS2 or PP7 labeling) and imaging of complexes/enzymes recruitment at a given locus (transcription factors, Pol II, enhancers, etc.), as well as the time course of protein synthesis observed from a single RNA in the cytoplasm (Morisaki et al., 2016; Wu, Eliscovich, Yoon, & Singer, 2016). In many contexts, this powerful analysis technique will help to dissect complex biological mechanisms, by building upon basic concepts of signal theory.

REFERENCES

Bentley, D. L. (2014). Coupling mRNA processing with transcription in time and space. *Nature Reviews. Genetics*, *15*(3), 163–175. http://doi.org/10.1038/nrg3662.

Bertrand, E., Chartrand, P., Schaefer, M., Shenoy, S. M., Singer, R. H., & Long, R. M. (1998). Localization of ASH1 mRNA particles in living yeast. *Molecular Cell*, *2*(4), 437–445.

Bothma, J. P., Garcia, H. G., Esposito, E., Schlissel, G., Gregor, T., & Levine, M. (2014). Dynamic regulation of eve stripe 2 expression reveals transcriptional bursts in living Drosophila embryos. *Proceedings of the National Academy of Sciences of the United States of America*, *111*(29), 10598–10603. http://doi.org/10.1073/pnas.1410022111.

Chao, J. A., Patskovsky, Y., Almo, S. C., & Singer, R. H. (2007). Structural basis for the coevolution of a viral RNA–protein complex. *Nature Structural & Molecular Biology*, *15*(1), 103–105. http://doi.org/10.1038/nsmb1327.

Coulon, A., Chow, C. C., Singer, R. H., & Larson, D. R. (2013). Eukaryotic transcriptional dynamics: From single molecules to cell populations. *Nature Reviews. Genetics*, *14*(8), 572–584. http://doi.org/10.1038/nrg3484.

Coulon, A., Ferguson, M. L., de Turris, V., Palangat, M., Chow, C. C., & Larson, D. R. (2014). Kinetic competition during the transcription cycle results in stochastic RNA processing. *eLife*, *3*. http://doi.org/10.7554/eLife.03939.

Craig, N., Green, R., Greider, C., Cohen-Fix, O., Storz, G., & Wolberger, C. (2014). *Molecular biology—Principles of genome function* (2nd ed.). Oxford, United Kingdom: Oxford University Press.

Femino, A. M., Fay, F. S., Fogarty, K., & Singer, R. H. (1998). Visualization of single RNA transcripts in situ. *Science, 280*(5363), 585–590.

Ferguson, M. L., & Larson, D. R. (2013). In *Measuring transcription dynamics in living cells using fluctuation analysis: Vol. 1042.* (pp. 47–60). Totowa, NJ: Humana Press. http://doi.org/10.1007/978-1-62703-526-2_4.

Gillespie, D. T. (1976). A general method for numerically simulating the stochastic time evolution of coupled chemical reactions. *Journal of Computational Physics, 22,* 403–434.

Guo, S.-M., He, J., Monnier, N., Sun, G., Wohland, T., & Bathe, M. (2012). Bayesian approach to the analysis of fluorescence correlation spectroscopy data II: Application to simulated and in vitro data. *Analytical Chemistry, 84*(9), 3880–3888. http://doi.org/10.1021/ac2034375.

Halstead, J. M., Lionnet, T., Wilbertz, J. H., Wippich, F., Ephrussi, A., Singer, R. H., et al. (2015). An RNA biosensor for imaging the first round of translation from single cells to living animals. *Science, 347*(6228), 1367–1671. http://doi.org/10.1126/science.aaa3380.

Konishi, S., & Kitagawa, G. (2008). *Information criteria and statistical modeling.* New York: Springer.

Larson, D. R., Zenklusen, D., Wu, B., Chao, J. A., & Singer, R. H. (2011). Real-time observation of transcription initiation and elongation on an endogenous yeast gene. *Science, 332*(6028), 475–478. http://doi.org/10.1126/science.1202142.

Lenstra, T. L., Coulon, A., Chow, C. C., & Larson, D. R. (2015). Single-molecule imaging reveals a switch between spurious and functional ncRNA transcription. *Molecular Cell, 60,* 597–610. http://doi.org/10.1016/j.molcel.2015.09.028.

Martin, R. M., Rino, J., Carvalho, C., Kirchhausen, T., & Carmo-Fonseca, M. (2013). Live-cell visualization of pre-mRNA splicing with single-molecule sensitivity. *Cell Reports, 4*(6), 1144–1155. http://doi.org/10.1016/j.celrep.2013.08.013.

Morisaki, T., Lyon, K., DeLuca, K. F., DeLuca, J. G., English, B. P., Zhang, Z., et al. (2016). *Real-time quantification of single RNA translation dynamics in living cells,* manuscript in preparation.

Porrua, O., & Libri, D. (2015). Transcription termination and the control of the transcriptome: Why, where and how to stop. *Nature Reviews. Molecular Cell Biology, 16*(3), 190–202. http://doi.org/10.1038/nrm3943.

Schatzel, K. (1990). Noise on photon correlation data. I. Autocorrelation functions. *Quantum Optics: Journal of the European Optical Society Part B, 2*(4), 287–305. http://doi.org/10.1088/0954-8998/2/4/002.

Segel, I. H. (1993). *Enzyme kinetics.* New York: Wiley-Interscience.

Van Etten, W. C. (2006). *Introduction to random signals and noise.* Chichester, England: John Wiley & Sons.

Wohland, T., Rigler, R., & Vogel, H. (2001). The standard deviation in fluorescence correlation spectroscopy. *Biophysical Journal, 80*(6), 2987–2999. http://doi.org/10.1016/S0006-3495(01)76264-9.

Wu, B., Eliscovich, C., Yoon, Y. J., & Singer, R. H. (2016). *Translation dynamics of single mRNAs in live cells and neurons,* manuscript in preparation.

CHAPTER EIGHT

IMAGEtags: Quantifying mRNA Transcription in Real Time with Multiaptamer Reporters

J. Ray[*,1], I. Shin[†], M. Ilgu[‡], L. Bendickson[§,¶], V. Gupta[∥],
G.A. Kraus[§,¶], M. Nilsen-Hamilton[§,¶,‡,1]

[*]Cornell University, Ithaca, NY, United States
[†]National Forensic Service, Seoul, South Korea
[‡]Aptalogic Inc., Ames, IA, United States
[§]Ames Laboratory, US DOE, Ames, IA, United States
[¶]Iowa State University, Ames, IA, United States
[∥]The Scripps Research Institute, Jupiter, FL, United States
[1]Corresponding authors: e-mail address: jr839@cornell.edu; marit@iastate.edu

Contents

1. Introduction	194
2. IMAGEtags	196
3. Visualizing Gene Expression with IMAGEtags	199
3.1 Materials and Instrumentation	201
3.2 Transformation of Yeast Cells	202
3.3 Cell Preparation for FRET Measurements	202
3.4 Data Acquisition and Analysis	203
3.5 Acceptor Photobleaching to Validate FRET	205
4. Measurement of IMAGEtag RNA Level by RT-qPCR	205
5. Synthesis of Ligands	207
5.1 Synthesis of Cy5- and Cy3-Tobramycin	207
5.2 Synthesis of Cy5-PDC-Gly	208
5.3 Synthesis of Cy3-PDC-Gly	208
6. Cloning Repetitive Sequences	209
6.1 Cloning Initial Repetitive Aptamer Sequences from Synthetic Oligonucleotides	209
6.2 Constructing Longer Repetitive Sequences	210
6.3 Multiplying the Original Repetitive Sequences	211
7. Conclusions	211
Acknowledgments	212
References	212

Abstract

Cell communications are essential to the organization, development, and maintenance of multicellular organisms. Much of this communication involves changes in RNA transcription and is dynamic. Most methods for studying transcription require interrupting the continuity of cellular function by sacrificing the communicating cells and capturing gene expression information by periodic sampling of individual cells or the population. The IMAGEtag technology to quantify RNA levels in living cells, demonstrated here in yeast, allows individual cells to be tracked over time as they respond to different environmental cues. IMAGEtags are short RNAs consisting of strings of a variable number of tandem aptamers that bind small-molecule ligands. The aptamer strings can vary in length and in configuration of aptamer constituents, such as to contain multiples of the same aptamer or two or more different aptamers that alternate in their occurrence. A minimum effective length is about five aptamers. The maximum length is undefined. The small-molecule ligands are enabled for imaging as fluorophore conjugates. For each IMAGEtag, two fluorophore conjugates are provided, which are FRET pairs. When a cell expresses an RNA containing an IMAGEtag sequence, the aptamers bind their ligands and bring the fluorophores into sufficiently close proximity to allow FRET. The background fluorescence of both fluorophores is minimal in the FRET channel. These features endow IMAGEtags with the sensitivity to report on mRNA expression levels in living cells.

1. INTRODUCTION

Transcriptional regulation is central to cellular function and to the abilities of cells to respond to their environments. Cellular responses are rapid and cells in the same environment can respond differently. Thus, methods of tracking transcription in individual cells in real time are required for developing a better understanding of the processes involved in cell regulation at the transcriptional level. Several RNA reporters have been developed for monitoring transcription in living cells. These RNA reporters fall in two classes. The first to be developed were the MS2-based reporters in which a series of identical viral RNA elements are present in the RNA sequence and the MS2 peptide to which they bind is expressed as a peptide tag fused to a fluorescent protein such as GFP. The RNA reporters are detected by their ability to concentrate the MS2-tagged GFP (Bertrand et al., 1998; Hocine, Raymond, Zenklusen, Chao, & Singer, 2013). This method requires computational deconvolution of the images and allows the analysis of RNA transcription and trafficking in single cells. The second class of aptamer applied as RNA reporters is a group of light-up aptamers for which the ligands have greatly elevated fluorescence yield in the aptamer–ligand complex than when free in solution. The Spinach and Broccoli aptamers are examples of light-up aptamers that have recently been applied to visualizing RNA transcription in individual cells (Filonov, Moon,

Svensen, & Jaffrey, 2014; Paige, Wu, & Jaffrey, 2011; Strack, Disney, & Jaffrey, 2013). These aptamers have been used to image rRNA transcription and to localize trinucleotide RNA sequences but have not been used to image mRNAs. The latter RNAs provide a big challenge for imaging because of their low levels of expression and the consequent requirement for very sensitive means of detection. To address this problem, we have used FRET (Förster resonance energy transfer) for mRNA detection. FRET only occurs when two molecules capable of interacting with energy transfer are colocated within 100 Å. These FRET pairs interact infrequently when moving randomly in a volume such as the cell cytoplasm. However, energy transfer is efficient when they are held within 100 Å of each other for extended periods. Thus, the major advantage of FRET is that the signal from appropriately tethered FRET pairs is much stronger than the signal from these molecules when they are moving randomly in solution.

IMAGEtag (*I*ntracellular *M*ulti*a*ptamer *G*enetic *tag*) reporters consist of strings of tandem aptamers, which bind small-molecule ligands that are conjugated with one or the other of a FRET pair. The FRET signal is produced by interaction of the FRET ligands with aptamers in the IMAGEtag strings. When bound by the aptamers the fluorophore-conjugated ligands are in sufficiently close proximity to participate in FRET, thereby creating a fluorescent signal that reports on the presence of the IMAGEtags (Fig. 1).

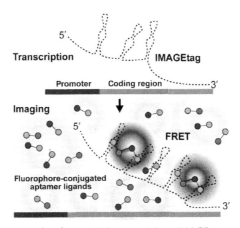

Fig. 1 The IMAGEtag technology. RNAs-containing IMAGEtag sequences are transcribed by the cell (*top panel*) and, when presented with their fluorescently conjugated ligands (o–o), will bind their ligand allowing the fluorophores to interact in FRET (*lower panel*) that enables a unique image of the IMAGEtags over the background fluorescence.

The IMAGEtag can consist of aptamers that recognize the same or different ligands. Two aptamers with different specificities would be coded in the IMAGEtag as alternating tandem structures. However, the combinations of aptamers that we have so far tested for proper folding in alternating series have not given promising results. Instead, we created IMAGEtags consisting of multiple copies of the same aptamer sequence in strings of at least five units. These IMAGEtags function well for imaging promoter activity when the aptamer ligand is presented in two forms, each conjugated with one of a FRET pair (Ilgu et al., 2016; Shin et al., 2014). The effectiveness of these single-aptamer IMAGEtags can be understood when it is considered that, when an internal aptamer of the string binds one of the two ligands, there is a 50% chance that one or the other adjacent aptamer will bind the FRET pair and therefore a 75% chance that the central aptamer will have an adjacent FRET partner. Although homo-FRET can occur if two of the same ligand are closely located by the IMAGEtags, this output is much lower in intensity than from the hetero-FRET pairs. Thus, an IMAGEtag containing a single-aptamer type will provide a signal that increases in efficiency with the number of tandem aptamers in the string. We have found that strings of five aptamers or more of a single type provide sufficient signal to detect transcribed mRNA (Ilgu et al., 2016; Shin et al., 2014).

2. IMAGEtags

We have so far made three IMAGEtag reporters, each consisting of a tandem series of the same aptamer (neomycin, tobramycin, or PDC aptamers). Any aptamer that binds a target with the ability to enter a cell is a potential component of IMAGEtags. However, the following practical considerations limit the choice of aptamers. (1) The ligand needs to be nontoxic in the concentration range used for imaging. (2) The aptamer components of IMAGEtags must be capable of binding their targets in the cellular cytoplasmic environment. (3) The aptamers must fold appropriately for binding their targets when in the context of the IMAGEtag that includes other aptamers and additional RNA sequence. The ability to fold appropriately depends on the RNA sequence including the spacing between aptamers. We have found that a space of four nucleotides is sufficient for small aptamers such as those described here, which all fold as stem-loops and are in the range of 20–40 nt long. It is preferable that the

IMAGEtags are as short as possible to minimize their effects on transcription rate. The ability to fold appropriately under intracellular conditions can be predicted for IMAGEtags by cotranscriptional folding programs such as KineFold (Xayaphoummine, Bucher, & Isambert, 2005). (4) The aptamers should be tested for their abilities to bind ligand under the salt conditions present in the cytoplasm in which the free Mg^{2+} concentration is much lower than is used for selecting most aptamers and where the major monovalent cation is K^+ rather than Na^+. We have found that aptamer affinities and ligand specificities can be quite different in buffers that approximate intracellular conditions compared with the buffers in which they were selected (Ilgu et al., 2016; Ilgu, Wang, Lamm, & Nilsen-Hamilton, 2013).

When used as reporters of promoter activity as we have done, IMAGEtags can form the majority of the RNA sequence and should have a short half-life so that they report specifically on promoter activity. The IMAGEtags described here fit those characteristics, with measured half-lives in yeast of 6–8 min. If IMAGEtags are applied to measure longer term steady-state RNA levels in the cell or to track RNA trafficking, the optimal position of the IMAGEtags in the RNA may vary. Additional RNA sequence can change the transcribed reporter RNA abundance by altering characteristics such as RNA stability and transcriptional elongation rate. Therefore, the tagged RNA should be demonstrated to be regulated similarly and have the same rate of turnover as the endogenous RNA.

The IMAGEtag reporters have been demonstrated to function in yeast in which they have been tested for reporting transcription from the GAL1, ADH1, and ACT1 promoters (Shin et al., 2014). The ligands for each aptamer were conjugated with cyanine3 (Cy3) or cyanine5 (Cy5) as the FRET pair. With the GAL1 promoter driving RNA expression and in the presence of the inducer, galactose, or with the constitutive ADH1 and ACT1 promoters driving RNA expression, FRET signals are generated when the cells are incubated with the appropriate ligand pair (Fig. 2). An important aspect of this method is that the background created by each fluorophore is absent in the FRET channel (Fig. 2A,B).

The high specificity of the aptamers for their ligands can be used to validate that the IMAGEtag is the source of the FRET signal. Thus, when yeast cells expressing the 6×PDC IMAGEtags (six copies of the PDC aptamer) were incubated with Cy3- or Cy5-conjugated PDC ligands, the FRET signals from cells that expressed the aptamer were

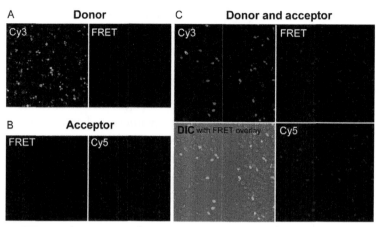

Fig. 2 FRET signal acquisition from IMAGEtags. (A) Yeast cells expressing 6×PDC IMAGEtag RNA were incubated with only Cy3-PDC (donor). Cells were excited at 550 nm with a white light laser and emissions were read with a PMT2 setting at 560–626 nm (Cy3) and with a PMT4 setting at 660–754 nm (Cy5/FRET channel). (B) Yeast cells expressing 6×PDC IMAGEtag RNAs were incubated with only Cy5-PDC (acceptor). The cells were excited at 550 nm and emission read at 660–754 nm followed by excitation at 650 nm and emission read at 660–754 nm (Cy5/FRET channel) (C) Yeast cells expressing 6×PDC IMAGEtag RNA were incubated with both Cy3-PDC (donor) and Cy5-PDC (acceptor). The cells were excited first at 550 nm with a white light laser and emissions are shown for the Cy3 (560–626 nm) and Cy5/FRET (660–754 nm) channels. The cells were then excited at 650 nm and emissions are shown for the Cy5/FRET channel (660–754 nm). The DIC image is shown with a FRET overlay.

significantly higher than obtained from cells that expressed a nonrelevant RNA from the same promoter (Fig. 3A). Cells expressing the 5×TOB IMAGEtags (five copies of the tobramycin aptamer) incubated with the PDC ligands gave FRET signals equal to the control level (Shin et al., 2014). By contrast, when tobramycin ligands were provided, an increase in FRET was observed in cells that express the 5×TOB IMAGEtags from the GAL1 promoter compared with the control RNA (Fig. 4A).

The importance of imaging live cells is that real-time changes in gene expression can be observed upon promoter induction. The increased transcriptional activity of the GAL1 promoter in response to galactose was monitored by the appearance of the IMAGEtag RNA with time in the cell population (Fig. 3B) and in individual cells (Fig. 3C). By comparison, only a decrease in FRET due to fluorophore quenching was observed in cells expressing the control RNA (Fig. 3B,C).

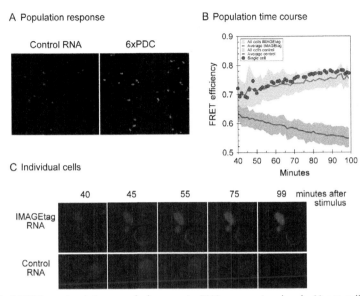

Fig. 3 IMAGEtags track temporal changes in RNA expression levels. Yeast cells were grown in 2% raffinose containing SD-Uracil and induced with 2% galactose. (A) FRET images are shown of two populations of yeast expressing either control RNA or 6×PDC IMAGEtags under the control of the GAL1 promoter. (B) Quantification of the FRET from 8 cells expressing control RNA and 13 cells expressing IMAGEtag RNA. The control cell FRET decreased with time due to cumulative exposure to laser light. To obtain the FRET efficiency values for IMAGEtag-expressing cells, the average control FRET emissions per cell at each time point were subtracted from the FRET emissions of individual IMAGEtag-expressing cells. The data are represented as the range of values (areas of *color*) with the mean value as a *line* within the range of values. The *circles* show the time course of increase in FRET for the single cell-expressing IMAGEtags shown in (C). (C) Images of a budding yeast cell that expresses IMAGEtags (*upper panel*) or that expresses control RNA (*lower panel*) are shown as a function of time after addition of galactose to the culture. Shown are merged DIC and FRET images. (See the color plate.)

3. VISUALIZING GENE EXPRESSION WITH IMAGEtags

To ensure that IMAGEtag expression is responsible for the observed FRET signals, it is important to have a series of control experiments and conditions, which include: (1) Control cells that are imaged under the same conditions as the IMAGEtag-expressing cells, which express (from the same vector as used for the IMAGEtag RNA) a nonrelevant RNA that does not bind the aptamer ligand. (2) Cells harboring the IMAGEtag-expressing

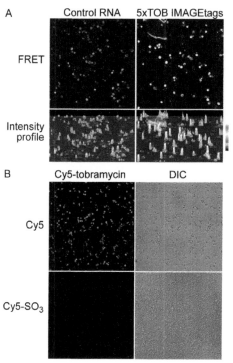

Fig. 4 Tobramycin IMAGEtags and the impact of ligand conjugate charge on cell uptake. (A) Yeast cells expressing either the control RNA or 5×TOB IMAGEtags were grown in 1% raffinose containing SD-Uracil and incubated for 30 min with Cy3-tobramycin and Cy5-tobramycin and 2% galactose. The cells were washed and FRET images were acquired. Quantification of the images is shown by the intensity profiles below each image. (B) Yeast cells expressing TOB IMAGEtags were incubated with either Cy5-tobramycin or Cy5-SO_3-tobramycin for 90 min, washed and then imaged in the Cy5 channel. The DIC images show the locations of the cells in the field. The DIC for the Cy5-SO_3 has been overlaid by the image to its left from the Cy5 channel. (See the color plate.)

plasmid that are not induced to express the IMAGEtag RNA, (3) validation of FRET by acceptor photobleaching, and (4) validation of changes in mRNA levels in the cell population by RT-qPCR. In addition, it is very important in all studies that involve microscopy to ensure objective data collection and analysis. For example, the choice of field should be randomized and/or performed by a "blinded" operator. Similarly, the analysis should be performed by a "blinded" operator. Operator blinding can be readily achieved if another individual labels the tubes/plates of cells with a numerical or alphabetical system that can be disclosed after the images have been obtained and analyzed.

3.1 Materials and Instrumentation

1. PLATE solution [40% polyethylene glycol (PEG) 3350, 0.1 M lithium acetate, 1 mM EDTA, 10 mM Tris·HCl, pH 7.5].
2. Dimethyl sulfoxide (DMSO).
3. SD-Uracil media: 6.7 g/L Difco™ Yeast nitrogen base without amino acids, 1.92 g/L Yeast synthetic dropout medium supplement without uracil (add 20 g agar/L for preparing plates). Autoclave then add filter-sterilized glucose or raffinose to a final concentration of 2% (w/v).
4. 20% galactose in water.
5. RNA extraction buffer (0.1 M NaCl, 5% SDS, 10 mM EDTA, 50 mM Tris·HCl, pH 7.5).
6. Phenol:chloroform:isoamyl alcohol (49.5:49.5:1). Note that the pH of this solution should be acidic (pH 4.5). Unlike dsDNA, which has a tendency to precipitate at lower pH and partitions into the organic phase (Sambrook & Russell, 2006), RNA remains water soluble due to H bonding between water and the protonated bases. Thus extraction at low pH minimizes contamination of the RNA preparation by chromosomal DNA.
7. Chloroform:isoamyl alcohol (24:1).
8. Fluorescent aptamer ligand conjugates. These will depend on the IMAGEtags that are used (see Fig. 5 for the fluorophore-conjugated ligands described here).

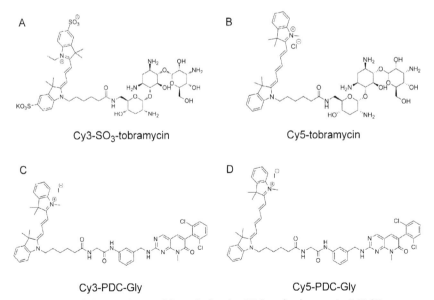

Fig. 5 Cy3- and Cy5-conjugated ligands for the PDC and tobramycin IMAGEtags.

9. The plasmids used in this study can be obtained from the corresponding author, M.N.H.
10. In these studies, we have used yeast strain BY4735, Genotype: MATα ade2Δ::hisGhis3Δ200leu2Δ0met15Δ0trp1Δ63ura3Δ0.
11. Confocal microscopy: In these studies we used a white light laser to excite and PMT for recording emissions from the fluorophores.

3.2 Transformation of Yeast Cells

1. Yeast expression plasmid pYES2 was used as the backbone to construct vectors from which RNA containing IMAGEtags could be expressed in yeast.
2. Grow *Saccharomyces cerevisiae* in YPD (1% yeast extract, 1% peptone, and 2% glucose) medium at 30°C overnight with shaking at 250 rpm.
3. Take 0.5 mL of the culture and spin for 30 s at $13,200 \times g$. Remove the supernatant and resuspend the pellet in 0.5 mL PLATE solution. Add 10 μL carrier DNA (100 μg salmon testes DNA) and 10 μL transforming plasmid DNA (1 μg) and vortex. Add 57 μL DMSO and vortex briefly. Leave for at least 30 min at room temperature (can be left on bench overnight).
4. Heat shock at 42°C for 15 min. Centrifuge the cells at $13,200 \times g$ for 30 s and carefully remove the supernatant. Resuspend the cells gently in 200 μL of sterile distilled and deionized water by aspirating up and down with the pipette tip. Immediately spread the cells on SD-Uracil plates. Incubate the plates at 30°C until colonies appear.

3.3 Cell Preparation for FRET Measurements

Grow yeast cells containing the plasmids with expression capability for either the control RNA or the IMAGEtag RNA.

1. Pick a colony from the SD-Uracil plates containing the transformed cells. Grow the cells in SD-Uracil medium with 2% (w/v) glucose at 30°C with shaking at 250 rpm until they reach OD600 of 1.0.
2. Inoculate the culture to a fresh SD-Uracil medium with 2% (w/v) raffinose to an OD600 of 0.02. Grow the cells at 30°C with shaking at 250 rpm until the culture reaches an OD600 of 0.4–0.6.
3. Centrifuge the cells at $1000 \times g$ for 5 min. Remove supernatant and resuspend the pellet in SD-Uracil medium with 2% raffinose. Divide the cell culture into three tubes and add one or the other or both fluorescent ligand conjugates (Shin et al., 2014). The appropriate

concentration of each ligand must be determined experimentally to account for differential uptake by the cells. This is done empirically by testing for optimal FRET signal from cells that express IMAGEtags and have been incubated with different ratios and concentrations of ligand pairs. For the PDC ligands, we have used final concentrations in the cell culture of 4 μM Cy5-PDC and 10 μM Cy3-PDC. For the tobramycin IMAGEtags, we have used final concentrations in the cell culture of 25 μM of each ligand conjugate and up to 40 μM for the sulfonated ligands, which do not enter the cells as readily as ligands lacking sulfonates on the fluorophore due to their net negative charge. For this reason, we prefer to use nonsulfonated ligands although the cost of purchasing the precursors is higher for their synthesis than for the sulfonated derivatives.

4. Incubate the cells for 1 h at 30°C.
5. For a time course experiment, in which the promoter driving IMAGEtag RNA expression is GAL1, plate 45 μL of cells on a poly-D-lysine-coated 35 mm glass bottom dish from MatTek Corporation (Ashland, MA). Place a lid over the dish and put it on a heated stage at 30°C under the confocal microscope. Allow the cells to settle down for about 10 min. Add 5 μL of 20% (w/v) galactose very slowly (to avoid disturbing the cells) to a final concentration of 2%. Focus the cells and start collecting data.
6. For a single time-point experiment in which the promoter driving IMAGEtag expression is GAL1, add 20% (w/v) galactose to a final concentration of 2% in each tube from step 4. Incubate for 60–90 min at 30°C. Place 50 μL of cells on a 35 mm glass bottom MatTek dish and put it over a heated stage at 30°C under the confocal microscope. Put a lid over the dish to avoid evaporation. Allow the cells to settle for 10 min and then collect data.

3.4 Data Acquisition and Analysis

Sensitized emission data are collected from the control RNA-expressing cells or the IMAGEtag-expressing cells from samples containing individual ligands (either Cy3-PDC or Cy5-PDC) and both the ligands together. For these experiments we have used a Leica SP5X laser scanning confocal microscope (Figs. 2–4B) or a Nikon Eclipse 200 confocal microscope (Fig. 4A). The following instructions are for the Leica microscope, but similar protocols can be used with other confocal microscopes. Although a wide-field microscope could be used for measuring FRET, it is not a good choice as it suffers the problem of signals being averaged from many focal planes.

1. Place the dish with cells and medium containing either Cy3 ligand conjugate (donor) or Cy5 ligand conjugate (acceptor) under the confocal microscope. Image acquisition should be done in a sequential mode. In the first sequence illuminate the cells with a 550 nm white light laser and collect images from the donor emission channel (Cy3; 560–626 nm) and the FRET channel (Cy5; 660–754 nm). In the second sequence illuminate the cells with a 650 nm laser and collect images at the acceptor emission channel (Cy5; 660–754 nm). Collect the DIC image of the cells simultaneously with the fluorescent images. In our experiments, focus on the cells was achieved with a 63 × oil immersion objective. The choice of field and initial focus is based on the bright-field image as seen through the eyepiece. Additional fine Z-focusing is based on the fluorescence/confocal signal on the screen.
2. Four channels should be visible on the screen including the DIC image (Fig. 2C). Adjust the gains of each emission channel so that signals are below saturation. The donor-only or acceptor-only control samples are essential to correct for cross talk.
3. Using identical settings from steps 1 and 2, collect data from cells in medium containing the combination of donor and acceptor ligands.
4. Data collection can be done from a single focal plane or by Z-stacking. For time course experiments, data can be collected manually or the microscope can be programmed to collect images at regular intervals (Fig. 3).
5. For FRET experiments, donor-only and acceptor-only controls are necessary (Fig. 2A and B) to calculate the cross-talk signals. FRET signals should be calculated by taking into consideration the emissions measured in the Cy3, FRET, and Cy5 channels (Fig. 2C) using the equation $FRET = (B - A \times b - c \times C)/C$, where B = FRET emission, A = donor emission, b = donor emission cross talk ratio (B in donor-only sample/A in donor-only sample), c = acceptor excitation cross talk ratio (B in acceptor-only sample/C in acceptor-only sample), C = acceptor emission (Wouters, Verveer, & Bastiaens, 2001). In our experiments we have found the FRET efficiency to be in a range of 0.1–0.8.

As an example, the increase in FRET signals of cells expressing 6×PDC IMAGEtag RNAs and the FRET in control cells over a period of 40–99 min after induction of the GAL1 promoter with galactose are shown in Fig. 3 after quantification as described here.
6. Although the cells are attached to the dish by way of poly-D-lysine, some can still move in and out of the confocal field and in the horizontal

direction. These movements are small and visual examination of consecutive images allows tracking of single cells. To analyze many cells in an image is laborious, but can be achieved automatically with the application of an algorithm implemented in Matlab™ (Mitra, Bouthcko, Ray, & Nilsen-Hamilton, 2015).

3.5 Acceptor Photobleaching to Validate FRET

FRET measurements should also be validated by the acceptor photobleaching method in which the acceptor is photobleached using a high-intensity laser power. The donor emission is measured before and after photobleaching. This procedure can be performed using the Leica FRET-AB acceptor photobleaching wizard or equivalent software associated with other instruments.

1. Place the dish with cells containing both the donor and the acceptor ligands. Focus and illuminate the identified cells with a 550 nm white light laser and record the emission from the donor channel (560–626 nm). Excite the same cells with a 650 nm laser and record emissions from the acceptor channel (660–754 nm). Adjust the gains of the channels to ensure that they are below saturation.
2. Select another cell as a region of interest (ROI) and illuminate the cell with 100% power of the 650 nm laser to bleach the acceptor. Record the emissions from the donor and the acceptor channels. Repeat for other cells. The FRET efficiencies will be calculated using the formula:
3. FRET efficiency $= \left(\text{Donor}^{\text{postbleach}} - \text{Donor}^{\text{prebleach}}\right)/\text{Donor}^{\text{postbleach}}$
4. Perform the same exercise for other cells and ROIs that do not include cells to measure background FRET efficiency.

4. MEASUREMENT OF IMAGEtag RNA LEVEL BY RT-qPCR

As an additional control, the IMAGEtag RNA levels can be quantified in the same cell populations as are imaged. This control should be done when evaluating new IMAGEtags or IMAGEtag expression in different cell types. The observation of parallel increases in FRET of individual cells with an increase in IMAGEtag RNA in the cell population supports the interpretation that the FRET signals emanate from the IMAGEtags. This control can be done readily with yeast because

they are liquid cultures from which some portion can be set aside for later analysis.
1. Harvest the cells in 0.6 mL RNA extraction buffer and 0.6 mL of phenol:chloroform:isoamyl alcohol (49.5:49.5:1) for 6 min at room temperature.
2. Add 0.2 g glass beads (0.45 mm diameter) and lyse the cells with vigorous agitation for 2 min.
3. Centrifuge at $2200 \times g$ for 5 min and transfer the upper aqueous phase to a new tube.
4. Add 1 volume of phenol:chloroform:isoamyl alcohol (49.5:49.5:1).
5. Vortex for 2 min and centrifuge at $2200 \times g$ for 5 min.
6. Transfer the aqueous phase to a new tube. Repeat steps 4–6.
7. Add 1 volume of chloroform:isoamyl alcohol (24:1).
8. Vortex for 2 min and centrifuge at $2200 \times g$ for 5 min.
9. Transfer the aqueous phase to a new tube.
10. Precipitate with 3 volumes of 100% ethanol (final=73%) and 0.1 volume of 3 M sodium acetate, pH 5.2 (final=73 mM).
11. Resuspend the RNA in DEPC-treated water.
12. Treat 1 μg RNA with RNAse-free DNase I (Life Technologies, Carlsbad, CA, Cat# AM2222) at room temperature for 15 min and stop the reaction by bringing the solution to 2.5 mM EDTA.
13. Reverse transcribe with Superscript III reverse transcriptase (Life Technologies, Cat# 18068-015) with oligodT (18mer, IDT, Coralville, IA, Cat# 1937) in a final volume of 20 μL.
14. Dilute the reaction to 50% and use 2 μL of the diluted cDNA from the previous reverse transcription step for qPCR with SYBR green (Life Technologies, Cat# S7563), GoTaq Flexi polymerase (Promega, Madison, WI, Cat# M8291), forward primer 5′-CTAGAGGGCCGCATCATGTAATTAG-3′, and reverse primer 5′-CCTAGACTTCAGGTTGTCTAACTCC-3′. These primers are located at the 3′ end of the IMAGEtag sequence and part of the pYES2 vector. The endogenous ACT1 mRNA can be used as an internal control, which is amplified by the forward primer 5′-ATTCTGAGGTTGCTGCTTT-3′ and reverse primer 5′-GTCCCAGTTGGTGACAATAC-3′. The qPCR was performed in an MJ Research Opticon 2 PCR machine with a protocol of 94°C for 5 min followed by 40 cycles (94°C for 15 s, 60°C for 15 s, 72°C for 15 s) then a plate read. Standards consisting of each IMAGEtag sequence to be amplified or the ACT1 cDNA fragment

defined by the stated primers were on the same plate as the samples and subjected to the same PCR protocol. The amount of RNA of each type was calculated from the standard curves.

5. SYNTHESIS OF LIGANDS

The aptamer ligand conjugates discussed in this chapter are shown in Fig. 5. The choice of ligand is important as the ligand must not be toxic to the cells under study in the range of ligand concentrations required for imaging. The nature of the ligand–fluorophore conjugate is also important because it should ideally readily enter the cells in order to be present in high enough concentrations in the majority of cells for achieving a FRET signal. A barrier to cell entry can come from the ligand or the conjugated fluorophore. For example, the Cy3 and Cy5 precursors for conjugate formation can be purchased as the sulfonated (Fig. 5A) or nonsulfonated (Fig. 5B–D) forms. The additional negative charge on the sulfonated ligands will reduce cell uptake (Fig. 4B) such that fewer cells in the population can be analyzed for FRET. Once in the cell, the ligand conjugate should interact as little as possible with cellular components other than the aptamers as this interaction may increase the background FRET and/or compete with the IMAGEtags for binding their ligands. Another very important property of aptamer ligand conjugates is their toxicity. Some aptamers must be ruled out as IMAGEtags due to the toxicity of their ligands (Ilgu et al., 2016; Kraus et al., 2008). We have found that Cy3- and Cy5-conjugated ligands can be effectively used in a concentration range in which they are not toxic. The fluorophores are commercially available in forms that can be readily linked to a variety of aptamer ligands. Other FRET pairs could be incorporated into the IMAGEtag protocol. However, it is important that the chosen fluorophores do not interact with the RNA independently such that their fluorescence is quenched. For example, guanine quenches the fluorescence of many fluorophores, but not Cy3 and Cy5 (Torimura et al., 2001).

The commercial sources of some reagents are listed for the syntheses described next. All other reagents were of reagent grade or higher and obtained from Sigma-Aldrich.

5.1 Synthesis of Cy5- and Cy3-Tobramycin

1. Prepare a 20 mL vial equipped with a stirring bar and wrapped in aluminum foil to ensure complete darkness during the procedure

2. In the vial mix Cy3 NHS ester (1 equiv.), aminoglycoside (1.1 equiv.), and N,N-diisopropylethylamine (20 equiv.), and DMSO (1.0 mL). Tobramycin was from Sigma (Cat# T4014). The sources of NHS esters were: GE Healthcare Life sciences (Cy3 NHS ester, Cat# PA13105 and Cy5 NHS ester, Cat# PA15100, both sulfonated) and Lumiprobe (Cy3 NHS ester, Cat# 41020 and Cy5 NHS ester, # 43020, both not sulfonated).
3. Stir at room temperature under an argon atmosphere in the dark for 12 h.
4. Purify by preparative TLC using a 5:4:1 mixture of DCM:MeOH:NH$_4$OH to resolve the product (R_f=0.4).
5. Evaluate by mass spectrometry with the following results expected: (1) *Cy5-Tobramycin:* HRMS: calculated for $C_{50}H_{74}ClN_7O_{10}$=967.52, found: 967.5867 (M^+-H); (2) *Cy3-Tobramycin:* HRMS: calculated for $C_{47}H_{70}N_7O_{10}$=892.52, found: 875.5918 (M^+-H_2O); (3) *Cy5-SO$_3$-Tobramycin:* MS (ESI): m/z=1104.46 (M^-); HRMS: calculated for $C_{51}H_{74}N_7O_{16}S_2$=1104.4633, found: 1104.4606; (4) *Cy3-SO$_3$-Tobramycin:* MS (ESI): m/z=1078.44 (M^-); HRMS: calculated for $C_{49}H_{72}N_7O_{16}S_2$=1078.4477, found: 1078.4447.

5.2 Synthesis of Cy5-PDC-Gly

1. Prepare a 20 mL vial equipped with a stirring bar and wrapped in aluminum foil to ensure complete darkness during the procedure
2. In the vial combine Cy5-NHS ester (Jung & Kim, 2006) (50.0 mg, 71.0 μmol), PDC-Gly (Kraus, Gupta, Mokhtarian, Mehanovic, & Nilsen-Hamilton, 2010) (34.0 mg, 71.0 μmol), and triethylamine (0.5 mL) and DMF (5.0 mL).
3. Stir at room temperature under an argon atmosphere in the dark for 12 h
4. Purify by preparative TLC using 10% MeOH:DCM to give the pure product in 66% yield (R_f=0.3).
5. Evaluate by mass spectrometry with the following results expected: MS for $C_{55}H_{57}Cl_2N_8O_3$: MS (m/z): 947.5 (M^+-1), 497.4, 383.3.

5.3 Synthesis of Cy3-PDC-Gly

1. Prepare a 20 mL vial equipped with a stirring bar and wrapped in aluminum foil to ensure complete darkness during the procedure.
2. In the vial combine Cy3-NHS ester (Jung & Kim, 2006) (30.0 mg, 43.0 μmol), PDC-Gly (Kraus et al., 2010) (25.0 mg, 52.0 μmol), and N,N-diisopropylethylamine (0.1 mL) in DMF (5.0 mL).

3. Stir at room temperature under an argon atmosphere in the dark for 12 h.
4. Purify by preparative TLC using 10% MeOH:DCM to give the pure product in 72% yield ($R_f = 0.3$).
5. Evaluate by mass spectrometry with the following results expected: Cy3-PDC-Gly: MS (m/z): 923.3910 ($M^+ + 1$), 471.2992; HRMS: calculated for $C_{53}H_{55}Cl_2N_8O_3^+$: 923.3931, found: 923.3915.

6. CLONING REPETITIVE SEQUENCES

IMAGEtags can be created from many different aptamers for which the ligand is small and can enter the cell. However, cloning repetitive sequences can be challenging, in part because of the probability of circularization of the enlarging inserts. We have developed a successful method of cloning repeated sequences that is based on previously reported methods (Meyer & Chilkoti, 2002; Won & Barron, 2002). Although repeated sequences have been reported to be genetically unstable, we have found that most of the sequences cloned using the method described here were genetically stable in *Escherichia coli* in which the plasmids are maintained (Shin et al., 2014).

Prior to cloning a repetitive sequence of aptamers as an IMAGEtag, it is important to check the likelihood of the aptamers folding appropriately in the string to enable each aptamer to retain a structure capable of binding ligand. A cotranscriptional folding program should be used to simulate the folding of an IMAGEtag in vivo (Fig. 6A).

6.1 Cloning Initial Repetitive Aptamer Sequences from Synthetic Oligonucleotides

1. Digest pZErO-2 vector with *Afl*III (New England Biolabs-NEB, Ipswich, MA, Cat# R0541S) and *Apo*I (NEB, R0566S) restriction enzymes and isolate the 1854 bp fragment using gel electrophoresis followed by extraction and purification by a method such as the QIAquick gel extraction kit (QIAGEN, Valencia, CA, Cat# 28704).
2. Use two pairs of complementary oligonucleotides to create inserts containing between two and seven repeats (150–250 bp) of the sequences flanked by restriction sites for *Bsa*I (NEB, R0535S) and *Bsm*AI (NEB, R0529S) for subsequent multiplication (Fig. 6B).
3. Anneal the complementary oligonucleotides by slow cooling from 95°C to 25°C.

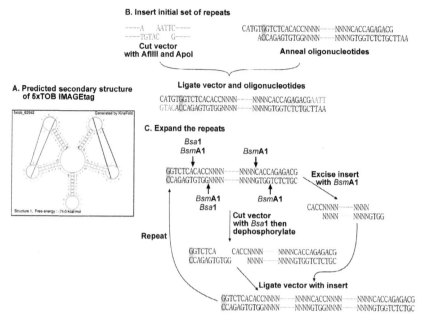

Fig. 6 Cloning multiple tandem sequences. (A) IMAGEtag sequence design involves checking for folding during transcription. The result of folding the 5×TOB IMAGEtag RNA by KineFold is shown. (B) The method for cloning multimers to create IMAGEtags starts with oligonucleotide inserts comprising the sequence of a series of four or more aptamers that are inserted by ligation into the pZERO vector, which has been minimized in size to allow for maximum insert sizes. (C) Aptamer repeats are expanded using *Bsa*1 to cut the vector and *Bsm*A1 to create a new fragment for insertion.

4. Ligate the double-stranded inserts into the digested vector with T4 DNA ligase (Promega, Cat# M1801) and transform into the *E. coli* chemically competent strain NEB 5-alpha (NEB, Cat# C1987I) by heat shock for 1 min at 42°C.

6.2 Constructing Longer Repetitive Sequences

1. Use *Bsm*AI to cut the initial repetitive sequences from the vector and purify the excised fragment by agarose gel electrophoresis using the QIAquick gel extraction kit (QIAGEN, Valencia, CA, cat #28704 or 28706).

2. *Bsa*I cuts at a single site at the 5′ end of the multiaptamer, leaving the aptamers in the vector. Dephosphorylate the *Bsa*I-digested vector with calf intestinal alkaline phosphatase (Promega, Cat# M1821) to eliminate self-ligation, and ligate the multiaptamer fragment excised with *Bsm*AI (created in step 1 in Section 6.2) into the cut vector.

3. Analyze the clones for inserts by sequencing in each direction with primers 2819 (5′-GGCCTTTTTACGGTTCCTGG-3′) and 2820 (5′-GCCGCTCCCGATTCGCAG-3′). Clones can also be analyzed by digestion with *Bsm*AI followed by agarose gel electrophoresis to establish the size of the insert.
4. In general, several plasmids are obtained that appear, by electrophoretic mobility, to have the expected insert sizes. The number of repeats in each plasmid is confirmed by sequencing.

6.3 Multiplying the Original Repetitive Sequences

The restriction enzymes *Bsa*I and *Bsm*AI are used to multiply the initial set of aptamer repeats. The strategy is based on the fact that the recognition sequences of these restriction enzymes are not palindromic and they cut DNA beyond their recognition sequences. The recognition sequence for *Bsm*AI is part of the longer *Bsa*I recognition sequence. Thus, a sequence cut by *Bsa*I is also cut by *Bsm*AI but not vice versa. Consequently, a region between the two restriction enzyme sites can be cut from the vector by *Bsm*AI, while digestion with *Bsa*I only cuts the vector at one end of the insert. This allows the *Bsm*AI fragment to be ligated back into the same vector but that has been digested at a single site with *Bsa*I. The result is multiplication of the number of repeats in the insert from the first round of cloning (Fig. 6C).

With each round of expansion we can expect to obtain cloned inserts with double the number of repeats of the previous insert. In addition to the expected inserts, some clones contained inserts with 3, 4, and up to 8 multiples of the inserts.

Additional rounds of cloning as just described can be used to further multiply the number of repeated sequences. As for the first round, the number of repeat sequences in the additional round products can be even multipliers of the original number of repeats. However, more than the expected repeats are sometimes found because this strategy allows the insertion of multiple fragments in a single ligation reaction. In addition, odd numbers of repeated sequences are occasionally obtained when an even number was expected.

7. CONCLUSIONS

The IMAGEtag technique provides a relatively simple way of monitoring promoter activity in living cells without requiring the expression of exogenous fluorescent proteins. The IMAGEtag reporters do not influence cell physiology as evidenced by the lack of change in cell growth rate or housekeeping mRNA levels with expression of RNAs containing IMAGEtags (Shin et al., 2014). Currently optimized for monitoring yeast gene expression as described here and previously reported (Shin et al., 2014), the IMAGEtag method is being developed for application in other eukaryotic cells. The considerations in adapting the method to other cell types for optimal signal/noise include: (1) ligand permeability, (2) ligand toxicity, (3) the promoter employed for aptamer expression, (4) 3′ UTR modifications that promote aptamer stability, and (5) the effect of the IMAGEtag and its position in the RNA on the steady-state level of the cellular RNA in which it is incorporated (Hogg & Collins, 2007; Said et al., 2009).

The IMAGEtag approach to tracking gene expression in living cells in real time provides an opportunity to observe rapid changes in gene expression in individual cells as they interact with their neighbors or in cell populations as they respond to environmental cues. The tags are small and readily integrated into any RNA. Application of FRET provides the sensitivity required to detect mRNAs in individual cells using standard confocal microscope equipment. The analysis of FRET is straightforward and can be automated for analyzing FRET associated with many cells in a field, even when the cells are moving.

ACKNOWLEDGMENTS
The National Institute of Health funded the development of the IMAGEtags (R01 EB005075), and selection and optimization of the PDC and tobramycin aptamers (R43CA-110222; 5R21AI-114283). The U.S. Department of Energy, Office of Biological and Environmental Research through the Ames Laboratory funded the implementation of the IMAGEtags in yeast. The Ames Laboratory is operated for the U.S. Department of Energy by Iowa State University under Contract No. DE-AC02-07CH11358. We thank Margie Carter in the ISU Confocal Microscopy and Multiphoton Facility for help with fluorescence microscopy, H. Eirik Haarberg for early work on designing vectors and modifying the method for cloning multiple aptamers, and Ivan Geraskin for providing data relating to the synthesis of the nonsulfonated Cy3 and Cy5 adducts.

REFERENCES
Bertrand, E., Chartrand, P., Schaefer, M., Shenoy, S. M., Singer, R. H., & Long, R. M. (1998). Localization of ASH1 mRNA particles in living yeast. *Molecular Cell, 2*(4), 437–445.

Filonov, G. S., Moon, J. D., Svensen, N., & Jaffrey, S. R. (2014). Broccoli: Rapid selection of an RNA mimic of green fluorescent protein by fluorescence-based selection and directed evolution. *Journal of the American Chemical Society, 136*(46), 16299–16308.

Hocine, S., Raymond, P., Zenklusen, D., Chao, J. A., & Singer, R. H. (2013). Single-molecule analysis of gene expression using two-color RNA labeling in live yeast. *Nature Methods*, *10*(2), 119–121.

Hogg, J. R., & Collins, K. (2007). RNA-based affinity purification reveals 7SK RNPs with distinct composition and regulation. *RNA*, *13*(6), 868–880.

Ilgu, M., Ray, J., Bendickson, L., Wang, T., Geraskin, I. M., Kraus, G. A., & Nilsen-Hamilton, M. (2016). Light-up and FRET aptamer reporters; evaluating their applications for imaging transcription in eukaryotic cells. *Methods*, *98*, 26–33.

Ilgu, M., Wang, T., Lamm, M. H., & Nilsen-Hamilton, M. (2013). Investigating the malleability of RNA aptamers. *Methods*, *63*(2), 178–187.

Jung, M. E., & Kim, W. J. (2006). Practical syntheses of dyes for difference gel electrophoresis. *Bioorganic & Medicinal Chemistry*, *14*(1), 92–97.

Kraus, G. A., Gupta, V., Mokhtarian, M., Mehanovic, S., & Nilsen-Hamilton, M. (2010). New effective inhibitors of the Abelson kinase. *Bioorganic & Medicinal Chemistry*, *18*(17), 6316–6321.

Kraus, G. A., Jeon, I., Nilsen-Hamilton, M., Awad, A. M., Banerjee, J., & Parvin, B. (2008). Fluorinated analogs of malachite green: Synthesis and toxicity. *Molecules*, *13*(4), 986–994.

Meyer, D. E., & Chilkoti, A. (2002). Genetically encoded synthesis of protein-based polymers with precisely specified molecular weight and sequence by recursive directional ligation: Examples from the elastin-like polypeptide system. *Biomacromolecules*, *3*(2), 357–367.

Mitra, D., Bouthcko, R., Ray, J., & Nilsen-Hamilton, M. (2015). Detecting cells in time varying intensity images in confocal microscopy for gene expression studies in living cells. In *Paper presented at the SPIE, Medical Imaging 2015*.

Paige, J. S., Wu, K. Y., & Jaffrey, S. R. (2011). RNA mimics of green fluorescent protein. *Science*, *333*(6042), 642–646.

Said, N., Rieder, R., Hurwitz, R., Deckert, J., Urlaub, H., & Vogel, J. (2009). In vivo expression and purification of aptamer-tagged small RNA regulators. *Nucleic Acids Research*, *37*(20), e133.

Sambrook, J., & Russell, D. W. (2006). Purification of nucleic acids by extraction with phenol:chloroform. *Cold Spring Harbor Protocols*, *2006*(1)pdb.prot4455.

Shin, I., Ray, J., Gupta, V., Ilgu, M., Beasley, J., Bendickson, L., ... Nilsen-Hamilton, M. (2014). Live-cell imaging of Pol II promoter activity to monitor gene expression with RNA IMAGEtag reporters. *Nucleic Acids Research*, *42*(11), e90.

Strack, R. L., Disney, M. D., & Jaffrey, S. R. (2013). A superfolding Spinach2 reveals the dynamic nature of trinucleotide repeat-containing RNA. *Nature Methods*, *10*(12), 1219–1224.

Torimura, M., Kurata, S., Yamada, K., Yokomaku, T., Kamagata, Y., Kanagawa, T., & Kurane, R. (2001). Fluorescence-quenching phenomenon by photoinduced electron transfer between a fluorescent dye and a nucleotide base. *Analytical Sciences*, *17*(1), 155–160.

Won, J. I., & Barron, A. E. (2002). A new cloning method for the preparation of long repetitive polypeptides without a sequence requirement. *Macromolecules*, *35*(22), 8281–8287.

Wouters, F. S., Verveer, P. J., & Bastiaens, P. I. (2001). Imaging biochemistry inside cells. *Trends in Cell Biology*, *11*(5), 203–211.

Xayaphoummine, A., Bucher, T., & Isambert, H. (2005). Kinefold web server for RNA/DNA folding path and structure prediction including pseudoknots and knots. *Nucleic Acids Research*, *33*, W605–W610. Web Server issue.

CHAPTER NINE

A Method for Expressing and Imaging Abundant, Stable, Circular RNAs In Vivo Using tRNA Splicing

C.A. Schmidt*, J.J. Noto*, G.S. Filonov[†], A.G. Matera*,[1]

*Integrative Program for Biological and Genome Sciences, Curriculum in Genetics and Molecular Biology, University of North Carolina, Chapel Hill, NC, United States
[†]Weill Cornell Medical College, New York, NY, United States
[1]Corresponding author: e-mail address: matera@unc.edu

Contents

1. Introduction	216
1.1 Circular RNAs	216
1.2 Engineering and Imaging circRNAs	216
2. Design and Generation of tricRNA Vectors	218
2.1 Isolation of Parental tRNA Gene	218
2.2 Mutagenesis of Parental tRNA Gene	221
2.3 Addition of External Promoters	222
3. In Vivo Expression of tricRNAs	226
3.1 Transfection	226
3.2 RNA Isolation	228
3.3 Analysis of Products	228
4. In-Gel Imaging of tricRNAs	229
5. Cellular Imaging of tricRNAs	233
6. Concluding Remarks	234
Acknowledgments	235
References	235

Abstract

Recent improvements in high-throughput sequencing technologies underscore the pervasiveness of circular RNA (circRNA) expression in animal cells. CircRNAs are distinct from their linear counterparts because they lack the 5′ caps and 3′ tails that typically help determine the cellular fate of a transcript. However, due to the lack of free ends, circRNAs are impervious to exonucleases and thus can evade normal RNA turnover mechanisms. Most circRNAs are derived from protein-coding pre-mRNAs, via a mechanism called "back-splicing." Existing methods of circRNA expression thus typically involve genes that have been engineered to contain sequence elements that promote back-splicing.

We recently uncovered an anciently conserved mechanism of RNA circularization in metazoans that involves splicing of tRNA introns. This splicing mechanism is completely independent from that of pre-mRNAs. In this chapter, we detail an orthogonal method that involves splicing of intron-containing tRNAs in order to produce circRNAs in vivo. We utilize fluorescence-based RNA reporters to characterize the expression, localization, and stability of these so-called tRNA intronic circular RNAs. Because tRNA biogenesis is essential for all cellular life, this method provides a means to express ultrastable, high-copy, circRNA effectors in a wide variety of metazoan cell types.

1. INTRODUCTION

1.1 Circular RNAs

The subject of circular RNAs (circRNAs) is emerging as an important and diverse field of biological study, due in large part to advances in high-throughput sequencing. Once thought to exist merely as rare or aberrantly generated artifacts of premessenger RNA splicing, circRNAs have recently been shown to be an abundant and evolutionarily conserved class of RNAs (Jeck et al., 2013; Salzman, Gawad, Wang, Lacayo, & Brown, 2012). A range of cellular roles have been attributed to circRNAs including regulation of transcription and the ability to behave as molecular sponges, competing with endogenous targets for binding of RNA-binding proteins (Ashwal-Fluss et al., 2014) or microRNAs (Hansen et al., 2013; Memczak et al., 2013). In addition, circRNA expression is associated with disease risk in humans (Burd et al., 2010). Expression of circRNAs can be tightly regulated by *cis*- and *trans*-factors and in a tissue-specific and developmentally timed manner (Conn et al., 2015; Kramer et al., 2015; Salzman, Chen, Olsen, Wang, & Brown, 2013), suggesting further functional relevance. CircRNAs can be generated through a number of biogenesis pathways, the most well known of which involves ligation of a splice donor to an upstream splice acceptor during mRNA splicing by a process termed "back-splicing."

1.2 Engineering and Imaging circRNAs

Few tools have been specifically developed to facilitate circRNA research. Although several methods have been used to circularize RNAs in vitro, directed generation of circRNAs in vivo has only recently been accomplished (for a review of RNA circularization strategies, see Petkovic & Müller, 2015). Engineered circRNAs have already been shown to be capable of serving as a template for translation (Chen & Sarnow, 1995; Wang & Wang, 2015) given the presence of an internal ribosome entry site. Several groups have engineered minigenes consisting of minimal elements optimized to promote circularization of a specific sequence via back-splicing (Ashwal-Fluss et al., 2014; Wang & Wang, 2015). Despite these advances, there remains a need for an in vivo, high-copy, circRNA expression system. We have recently developed an orthogonal approach to direct circRNA expression (Lu et al., 2015), utilizing the process of tRNA splicing to produce tRNA intron-derived circular RNAs (tricRNAs). First identified in archaea (Salgia, Singh, Gurha, & Gupta, 2003) tricRNA generation seems to be a conserved feature of tRNA splicing in metazoans (Lu et al., 2015). tRNAs undergo extensive posttranscriptional RNA processing (for a review, see Yoshihisa, 2014), including recognition and cleavage of intron-containing tRNAs by the tRNA splicing endonuclease (TSEN) followed by ligation of exon halves by the RNA ligase RtcB. During this process, the tRNA intron is also ligated by RtcB in a head-to-tail fashion to form a stable circRNA (Fig. 1). In our engineered tricRNA system, tRNA-intronic sequences are replaced with the sequence of the desired circRNA. Expression and splicing of these tRNAs results in robust production of stable "designer" tricRNAs. circRNAs made from this platform are

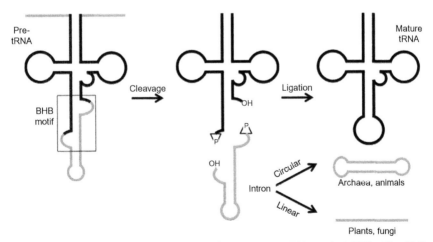

Fig. 1 pre-tRNAs with introns are processed into mature tRNAs and tricRNAs. The tRNA splicing endonuclease (*TSEN*) complex cleaves at the bulge-helix-bulge (*BHB*) motif of the pre-tRNA, generating 5′-OH and 2′,3′-cyclic phosphate ends. Processing also includes cleavage of the 5′ leader and the 3′ trailer sequences by RNaseP and RNaseZ, respectively. Ligation of the tRNA exons and circularization of the intron are carried out by the RtcB ligase.

expressed at the levels comparable to those of other high-copy housekeeping genes transcribed by RNA polymerase III (pol III).

To create tricRNA reporter constructs that monitor expression, we have taken advantage of fluorescent RNA aptamer technology. We inserted sequences corresponding to the RNA aptamers Broccoli and Spinach2 (Filonov, Kam, Song, & Jaffrey, 2015; Filonov, Moon, Svensen, & Jaffrey, 2014; Paige, Wu, & Jaffrey, 2011; Strack, Disney, & Jaffrey, 2013) into tricRNA expression vectors. These aptamers bind to the fluorophore DFHBI-1T, which has negligible toxicity when applied to living cells. DFHBI-1T is a chromophore that mimics green fluorescent protein (GFP) and has very low background fluorescence. When bound by Broccoli or Spinach2, DFHBI-1T fluorescence is greatly enhanced, allowing us to use cellular and electrophoretic imaging techniques to characterize tricRNA production, localization, and stability.

2. DESIGN AND GENERATION OF tricRNA VECTORS

Here we describe the cloning steps that are necessary to isolate a parental tRNA gene, place it into an appropriate cloning vector, mutagenize the intron, and add an external promoter.

2.1 Isolation of Parental tRNA Gene

To generate a tricRNA expression vector, a tRNA gene with an intron must be chosen. One good resource for this criterion is the genomic tRNA database (Chan & Lowe, 2009). For our purposes we used *Drosophila CR31905*, a Tyr:GTA gene with a 113 nucleotide intron. We have also had success with *CR31143*, a Leu:CAA gene that bears a 40 nucleotide intron. We chose these tRNAs based on their intron size and on the number of RNA-seq reads mapping to the intron. In addition, we have used *TRY-GTA3-1*, a human Tyr:GTA tRNA gene that contains a 16 nucleotide intron.

2.1.1 Notes on Cloning Method

Cloning a given tRNA gene into a suitable vector can be achieved in a variety of ways. For example, traditional "cut and paste" restriction enzyme cloning can be used for this task. Alternatively, several commercial kits such as Gateway, In-Fusion, and TOPO-TA avoid the use of restriction enzymes.

2.1.2 Notes on Primer Design

Primers should be designed according to the specific cloning protocol (eg, adding the attB/attP sites to primers for Gateway cloning). For the annealing portion of the primer, we recommend amplifying ~20 nucleotides upstream of the 5′ tRNA exon and ~35 nucleotides downstream of the 3′ tRNA exon. The region downstream of the 3′ tRNA exon must contain a run of at least four T residues, which, when transcribed, serve as a termination signal for pol III. Conversely, the tricRNA you are trying to express should not contain such a run of T residues. If restriction enzyme cloning is utilized, ensure that the chosen enzymes are buffer compatible for ease of cloning.

2.1.3 Notes on the Vector

For the initial cloning, we recommend using a relatively compact vector, such as pGEM, pBluescript, or pDONR. These vectors are more amenable to downstream site-directed mutagenesis that is necessary for generating a designer tricRNA. Larger vectors, such as pAV, are more difficult to mutagenize. Thus, we recommend using larger vectors for the final cloning step (see Fig. 4).

Materials and equipment
- Micropipette and tips
- Vector (pGEM and pBlueScript variants can be purchased from Addgene (http://www.addgene.org); pDONR can be purchased from ThermoFisher Scientific (Waltham, MA), cat# 12536–017)
- Template DNA (genomic DNA)
- PCR primers
- Thin-walled PCR tubes
- Thermal cycler
- PCR reagents
    - Gateway: ThermoFisher Scientific cat# 11789013 for BP reaction; cat# 11791019 for LR reaction
    - In-Fusion: Clontech (Mountain View, CA) cat# 638909
    - TOPO-TA: ThermoFisher Scientific cat# 451641
    - "Cut and paste" restriction enzyme cloning supplies: heat-stable DNA polymerase, dNTPs, PCR buffer, $MgCl_2$, restriction enzymes and corresponding buffer, DNA ligase and corresponding buffer
- Agarose gel casting supplies
- Gel box
- Power supply
- DNA size ladder

- Loading buffer
- PCR purification kit (*optional*—we recommend QIAquick PCR Purification Kit; Qiagen (Hilden, Germany) cat#28104)
- Chemically competent cells
- 42°C heat source
- Agar plates with appropriate selection agent
- Plate spreader
- 37°C incubator
- Culture tubes
- LB broth with appropriate selection agent
- Miniprep kit (we recommend QIAprep Spin Miniprep Kit; Qiagen cat# 27104)
- DNA sequencing facility or service

Workflow (Fig. 2)

1. Design primers appropriate for the chosen cloning method using the above criteria.
2. Use PCR to isolate the tRNA gene from genomic DNA. Ensure that the product is the correct size by agarose gel electrophoresis.
3. Optional: Column purify the PCR product to remove unincorporated nucleotides.
4. Follow the manufacturer's protocol to insert the PCR product into the vector of choice.

   We used Gateway cloning to insert *CR31905* into pDONR.

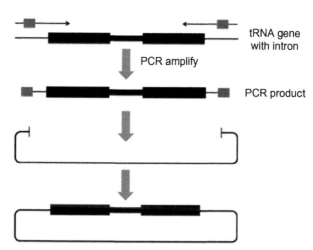

Fig. 2 tRNA genes with introns can be cloned into a small vector using a PCR-based cloning method.

5. Screen transformed colonies by sequencing to ensure accuracy.
   We termed our resultant construct, pTRIC31905.
6. Optional: Transfect the construct into an appropriate cell line, isolate RNA, and determine if tricRNAs are made using a method described in Fig. 5.
   We used RT-PCR to show that pTRIC31905 makes a tricRNA in HeLa cells (Lu et al., 2015). See Table 1.

2.2 Mutagenesis of Parental tRNA Gene

The next step is to modify the intron of the cloned tRNA by adding restriction sites for future subcloning. We have had good success using NEB's Q5 Site-Directed Mutagenesis Kit. The NEB website includes a particularly useful tool, called NEBaseChanger, which helps design the primers for site-directed mutagenesis and suggests PCR conditions (http://nebasechanger.neb.com/).

Table 1 Precursor tRNAs and Their Resultant tricRNAs

pTRIC31905	pTRIC-Y	pTRIC-Y:Broccoli	Name
			Precursor
			Product
Endogenous tricRNA (tric31905)	tricRNA with filler sequence	Contains Broccoli fluorescent RNA aptamer	Notes

2.2.1 Notes on Choosing Restriction Sites

The bulge-helix-bulge (BHB) motif within the pre-tRNA is necessary for proper splicing (Fig. 1). Thus, it is important to maintain this structure. We chose *Not*I and *Sac*II, which partially pair in a helix below the BHB of Tyr:GTA. We retained several nucleotides of the native intron to maintain the tRNA structure. We recommend using an RNA drawing application, such as VARNA (Darty, Denise, & Ponty, 2009), to assist in the visualization of these changes. Again, it is important to avoid introducing a run of four T residues, which will terminate transcription.

Materials and equipment
- Micropipette and tips
- PCR primers
- Template DNA (created in previous step)
- Thin-walled PCR tubes
- Thermal cycler
- Q5 Site-Directed Mutagenesis Kit (NEB (Ipswich, MA) cat# E0554S)
- 42°C heat source
- Agar plates with appropriate selection agent
- Plate spreader
- 37°C incubator
- Culture tubes
- LB broth with appropriate selection agent
- Miniprep kit
- DNA sequencing facility or service

Workflow (Fig. 3)
1. Design primers using NEBaseChanger.
2. Follow the manufacturer's protocol for the Site-Directed Mutagenesis Kit.
3. Screen colonies by sequencing to ensure accuracy.
    We termed our resultant construct, pTRIC-Y (Y for tyrosine).
4. Optional: Transfect construct into an appropriate cell line, isolate RNA, and determine if tricRNAs are made using a method described below. See Table 1.

2.3 Addition of External Promoters

The expression of tricRNAs is limited by the strength of the internal tRNA promoter. Expression can be increased by the addition of external

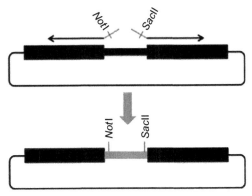

Fig. 3 The tRNA gene can be modified by site-directed mutagenesis to place restriction enzymes in the intron for convenient subcloning.

RNA pol III promoters, such as 5S, 7SK, or U6. We used a three-way cloning approach to put pTRIC-Y under the control of the human 5S and U6 promoters.

2.3.1 Notes on pol III Promoters

We have observed that addition of proximal nucleotides from the poll III-transcribed gene can have a positive effect on transcription. For example, it has been shown that adding the first 27 nucleotides of the human U6 snRNA promotes capping of the transcript, greatly increasing its stability (Good et al., 1997). Furthermore, we used the human 5S promoter with 117 nucleotides of the rRNA, leading to a much longer pre-tRNA but also a more highly expressed tricRNA (see Fig. 6A). Thus, in choosing a promoter, consider making two constructs: one with just the promoter, and one with the promoter plus several nucleotides of the downstream RNA.

2.3.2 Notes on Cloning

We took a three-way restriction enzyme cloning approach for this step. A schematic can be seen in Fig. 4. When selecting a vector, ensure that the restriction sites that are now in the intron do not appear anywhere else in the vector. Additionally, a vector more suited to specific applications can be used. For example, we chose pAV because we frequently transfect our constructs into HEK293T cells.

Materials and equipment
- Micropipette and tips
- Template DNA

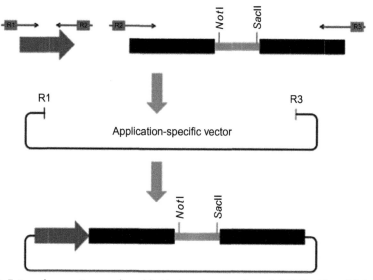

Fig. 4 External promoters can be added to boost expression of tricRNAs. The addition of these elements can be carried out by three-way restriction enzyme cloning.

- tRNA template—created in previous step
- pol III promoter template–vector that contains the desired promoter (check Addgene) or genomic DNA
- PCR primers
- Thin-walled PCR tubes
- Thermal cycler
- Heat-stable DNA polymerase
- dNTPs
- PCR buffer
- $MgCl_2$
- Restriction enzymes and corresponding buffer
- Agarose gel casting supplies
- Gel box
- Power supply
- DNA size ladder
- Loading buffer
- PCR purification kit
- Gel purification kit (we recommend QIAquick Gel Extraction Kit; Qiagen cat# 28704)

- DNA ligase and corresponding buffer
- Chemically competent cells
- 42°C heat source
- Agar plates with appropriate selection agent
- Plate spreader
- 37°C incubator
- Culture tubes
- LB broth with appropriate selection agent
- Miniprep kit
- DNA sequencing facility or service

Workflow (Fig. 4)

1. Design primers to isolate both the intron-modified tRNA and the pol III promoter of choice (including additional nucleotides, if desired) with appropriate restriction enzymes. Ensure the enzymes are buffer compatible.

    We used *Bam*HI and *Sal*I to isolate the promoters, and *Sal*I and *Xba*I to isolate the tRNA.

2. Use PCR to isolate the tRNA and the promoter from appropriate templates. For the intron-modified tRNA, use the construct created by site-directed mutagenesis. For the promoter, use genomic DNA or a plasmid that contains the desired promoter sequence.

    We used pAV U6 + 27 from Addgene for the U6* promoter.

3. Optional: Column purify the PCR products to remove unincorporated nucleotides.

4. Digest an application-specific vector with the appropriate restriction enzymes.

    We typically digest 1 μg of vector for 1 h at 37°C and purify on a 0.5% agarose gel.

5. Ligate the digested PCR products and vector, using a 3:1 molar ratio of each insert to vector.

6. Transform the ligation reaction into chemically competent cells. Plate the transformation reaction onto selection media.

7. Screen transformed colonies by sequencing to ensure accuracy.

    We termed out resultant constructs pTRIC-Y_5S, pTRIC-Y_U6, and pTRIC-Y_U6*.

8. Optional: Transfect the constructs into an appropriate cell line, isolate RNA, and determine if tricRNAs are made using a method described in Fig. 5.

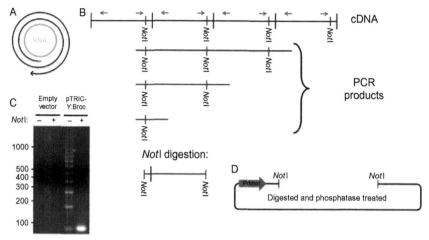

Fig. 5 tricRNA can be analyzed several ways. (A) The reverse transcriptase enzyme can transcribe around a circular RNA template many times. (B) The resulting cDNA is a concatamer, containing many repeats of the circle in a linear form. Thus, there are many binding sites for the RT-PCR primers. If there is a restriction site in the tricRNA, the PCR products can be digested with that restriction enzyme. (C) Running an RT-PCR reaction on an agarose gel results in a ladder of products. The digested PCR products collapse into one band. (D) The digested RT-PCR products can be ligated into a vector that has been cut with the same enzyme and phosphatase treated. It can then be sequenced to determine if the tricRNA is splicing as expected.

3. IN VIVO EXPRESSION OF tricRNAs

Typical RNA-seq cDNA libraries are size-excluded and thus we currently do not have evidence for endogenous human tricRNAs (which would be circular and only 16–21 nt long). However, we have found that human cells possess all the necessary machinery to generate tricRNAs (Lu et al., 2015). Using either *Drosophila* or human pre-tRNA constructs, we have observed tricRNA production in HeLa, HEK293T, and U87 cells. We have also expressed exogenous tricRNAs in *Drosophila* S2 cells.

3.1 Transfection

To assess tricRNA expression, constructs are transfected into an appropriate cell line. RNA is then isolated, and products are analyzed.

3.1.1 Notes on Cell Line

We have successfully expressed tricRNAs in several cell lines, including U87, HeLa, HEK293T, and *Drosophila* S2. The appropriate cell line depends on downstream applications. We have found that HEK293T cells produce tricRNAs most robustly, with HeLa cells still an appreciable choice.

3.1.2 Notes on Transfection Reagent
We have used Effectene, Lipofectamine 2000, and FuGENE HD and have had the most success with FuGENE.

Materials and equipment
- Cell line of choice
- Laminar flow hood
- CO_2 incubator
- Sterile cell culture plates or flasks
- Cell culture media (*we use DMEM supplemented with 10% FBS and 1% Pen/Strep*) and passaging reagents (*PBS, trypsin*)
- Micropipette and tips
- 1.5-mL tubes
- Transfection reagent
  - Effectene: Qiagen cat# 1054250
  - Lipofectamine 2000: ThermoFisher Scientific cat# 11668030
  - FuGENE HD: Promega (Madison, WI) cat# E2311
- Materials for seeding cells
  - Inverted microscope
  - Hemocytometer
  - Trypan blue
- DNA to transfect (tricRNA constructs and control)
- TRIzol: ThermoFisher Scientific cat# 15596026
- 3 *M* Sodium acetate
- 100% Ethanol
- Glycogen
- 70% Ethanol
- TE buffer
- Instrument to measure RNA concentration

Workflow
1. Follow the transfection reagent protocol for seeding cells.
   When using FuGENE, we seed 1.0×10^6 cells in T25 flasks.
2. Transfect cells according to the manufacturer's protocol.
   We use1 μg of DNA with 5 μL of FuGENE in 100 μL of serum-free media.
3. Incubate cells for 1–3 days. For a time course experiment, ensure that extra cells are seeded for the desired time points.
   We typically transfect for 3 days, changing the media every day.

3.2 RNA Isolation

For analysis of tricRNA expression, we isolate RNA using TRIzol reagent. While the manufacturer's protocol can be followed with good RNA yields, we have made the following modifications with better success:
1. Perform a second chloroform extraction.
2. Precipitate RNA with 2.5 volumes of ethanol, 0.1 volumes of sodium acetate, and 5 μL of glycogen instead of isopropanol.

We use 1 μL of TRIzol per T25 flask. After resuspending the RNA pellet in 100 μL of TE, we make a 1:10 dilution and quantify the diluted RNA using a NanoDrop. We then calculate RNA concentration based on the dilution. Typical concentrations of the dilutions are 100–500 ng/μL (with original samples thus being 1–5 μg/μL).

3.3 Analysis of Products

There are several ways to confirm tricRNA expression, including Northern blotting and RT-PCR. In addition, sequencing can be used to ensure that proper tRNA splicing is occurring.

3.3.1 Northern Blotting

We used Northern blotting to detect endogenous tricRNAs during *Drosophila* development, as well as in S2 cells (Lu et al., 2015). A similar approach can be taken to detect exogenous tricRNAs from a transfection experiment. The probe must be designed so that it spans the putative junction of the tricRNA. It should be noted that circRNAs migrate anomalously in polyacrylamide gels (Lu et al., 2015; Salgia et al., 2003); thus, the tricRNA band may not appear exactly where it is expected. As verification, a probe against the intron would detect the tricRNA, as well as unspliced pre-tRNA.

3.3.2 RT-PCR

This method is a straightforward way to confirm the presence of circRNAs. To perform PCR, cDNA must first be made. Several commercial kits are available for this purpose; we typically use SuperScriptIII (ThermoFisher Scientific cat# 18080051). When the reverse transcriptase enzyme encounters a circular template, such as a tricRNA, it can transcribe the circle many times before falling off (Fig. 5A), creating a linear cDNA concatamer with many repeats of the circle (Fig. 5B). The concatameric cDNA has many binding sites for RT-PCR primers, and thus many PCR products are possible. The formula for the potential PCR products is (length of PCR product) + (length of circle)n, where n is repeats of the circle. If the tricRNA has a restriction site, the PCR product can be digested with a restriction enzyme, and the ladder of products will collapse into one band (Fig. 5C).

> Note: the primers used for RT-PCR must be designed so that they would only amplify a circular product. The primers that bind to an individual repeat of the circle point away from each other, and amplification would not be possible if the RNA were linear (see Fig. 5).

3.3.3 Sequencing

To determine if the pre-tRNA is being spliced properly, the RT-PCR products can be sequenced. We took the restriction enzyme-digested PCR product from the above experiment and ligated it into a vector cut with the same enzyme and treated with a phosphatase (Fig. 5D). After transformation and harvesting DNA from cultures, we sequenced the plasmids using a sequencing primer in the vector. We were able to detect the insert in both directions, because the cloning was not directional. An alternative way to sequence the splice junction is to design PCR primers with a sequencing primer site on the 5′ end, perform PCR, and then send the PCR products to a sequencing facility.

4. IN-GEL IMAGING OF tricRNAs

The recent development of fluorescent RNA aptamers has enormous potential for live imaging of RNAs. Additionally, the small size of certain aptamers (such as Broccoli, which is 49 nt) allows for noninvasive tagging of cellular RNAs. In order to study the biogenesis and function of tricRNAs, we wanted to generate a fluorescent tricRNA reporter. To do this, we cloned the 49 nt Broccoli sequence into pTRIC-Y using *Not*I and *Sac*II restriction sites. We termed the resulting construct pTRIC-Y:Broccoli (see Table 1).

In addition to the internal promoter construct, we generated several external promoter constructs using the method described in Fig. 4:

pTRIC-Y:Broccoli_5S
pTRIC-Y:Broccoli_U6
pTRIC-Y:Broccoli_U6*

To analyze the products generated from transfection of these constructs, we utilized a novel in-gel staining protocol developed by the Jaffrey lab. This system allows for the visualization of all Broccoli-containing products; thus, both the pre-tRNA and the tricRNA are visible on the gel. We have confirmed the identity of both the circular and linear species using RNase R (Epicentre (Madison, WI) cat# RNR07250), an exonuclease that digests the majority of linear RNAs while leaving circRNAs intact (not shown).

Materials and equipment
- Polyacrylamide gel electrophoresis chamber
- Power supply
- 1X TBE
- TBE-urea gel
- Gel-loading tips
- RNA ladder
- Gel-loading buffer
- Trays to stain gels (tip box lids work well)
- DFHBI staining solution
    40 mM HEPES, pH 7.4
    100 mM KCl
    1 mM MgCl$_2$
    10 µM DFHBI-1T (Lucerna (New York, NY) cat# 400–1 mg).
        Prepare 50 mL of the staining solution. Store up to several months at 4°C. Can be reused several times.
- Imager (*we use a Typhoon TRIO+ Variable Mode Imager; a ChemiDoc MP can also be used*)
- Shaking platform
- SYBR gold or ethidium bromide
- Imaging processing software

Workflow
1. Transfect the desired cell line as described earlier with a Broccoli-containing tricRNA construct.
    We transfected HEK293T cells with several constructs using FuGENE HD, as described earlier.
2. Harvest the cells and isolate RNA after the desired amount of time. Quantify RNA yields.
3. Prepare an electrophoresis chamber with a TBE-urea polyacrylamide gel.
    We normally use 10% gels.
4. Run 5–10 µg of total RNA per well.
    We ran 10 µg of total RNA at 300 V for ~40 min in 1X TBE.
5. Wash the gel three times for 5 min each in deionized water on a shaking platform.

6. Incubate the gel in DFHBI-1T staining solution for 30 min on a shaking platform.
7. Image the gel with an appropriate instrument.

    A fairly wide variety of gel-documentation systems that have fluorescence capabilities can be used. We use a Typhoon laser scanner for imaging. In acquisition mode, choose fluorescence, then choose setup. Ensure that an appropriate excitation wavelength (488) is chosen, and that the correct emission filter (526 SP Fluorescein, Cy2, AlexaFluor488) is the only box selected.

8. Wash the gel three times for 5 min each in deionized water on a shaking platform.
9. To visualize total RNA, stain the gel with ethidium bromide or SYBR gold according to the manufacturer's protocol.

    Do not allow a total nucleic acid stain, like EtBr or SYBR Gold, to come into contact with the DFHBI-1T staining solution. Because DFHBI-1T imaging is extremely sensitive, the imager will pick up residual EtBr or SYBR Gold. Thus, ensure that separate staining trays are used.

10. Process images using software such as ImageJ or Photoshop.

    We have found that increasing both the brightness and the contrast from a Typhoon image gives the best visualization.

As a positive control, an in vitro transcript (IVT) can be generated and run on the gel along with total RNA. We have made an IVT of the pre-tRNA containing Broccoli in the intron. This transcript does not contain any leader or trailer sequences. We normally run 5 ng of this RNA on gels. It serves as a positive control for both pre-tRNA size and DFHBI-1T staining efficiency.

Fig. 6 shows the expression a variety of constructs that were transfected into HEK293T cells. Notably, the external pol III promoters greatly increase the expression of the tricRNA, seen in Fig. 6A. Each of the external promoters adds to the length of the pre-tRNA, but the circRNA is unchanged. The pre-tRNA IVT is shown in the last lane of the gel. The pre-tRNA bands from the total RNA lanes are slightly longer, which is due to the fact that they contain leader and trailer sequences. The doublet of bands in the U6* lane is likely due to transcription starting at both the internal and external promoter.

In addition to the chimeric human promoter/fly tRNA constructs we have generated, we have also made an all-human construct and transfected it into HEK293T cells for comparison (Fig. 6B). The expression of the

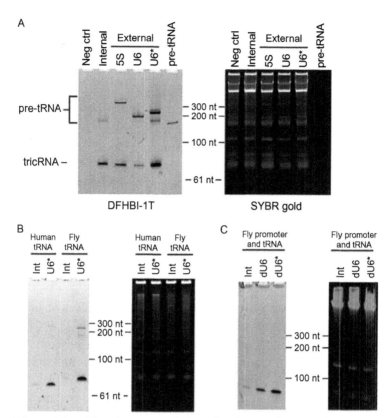

Fig. 6 A novel in-gel imaging protocol allows for the visualization of Broccoli-tagged reporter tricRNAs in vitro. For all images, the DFHBI-1T (Broccoli-specific) stain is on the *left*, whereas the SYBR gold (nonspecific) stain is on the *right*. The *top bands* are pre-tRNAs; the *lower bands* are tricRNAs. (A) The addition of external promoters greatly increases tricRNA expression. (B) In addition to fly tRNA constructs expressed under the control of human external promoters, we have also created an all-human construct with a human tRNA under the control of human U6*. (C) We have also created an all-fly construct with a fly tRNA under the control of a fly external pol III promoter. This construct and its variants were transfectecd into S2 cells.

human tricRNA is similar to that of the chimeric construct. The human tricRNA is also expected to be slightly shorter, which is reflected on the gel.

In a third experiment, we have created an all-fly construct, with a fly tRNA and the fly U6 promoter, and have transfected it into S2 cells (Fig. 6C). Interestingly, there is no visible pre-tRNA in any lane.

5. CELLULAR IMAGING OF tricRNAs

One of the advantages of the Broccoli system is that the fluorophore DFHBI-1T is not cytotoxic. Thus, it can be used for live-imaging experiments. Using this system, we wanted to determine the localization of our reporter in HEK293T cells.

Materials and equipment
- Micropipette and tips
- PBS
- Trypsin
- CO_2 incubator
- Clear cell culture media (DMEM without phenol red)
- Laminin-coated glass-bottomed dishes
- Hoechst
- DFHBI-1T
- Microscope with appropriate filters to detect both Broccoli and positive control (if used)
- Image processing software

Workflow
1. Transfect cells as described earlier for the desired amount of time.
    A positive control plasmid can be cotransfected, such as one that expresses mCherry.
2. When ready to begin, wash the cells with PBS.
3. Trypsinize the cells and resuspend in clear cell culture media (containing no phenol red) until no clumps are visible.
4. Plate the transfected cells at a 1:3 dilution in clear cell culture media onto laminin-coated glass-bottomed dishes and allow to incubate an additional day at 37°C.
    The remaining cells can be used for RNA isolation, if desired.
5. For live imaging, ensure that the environmental chamber on the microscope is at 37°C and 5% CO_2.
6. Add DFHBI-1T and Hoechst to final concentrations of 40 μM and 5 μg/mL, respectively, to the glass-bottomed dishes with transfected cells. Incubate for 30 min in a 37°C cell culture incubator.
7. Collect images on the microscope and process in an imaging software such as ImageJ.
    Use a FITC filter to detect Broccoli, a DAPI filter to detect Hoechst, and an appropriate filter to detect a positive control, if used.

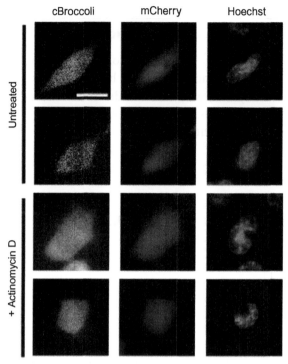

Fig. 7 HEK293T cells expressing Broccoli-containing tricRNAs can be imaged, with mCherry as a positive control for transfection. Scale bar is 20 μm. (See the color plate.)

Fig. 7 shows the data collected from an imaging experiment. We transfected pTRIC-Y:Broccoli_U6* into HEK293T cells to observe localization of our reporter. Circular Broccoli localizes primarily to the cytoplasm, while the positive control mCherry appears to be pan-cellular. Halting transcription with actinomycin D does not result in a noticeable drop in fluorescence intensity, indicating the incredible stability of tricRNAs.

6. CONCLUDING REMARKS

This manuscript delineates the process of generating a circRNA-expressing vector for use in visualizing RNA. Our method utilizes a conserved aspect of tRNA processing along with cutting-edge RNA imaging technology. While we have not explicitly tested the size limits of tricRNAs, we have detected circles up to 250 bp, which is well within the processivity of pol III (for comparison, *Drosophila* 7SK snRNA is over 400 nt). Future experiments will determine the upper limit of tricRNA size. We hope that our work will be helpful to the ever-growing fields of circRNAs and RNA imaging.

ACKNOWLEDGMENTS
The authors wish to thank S.R. Jaffrey for helpful discussions and for Spinach and Broccoli plasmids. C.A.S. was supported in part by a National Scientific Foundation Graduate Research Fellowship, DGE-1144081. This work was supported by a Grant from the National Institutes of Health, R01-GM053034-18 to A.G.M.

REFERENCES
Ashwal-Fluss, R., Meyer, M., Pamudurti, N. R., Ivanov, A., Bartok, O., et al. (2014). circRNA biogenesis competes with pre-mRNA splicing. *Molecular Cell, 56*(1), 55–66.

Burd, C. E., Jeck, W. R., Liu, Y., Sanoff, H. K., Wang, Z., & Sharpless, N. E. (2010). Expression of linear and novel circular forms of an INK4/ARF-associated non-coding RNA correlates with atherosclerosis risk. *PLoS Genetics, 6*(12), e1001233.

Chan, P. P., & Lowe, T. M. (2009). GtRNAdb: A database of transfer RNA genes detected in genomic sequence. *Nucleic Acids Research, 37*(Database issue), D93–D97.

Chen, C. Y., & Sarnow, P. (1995). Initiation of protein synthesis by the eukaryotic translational apparatus on circular RNAs. *Science, 268*(5209), 415–417.

Conn, S. J., Pillman, K. A., Toubia, J., Conn, V. M., Salmanidis, M., et al. (2015). The RNA binding protein quaking regulates formation of circRNAs. *Cell, 160*(6), 1125–1134.

Darty, K., Denise, A., & Ponty, Y. (2009). VARNA: Interactive drawing and editing of the RNA secondary structure. *Bioinformatics, 25*(15), 1974–1975.

Filonov, G. S., Kam, C. W., Song, W., & Jaffrey, S. R. (2015). In-gel imaging of RNA processing using Broccoli reveals optimal aptamer expression strategies. *Chemistry & Biology, 22*(5), 649–660.

Filonov, G. S., Moon, J. D., Svensen, N., & Jaffrey, S. R. (2014). Broccoli: Rapid selection of an RNA mimic of green fluorescent protein by fluorescence-based selection and directed evolution. *Journal of the American Chemical Society, 136*(46), 16299–16308.

Good, P. D., Krikos, A. J., Li, S. X. L., Bertrand, E., Lee, N. S., et al. (1997). Expression of small, therapeutic RNAs in human cell nuclei. *Gene Therapy, 4*(1), 45–54.

Hansen, T. B., Jensen, T. I., Clausen, B. H., Bramsen, J. B., Finsen, B., et al. (2013). Natural RNA circles function as efficient microRNA sponges. *Nature, 495*(7441), 384–388.

Jeck, W. R., Sorrentino, J. A., Wang, K., Slevin, M. K., Burd, C. E., et al. (2013). Circular RNAs are abundant, conserved, and associated with ALU repeats. *RNA, 19*(2), 141–157.

Kramer, M. C., Liang, D., Tatomer, D. C., Gold, B., March, Z. M., et al. (2015). Combinatorial control of Drosophila circular RNA expression by intronic repeats, hnRNPs, and SR proteins. *Genes & Development, 29*(20), 2168–2182.

Lu, Z., Filonov, G. S., Noto, J. J., Schmidt, C. A., Hatkevich, T. L., et al. (2015). Metazoan tRNA introns generate stable circular RNAs in vivo. *RNA, 21*(9), 1554–1565.

Memczak, S., Jens, M., Elefsinioti, A., Torti, F., Krueger, J., et al. (2013). Circular RNAs are a large class of animal RNAs with regulatory potency. *Nature, 495*(7441), 333–338.

Paige, J. S., Wu, K. Y., & Jaffrey, S. R. (2011). RNA mimics of green fluorescent protein. *Science, 333*(6042), 642–646.

Petkovic, S., & Müller, S. (2015). RNA circularization strategies in vivo and in vitro. *Nucleic Acids Research, 43*(4), 2454–2465.

Salgia, S. R., Singh, S. K., Gurha, P., & Gupta, R. (2003). Two reactions of Haloferax volcanii RNA splicing enzymes: Joining of exons and circularization of introns. *RNA*, *9*(3), 319–330.

Salzman, J., Chen, R. E., Olsen, M. N., Wang, P. L., & Brown, P. O. (2013). Cell-type specific features of circular RNA expression. *PLoS Genetics*, *9*(9), e1003777.

Salzman, J., Gawad, C., Wang, P. L., Lacayo, N., & Brown, P. O. (2012). Circular RNAs are the predominant transcript isoform from hundreds of human genes in diverse cell types. *PloS One*, *7*(2), e30733.

Strack, R. L., Disney, M. D., & Jaffrey, S. R. (2013). A superfolding Spinach2 reveals the dynamic nature of trinucleotide repeat-containing RNA. *Nature Methods*, *10*(12), 1219–1224.

Wang, Y., & Wang, Z. (2015). Efficient backsplicing produces translatable circular mRNAs. *RNA*, *21*(2), 172–179.

Yoshihisa, T. (2014). Handling tRNA introns, archaeal way and eukaryotic way. *Frontiers in Genetics*, *5*, e213.

CHAPTER TEN

RNA-ID, a Powerful Tool for Identifying and Characterizing Regulatory Sequences

C.E. Brule, K.M. Dean[2], E.J. Grayhack[1]
Center for RNA Biology, University of Rochester School of Medicine and Dentistry, Rochester, NY, United States
[1]Corresponding author: e-mail address: elizabeth_grayhack@urmc.rochester.edu

Contents

1. Introduction	238
2. Features of the RNA-ID System	239
3. Applications of the RNA-ID System	241
4. Analysis of *cis*-Regulatory Sequences in the RNA-ID Reporter	244
4.1 Cloning Sequences into the RNA-ID Reporter	245
4.2 Transformation of the RNA-ID Reporter into Yeast	248
4.3 Flow Cytometry	249
4.4 Tips for Analysis of a Library	250
5. Verification of Regulatory Sequences	250
6. Conclusions and Extensions of the RNA-ID Reporter System	251
Acknowledgments	252
References	252

Abstract

The identification and analysis of sequences that regulate gene expression is critical because regulated gene expression underlies biology. RNA-ID is an efficient and sensitive method to discover and investigate regulatory sequences in the yeast *Saccharomyces cerevisiae*, using fluorescence-based assays to detect green fluorescent protein (GFP) relative to a red fluorescent protein (RFP) control in individual cells. Putative regulatory sequences can be inserted either in-frame or upstream of a superfolder GFP fusion protein whose expression, like that of RFP, is driven by the bidirectional *GAL1,10* promoter. In this chapter, we describe the methodology to identify and study *cis*-regulatory sequences in the RNA-ID system, explaining features and variations of the RNA-ID reporter, as well as some applications of this system. We describe in detail the methods to analyze a single regulatory sequence, from construction of a single GFP variant to assay of variants by flow cytometry, as well as modifications required to screen libraries of different strains simultaneously. We also describe subsequent analyses of regulatory sequences.

---

[2] Current address: BD, 2350 Qume Drive, San Jose, CA 95131-1807, United States.

ABBREVIATIONS
FACS fluorescence-activated cell sorting
GFP green fluorescent protein
LIC ligation-independent cloning
RFP red fluorescent protein
RTD rapid tRNA decay
SD synthetic media with dextrose as the carbon source

1. INTRODUCTION

Identification of elements in DNA or RNA that regulate gene expression is crucial to mapping regulatory pathways and elucidating the mechanisms of such regulation. There is no comprehensive list of RNA or DNA regulatory sequences, despite the advances in technology that have made it easier to identify regulatory elements, such as high-throughput methods to identify motifs recognized by RNA binding proteins or sequences targeted by miRNAs (Campbell & Wickens, 2015; Hafner et al., 2010). However, methods to identify sequences that modulate expression in vivo are also needed, since comprehensive identification of sequences regulated by RNA binding proteins will require mapping the targets of over 700 RNA binding proteins even in the single-celled yeast *Saccharomyces cerevisiae* (Matia-Gonzalez, Laing, & Gerber, 2015), and since some regulatory sequences, such as synonymous codons, are not bound by proteins. RNA-ID is a method to identify and study *cis*-regulatory sequences, based on their effects on expression of superfolder green fluorescent protein (GFP; Fig. 1A; Dean & Grayhack, 2012); putative regulatory sequences are inserted either in-frame with GFP or upstream of GFP and are assayed by flow cytometry.

RNA-ID is an effective method that has been particularly useful in analysis of sequences whose effects are mediated by translation. Several features contribute to the efficacy of the RNA-ID system. First, the signal is robust, allowing differences in expression to be measured over a 250-fold range. Second, noise is minimized by normalization of GFP to a red fluorescent protein (RFP) that is controlled by the same promoter and by integration of the reporter into the chromosome (eliminating copy number effects). Third, results are reproducible over biological replicates (standard deviations are typically less than 5% for four independent biological replicates; Dean & Grayhack, 2012). This method can also be coupled to deep sequencing for a

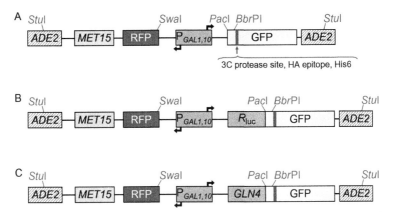

Fig. 1 The RNA-ID reporter and its derivatives. (A) In the RNA-ID vector, expression of two reporters GFP and RFP is driven by the bidirectional *GAL1,10* promoter. The GFP reporter is a fusion protein encoding a 3C protease site, an HA epitope, His6 (marked by *dark box*), followed by superfolder GFP. Putative regulatory sequences are inserted into the *Pac*I, *Bbr*PI sites (GFP), or into the *Swa*I site (RFP) using LIC cloning. The GFP lacks a start codon, allowing insertion of sequences upstream or within the coding sequence. (B) The *Renilla* luciferase–GFP fusion protein allows for analysis of the upstream nascent polypeptide and uses the same restriction sites for cloning. (C) In the *GLN4*-GFP fusion protein, varying lengths of *GLN4* can be PCR amplified to allow for the insertion of sequences at different locations relative to the start codon.

high-throughput method, which has been used to identify features of tRNAs required for a functional tRNA (Guy et al., 2014).

2. FEATURES OF THE RNA-ID SYSTEM

The RNA-ID system is a fluorescence-based assay where the bidirectional *GAL1,10* promoter drives expression of both a GFP fusion protein and RFP (Fig. 1A). Both the GFP and RFP genes were designed to carry unique restriction sites flanked by sequences that allow ligation-independent cloning (LIC; Aslanidis & de Jong, 1990) to insert novel sequences into either GFP or RFP. Furthermore, the reporter is integrated into the chromosome to eliminate copy number effects and expression of GFP or RFP variants is assayed using flow cytometry.

We used superfolder GFP to diminish expression effects from different amino acid insertions, since superfolder GFP has robust fluorescence when insoluble proteins are fused to its N-terminus (Pedelacq, Cabantous, Tran, Terwilliger, & Waldo, 2006). Thus, GFP expression from RNA-ID should be relatively insensitive to inserted amino acids. Indeed, the difference in

GFP expression from six different fusion reporters with varying lengths of *GLN4* (from 5 amino acids to 99 amino acids) fused to GFP is small, from 60 to 87 GFP/RFP units (GFP/RFP × 100) (Wolf & Grayhack, 2015). In the RNA-ID reporter, superfolder GFP is preceded by a site for 3C protease, HA epitope, and His6, all of which facilitate analysis and purification of the resulting protein (Quartley et al., 2009). Antibody detection of the HA epitope, which allows analysis of the polypeptide size in SDS-PAGE gels, was used to demonstrate the presence of a full-length frameshifted protein (Wolf & Grayhack, 2015). Moreover, sequences are inserted outside of the essential GFP regions just upstream of the HA epitope, and can even be inserted upstream of the start codon AUG, outside of the coding sequence.

Normalization of GFP to RFP reduces noise between cells of different sizes and reduces extrinsic noise that occurs due to heterogeneity or stochastic events (Raser & O'Shea, 2004), since the expression of these two proteins is driven by the same regulatory elements in the bidirectional *GAL1,10* promoter. Additionally, RFP is used to impose a cutoff for cells in which GAL induction is not effective, since RFP expression is presumably independent of the regulatory sequences in GFP. The RFP in the RNA-ID reporter is a version of mCherry that has been recoded with optimal codons for high expression in yeast (Keppler-Ross, Noffz, & Dean, 2008).

Derivatives of the RNA-ID reporter have been constructed to analyze the effects of regulatory sequences at internal locations (Fig. 1B and C). A specific reporter was constructed to allow analysis of effects on sequences both upstream and downstream of the regulatory sequence (Letzring, Wolf, Brule, & Grayhack, 2013). Thus, in a vector with a *Renilla* luciferase–GFP fusion protein, the regulatory sequence is inserted at amino acid 318 downstream of *Renilla* luciferase and upstream of GFP (Fig. 1B). Inhibitory CGA codons result in the recruitment of a quality control apparatus that targets the upstream nascent polypeptide, *Renilla* luciferase, for degradation (Letzring et al., 2013). Deletion of *LTN1*, which encodes the E3 ubiquitin ligase component of the quality control apparatus, restores *Renilla* luciferase activity, but does not affect readthrough of the codons and thus does not affect GFP (Letzring et al., 2013). The ability to measure both the upstream and downstream consequences of a regulatory sequence is invaluable. In another example, varying lengths of the N-terminal domain of *GLN4* were placed upstream of the regulatory sequence to measure the effects of internal gene location on GFP expression (Fig. 1C; Wolf & Grayhack, 2015).

LIC cloning in the RNA-ID system allows efficient and robust cloning of large numbers of different sequences without regard to the restriction sites

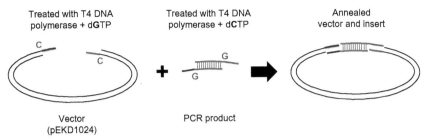

Fig. 2 An illustration of the general principles of LIC.

present in the inserted sequences. In LIC cloning (Fig. 2), the linearized vector is incubated with T4 DNA polymerase in the presence of a single dNTP, which inhibits its exonuclease activity at the complementary nucleotide, thus generating specific single-stranded ends of 12–18 nucleotides. A major advantage of using LIC cloning is that all PCR products generated with the appropriate LIC complementary ends can be efficiently cloned in this system, even if they do not contain any (or contain multiple) sites for the restriction endonucleases used to cleave the vector. Indeed, we used LIC cloning methods to clone and express more than 4000 ORFs from *Leishmania major*, *Trypanosoma brucei*, *Trypanosoma cruzi*, *Plasmodium falciparum*, and closely related organisms (Quartley et al., 2009) and developed a system to coexpress multiple genes without requirement for PCR amplification using LIC to join two vectors (Alexandrov et al., 2004). Furthermore, LIC cloning has been used to ligate multiple PCR products in a single reaction; these constructs also contained unique sequences (added during PCR) for identification of the particular construct (Quivey et al., 2015).

3. APPLICATIONS OF THE RNA-ID SYSTEM

There are several applications of the RNA-ID system; here we describe examples of three major applications and suggest potential applications. RNA-ID has been used to study translation (eg, inhibitory codons, chemical inhibitors, frameshifting), *trans*-acting factors (eg, tRNA functionality), and the effects of mutations on regulation (eg, $asc1\Delta$ on inhibitory codons and $met22\Delta$ on tRNAs).

The RNA-ID reporter has been used extensively to measure translation inhibition, particularly the effects of synonymous codons on expression (Dean & Grayhack, 2012). It has long been known that the choice of synonymous codons modulates translation efficiency, but the identity of codons or codon combinations that impede translation is unknown (Letzring, Dean, & Grayhack, 2010; Plotkin & Kudla, 2011). Such inhibitory codon-mediated signals are *cis*-acting regulators of translation that are almost certainly not recognized by RNA or DNA binding proteins. Thus, their detection is entirely dependent on a functional assay for expression, like RNA-ID.

In a systematic analysis, we had previously demonstrated that the Arg CGA codon is poorly translated due to a defect in I•A wobble decoding (Letzring et al., 2010). To examine effects of the Arg CGA codons in RNA-ID, GFP/RFP from reporters with CGA codons was compared to GFP/RFP from synonymous reporters with optimal AGA codons to control for amino acid effects. The sensitivity of the RNA-ID system is illustrated by the demonstration that the effects from a single CGA–CGA codon pair are easily detected by flow cytometry (Fig. 3). Either overproducing the native tRNA, tRNA$^{Arg(ICG)}$, or expressing a nonnative, exact base-pairing tRNA, tRNA$^{Arg(UCG)}*$, resulted in increased GFP/RFP from a reporter with CGA codons, but not from a synonymous reporter with AGA codons (Dean & Grayhack, 2012). This demonstrated that reduced expression was specifically due to translation of CGA codons. The RNA-ID system was also used to measure the effects of chemicals on translation of reporters with stop codons, as well as reporters with CGA codons. As expected based on previous analysis of stop codon readthrough (Salas-Marco & Bedwell, 2005), paromomycin enhanced readthrough of several stop codons in GFP, while cycloheximide had no effect (Dean & Grayhack, 2012). We also found that expression of CGA-containing reporters increased in the presence of paromomycin, suggesting that CGA codons are recognized as near-cognate mismatches in the A site of

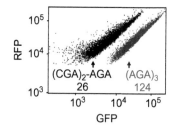

Fig. 3 Demonstration of the sensitivity of the RNA-ID system, showing a flow cytometry scatter plot in which each cell is represented by a *dot*. (AGA)$_3$ has a GFP/RFP value of 124 and (CGA)$_2$-AGA has a GFP/RFP value of 26.

the ribosome (Letzring et al., 2013). Frameshifting events can also be detected using the RNA-ID system because of the large dynamic range that allows detection of expression levels as low as 1% of the in-frame construct (Wolf & Grayhack, 2015).

RNA-ID has also been used to study *trans*-acting regulatory factors, specifically to identify the determinants of tRNA function in a high-throughput analysis of >25,000 tRNA variants (Guy et al., 2014). tRNA function requires a similarity in shape between all tRNAs for their role in translation, but differences in their individual sequence to allow specific aminoacylation by tRNA synthetases and unique modifications (Phizicky & Hopper, 2010). To study the features of tRNAs that determine tRNA function and susceptibility to rapid tRNA decay (RTD), Guy et al. (2014) generated a library of $SUP4_{oc}$ tRNA variants with a UUA anticodon that can readthrough a UAA stop codon. The stop codon is inserted into GFP; thus GFP expression reports on the functionality of a given tRNA variant. $SUP4_{oc}$ tRNA function was remarkably tolerant of mutations with 70 single mutations that were highly functional, although over half of these mutations resulted in RTD of the tRNA (Guy et al., 2014). Furthermore, this method of analyzing tRNA variants could be applied to any regulatory protein, where GFP expression reports on the functionality of protein variants.

RNA-ID system is a powerful genetic tool that can be used to find components in a regulatory system or to study the effects of mutations on regulation. Specifically, RNA-ID has been used to study the effects of mutations on codon-mediated regulation of translation (Letzring et al., 2013; Wolf & Grayhack, 2015) and to study the determinants of tRNA susceptibility to the RTD pathway (Guy et al., 2014). *ASC1* is a ribosomal small subunit protein involved in quality control pathways related to stalled ribosomes (Brandman et al., 2012; Kuroha et al., 2010). Since ribosomes are likely to stall at CGA codons (Letzring et al., 2013), we measured GFP/RFP in wild-type and *asc1Δ* strains from reporters with either CGA or AGA codons (Letzring et al., 2013; Wolf & Grayhack, 2015); indeed, deletion of *ASC1* improved readthrough of CGA codons at amino acid 318 (relative to the AGA control), but not of CGA codons at amino acid 6 (Letzring et al., 2013; Wolf & Grayhack, 2015). The synonymous control is very important in measurements in mutant strains because some mutations affect expression of the GFP or RFP reporters themselves. Similarly, Guy et al. (2014) defined RTD substrates by comparing tRNA function in wild-type and *met22Δ* strains and then verified these results by reconstruction of variants and direct measurement of levels of $SUP4_{oc}$ tRNA variants.

4. ANALYSIS OF *CIS*-REGULATORY SEQUENCES IN THE RNA-ID REPORTER

In this section, we describe the steps to construct RNA-ID reporters with specific sequences, integrate these reporters into the yeast genome, and analyze GFP/RFP expression by flow cytometry. A schematic of the steps involved in this entire process is shown in Fig. 4.

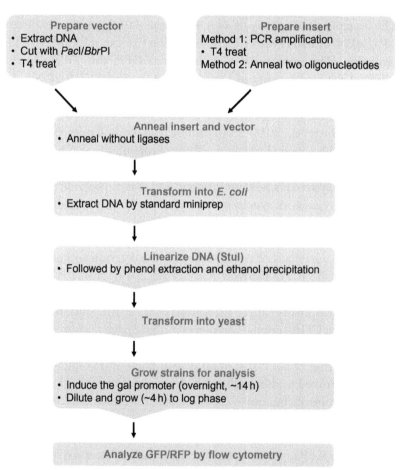

Fig. 4 Flowchart of the steps involved in analysis of *cis*-regulatory sequences in the RNA-ID system.

4.1 Cloning Sequences into the RNA-ID Reporter

To clone sequences into the RNA-ID reporter requires that both the vector and the insert have complementary single-stranded ends, and thus the first steps in this process are designed to generate those ends. After this preparation, annealing the insert and vector is sufficient for transformation. In the RNA-ID vector, GFP lacks an initiating AUG and thus there is no GFP expression from the base vector (pEKD1024). This allows insertion of inserts with the variable region either upstream or downstream of the start codon.

4.1.1 Preparation of Vector for Cloning into GFP

1. To obtain vector with appropriate ends, incubate 5–20 µg of pEKD1024 DNA with two restriction enzymes, *Pac*I (NEB) and *Bbr*PI (Roche), in a 100–400 µL reaction. Purify the vector by electrophoresis in a 1% LMP agarose (Invitrogen) gel in 1 × TAE buffer (40 mM Tris–acetate, 1 mM EDTA, pH 8.1, from a 50 × stock) at 4°C. Excise the 10,184 bp band and extract DNA from the gel using the QIAQuick Gel Extraction Kit (Qiagen). The concentration of the gel-purified DNA (ng/µL), measured on a NanoDrop Spectrophotometer at 260λ using an extinction coefficient of 50 ng cm/µL, is needed to calculate the pmol/µL, which is essential for the resection with T4 DNA polymerase. (*Note: use conversion factor of 660 g/mol bp).

2. To generate the single-stranded ends, incubate 0.03–0.10 pmol of gel-purified vector DNA with T4 DNA polymerase (Roche) in the presence of dGTP (Roche; Fig. 5A). On ice, mix the following in a 1.5 mL microcentrifuge tube: 2.0 µL of T4 buffer (Roche), 2.0 µL of 25 mM dGTP, 1.0 µL of 100 mM DTT, 0.03–0.10 pmol of gel-purified vector, 0.8 µL of T4 DNA polymerase (Roche, CAT# 70099), ddH$_2$O up to 20.4 µL. Incubate at 22°C (room temperature) for 40 min and then inactivate the T4 DNA polymerase by incubating the reaction at 75°C for 40 min. Store at −20°C.

4.1.2 Preparation of Inserts

Here, we describe three different inserts for LIC cloning into GFP: a PCR product, oligonucleotides, and a partially randomized sequence.

1. To insert a large DNA fragment, it is necessary to amplify the desired DNA by PCR using oligonucleotide primers with appropriate sequences added to their 5′ ends (Fig. 5B). Design oligonucleotides to PCR amplify

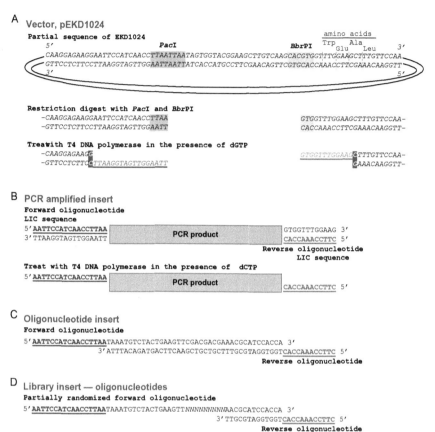

Fig. 5 Preparation of vector and inserts for LIC. (A) A schematic of the vector pEKD1024 showing the major steps to generate single-stranded DNA ends for LIC using T4 DNA polymerase and dGTP. Restriction enzyme sites are *shaded* and the nucleotide where the T4 DNA polymerase stops is indicated by *white text* over a *dark background*. (B–D) Schematics of various types of inserts that can be cloned into the vector pEKD1024. (B) A PCR amplified insert and the product after treatment with T4 DNA polymerase in the presence of dCTP. (C, D) Inserts formed with oligonucleotides. (C) Complete overlap of inserted sequences. (D) Partially randomized sequence with 13-nucleotide overlap.

the desired DNA fragment, to the 5′ end of the forward oligonucleotide add the following sequence: 5′-AATTCCATCAACCTTAAT, and to the 5′ end of the reverse oligonucleotide add the following sequence: 5′-CTTCCAAACCAC. After PCR amplification, purify the DNA on a 1% LMP agarose gel and incubate 0.12–0.48 pmol with T4 DNA polymerase (in a 20.4 μL reaction, as described for the vector) in the presence of *dCTP*, the dNTP complementary to that used on the vector.

2. Sequences up to 42 nucleotides can be easily inserted with two complementary oligonucleotides that contain the test sequence and the appropriate sequences for LIC added to their 5′ ends (5′-AATTCCATCAACCTTAAT to the forward oligonucleotide and 5′-CTTCCAAACCAC to the 5′ end of the reverse oligonucleotide: Fig. 5C). Make a 25 μM working stock of each of the forward and reverse oligonucleotides by diluting in TE 8 (10 mM Tris–HCl, pH 8.0, 1 mM EDTA) and then mix 4 μL of the forward oligonucleotide, 4 μL of the reverse oligonucleotide, 3 μL of 0.5 M NaCl, and 19 μL ddH$_2$O in a 1.5-mL microcentrifuge tube. Boil for 2–3 min using either a beaker of water over an open flame or a heat block (be sure to preboil the water or preheat the heat block before starting the time). Then, let the reaction cool slowly to 55–65°C (either by setting the entire beaker or the heat block on the bench). For each set of annealed oligonucleotides, make a 1:50 dilution in ddH$_2$O and store at −20°C.

3. To generate a library of variants or insert a longer sequence, we use two oligonucleotides, one of which only contains the nonrandomized portion of the sequence. We use a minimum of 13 nucleotides that form a complementary double-stranded region, but the remaining sequences can be single stranded (Fig. 5D). For a longer sequence, simply overlap the 2 oligonucleotides by 13 nucleotides and extend the insert with the reverse oligonucleotide. Subsequent preparation of the insert is the same as described for annealing two oligonucleotides.

4.1.3 Annealing the Insert and Vector

To complete the RNA-ID reporter, the vector and insert are annealed together without the need for ligases.

1. To a 1.5-mL microcentrifuge tube, add 1 μL of T4-treated vector (be sure the 1 μL drop is at the bottom of the microcentrifuge tube). Add 2 μL of the 1:50 dilution of the annealed oligonucleotides or T4-treated PCR reaction (slowly eject 2 μL into the 1 μL drop and physically mix with pipette tip). Incubate the reaction at 22°C (room temperature) for 5 min.

2. Add 1 μL of 25 mM EDTA (slowly eject into mix and physically mix with pipette tip). Incubate the reaction at 22°C (room temperature) for 5 min. Store at 4°C, or immediately transform into competent *Escherichia coli* cells.

4.1.4 Variation: Cloning into RFP

To clone into RFP, the vector is prepared for LIC cloning by incubating DNA with the restriction endonuclease *Swa*I, followed by gel purification and other steps described for the GFP vector. Inserts are prepared by adding common sequences (different from the GFP common sequences) to the 5′ ends of the forward oligonucleotide (5′-TAATCCATCAACCATTT) and the reverse oligonucleotide (5′-AAACCATTCTCCTATTT).

4.2 Transformation of the RNA-ID Reporter into Yeast
4.2.1 Linearization of DNA

To obtain a linear DNA fragment with the RNA-ID reporter flanked on each end by sequences homologous to *ADE2*, the DNA is incubated with *Stu*I (NEB) and then subjected to phenol extraction and ethanol precipitation. Specific and efficient integration into the *ADE2* locus is facilitated by the ∼400 bp of *ADE2* sequences on either end of the linear fragment of DNA (∼5700 nucleotides; will vary based on insertion length).

4.2.2 Transformation of the RNA-ID Reporter into Yeast

Each transformation requires 10 mL of cells, grown to OD_{600} between 1.3 and 2.5 (log phase), ∼300 ng of *Stu*I cut DNA, and yields 50–200 transformants. Directions are given for a single transformation, but Step 1 can be performed in bulk.

1. To prepare competent yeast cells for transformation, grow the desired yeast strain in YPD to OD_{600} between 1.3 and 2.5 (log phase). For each transformation, pellet 10 mL of cells by centrifugation at 3000 rpm for 5 min. Decant supernatant. Wash twice with half volume of 100 mM LiOAc. Resuspend cells in 100 μL of 100 mM LiOAc per transformation. Add 10 μL salmon sperm DNA per transformation.
2. Aliquot 110 μL of competent cells to one 2059 tube (13 mL conical tube; Falcon) per transformation. Add 300 ng of linearized DNA to each tube and tap mix. Incubate at 30°C in a water bath for 15 min (do not shake). After the first incubation, add 600 μL of a LiOAc-PEG solution (8 volumes of 60% PEG 3350 to 1 volume of 1 M LiOAc to 1 volume of ddH_2O) to each tube and tap mix. Incubate at 30°C in a water bath for 20 min (do not shake). After the second incubation, add 68 μL of DMSO to each tube and tap mix. Heat shock cells at 42°C in a water bath for 15 min, but do not shake. After the third incubation, pellet cells by centrifugation at 1500 rpm for 5 min. Carefully remove the LiOAc-PEG solution using a pipetteman.

3. To facilitate growth of transformed cells, include a recovery period where each tube receives 1 mL YPD (gently tap mix to resuspend cells) and is then incubated in a 30°C shaker at 250 rpm for 3–4 h. After this recovery period, plate 250 µL of each transformation on an SD-met plate and incubate at 30°C for 2 days. These plates have a background of nontransformed cells; thus, individual colonies are streaked onto SD-met plates to obtain single colonies and are incubated at 30°C for 2 days.

4.3 Flow Cytometry

Yeast strains carrying RNA-ID reporters are grown in YP + 2% raffinose + 2% galactose + 80 mg/L Ade (Sherman, Fink, & Hicks, 1986) for two reasons: (1) The bidirectional *GAL1,10* promoter is repressed in the presence of dextrose, but induced in the presence of galactose. (2) The RNA-ID reporter is integrated into the yeast chromosome at the *ADE2* locus; thus the cells are deficient in adenine biosynthesis causing them to turn pink; this pigment increases the background in the GFP channel.

4.3.1 Growth of Strains for Flow Cytometry

Single colonies from the transformation (or from saves) are used to inoculate 5 mL YP + 2 % raffinose + 2 % galactose + 80 mg/L Ade and grown overnight (~14 h) at 30°C. Then each culture is diluted into 5 mL of the same media to OD_{600} between 0.1 and 0.2 and grown for 4–6 h to OD_{600} of ~0.8 (Dean & Grayhack, 2012). For each strain, 500 µL of culture is transferred to a 5-mL polystyrene round-bottom tube (Falcon) and kept on ice (for up to an hour) until analysis by flow cytometry. Flow cytometry and downstream analysis are performed for at least three, preferably four, independent isolates of each strain (Dean & Grayhack, 2012).

4.3.2 Reference and Control Strains

A reference strain carrying the RNA-ID reporter with ATG inserted into GFP is used as the primary control in all flow cytometry experiments (Dean & Grayhack, 2012). GFP and RFP voltages are adjusted until fluorescence in both channels is 26,000, resulting in a 1:1 ratio of GFP:RFP. GFP and RFP from several control strains, that have been extensively analyzed, are measured in each experiment to assure reproducibility: $(AGA)_3$, a reporter with high GFP/RFP; $(CGA)_3$, a reporter with low GFP/RFP; JW132, the base vector for the RNA-ID reporter (does not contain GFP or RFP; Whipple, Lane, Chernyakov, D'Silva, & Phizicky, 2011); and an RNA-ID reporter lacking a GFP insert (GFP^- RFP^+).

4.4 Tips for Analysis of a Library
4.4.1 Generating a Large Library of Variants
To adequately cover the sequence scope of many libraries, it is important to obtain 50,000–100,000 yeast transformants. It is difficult to obtain this number of chromosomally integrated yeast transformants using many protocols. Yoshiko Kon and Eric Phizicky (personal communication) found that using ~2 μg of DNA to transform 10 mL of yeast at OD_{600} of 1.3–2.5 resulted in >50,000 yeast transformants; this protocol was used by Guy et al. (2014).

4.4.2 Purifying Transformants
Analyzing libraries of transformants presents a unique difficulty since each plate of yeast transformants also contains many nontransformed cells that are viable and will grow if transferred to rich media. Thus, it is necessary to impose an additional round of selection for the transformed yeast, but this must be done in bulk. We have resolved this issue in two ways, either by replica plating the transformation plates onto selective media (Dean & Grayhack, 2012) or by scraping and pooling transformation plates, followed by additional growth in media that selects for transformants (Guy et al., 2014), the latter of which seems to be the more effective approach.

4.4.3 Fluorescence-Activated Cell Sorting
The protocol for growing libraries of yeast for fluorescence-activated cell sorting (FACS) analysis is described by Guy et al. (2014). During the growth required to induce the GFP and RFP expression, cells need to be maintained in log phase, which necessitated a series of dilutions prior to sorting. To maintain the complex relative composition of the library, it is extremely important to inoculate each dilution series with $10 \times$ or more cells corresponding to the original number of transformants. FACS instrumentation and the gate setup is described by Dean and Grayhack (2012).

5. VERIFICATION OF REGULATORY SEQUENCES
RNA-ID is a useful tool for verifying the effects of regulatory sequences on translation and has been used extensively to distinguish between codon-mediated regulation and sequence-/structure-mediated regulation. For example, inhibition caused by CGA codons is suppressed

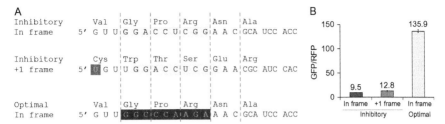

Fig. 6 Analysis to determine if reading frame is important for regulation. (A) The inhibitory sequence in frame and in the +1 frame and the optimal sequence in frame are shown. To frameshift the inhibitory sequence, a single nucleotide (*white text, shaded background*) is inserted upstream of the Val GUU and a single nucleotide is deleted downstream (the third nucleotide of the in-frame ACC codon). (B) GFP/RFP values for the three sequences described. *Error bars* represent ±SD.

by expressing either tRNA$^{Arg(ICG)}$ or tRNA$^{Arg(UCG)}$*, indicating that the inhibitory defect of CGA codons occurs in translation.

A second method to assess whether or not codon use is important for the observed regulation is to determine if the reading frame is required for inhibition. In this case, the inserted sequence is moved out of frame and strains carrying the reporter with the frameshifted insert are analyzed by flow cytometry. An example of this type of analysis is shown in Fig. 6. The inhibitory sequence GGA CCT CGG was identified by sequencing GFP variants isolated from a low-expression pool of cells obtained by FACS (Dean & Grayhack, 2012). To determine if this sequence or translation of these codons is responsible for low GFP, we reconstructed the original variant and a frameshifted derivative (Fig. 6A), both of which yield low GFP/RFP (Fig. 6B). Thus, inhibition by this sequence is not due to synonymous codon effects.

6. CONCLUSIONS AND EXTENSIONS OF THE RNA-ID REPORTER SYSTEM

While the methodology described earlier provides useful tools for identification and characterization of regulatory sequences, three limitations of the current RNA-ID system could be addressed in the future. First, reengineering the promoter and UTR regions of the RNA-ID vector would allow analysis of regulatory elements in different growth conditions, in different stresses, with different expression levels, and with different transcription start sites and could also allow analysis of uORFs. Use of the *GAL1,10* promoter to direct GFP and RFP expression allows good regulation of expression, but is not compatible with growth of yeast in its preferred carbon source, glucose. Since the *GAL1,10* regulatory region is flanked by unique restriction sites (*PacI* and *SpeI*), it would be relatively simple to replace the promoter and 5′ UTR sequences in the current vector with other bidirectional yeast promoters, such as the *RPL13A, RPL16B* promoter (Xu et al., 2009). This might facilitate detection of sequences that only regulate expression in particular conditions. Second, the vector could be converted to allow cloning with standard restriction enzyme technology, simply by introducing restriction sites with 5′ or 3′ overhangs at the *PacI* and *Bbr*PI sites. Third, since many regulatory signals are found in the 3′ UTR, we could modify the vector to allow cloning at the 3′ end of the GFP. Thus, RNA-ID is a robust and sensitive method to identify regulatory sequences that can be expanded in new directions.

ACKNOWLEDGMENTS

We thank Eric Phizicky, Andrew Wolf, Yoshiko Kon, and Michael Guy for their important contributions in developing the approach described in this chapter. This work was supported by NSF Grant MCB-1329545 to E.J.G. C.E.B. was partially supported by NIH Training Grant in Cellular, Biochemical, and Molecular Sciences 5T32 GM068411.

REFERENCES

Alexandrov, A., Vignali, M., LaCount, D. J., Quartley, E., de Vries, C., De Rosa, D., et al. (2004). A facile method for high-throughput co-expression of protein pairs. *Molecular & Cellular Proteomics, 3*, 934–938.

Aslanidis, C., & de Jong, P. J. (1990). Ligation-independent cloning of PCR products (LIC-PCR). *Nucleic Acids Research, 18*, 6069–6074.

Brandman, O., Stewart-Ornstein, J., Wong, D., Larson, A., Williams, C. C., Li, G. W., et al. (2012). A ribosome-bound quality control complex triggers degradation of nascent peptides and signals translation stress. *Cell, 151*, 1042–1054.

Campbell, Z. T., & Wickens, M. (2015). Probing RNA-protein networks: Biochemistry meets genomics. *Trends in Biochemical Sciences, 40*, 157–164.

Dean, K. M., & Grayhack, E. J. (2012). RNA-ID, a highly sensitive and robust method to identify *cis*-regulatory sequences using superfolder GFP and a fluorescence-based assay. *RNA, 18*, 2335–2344.

Guy, M. P., Young, D. L., Payea, M. J., Zhang, X., Kon, Y., Dean, K. M., et al. (2014). Identification of the determinants of tRNA function and susceptibility to rapid tRNA decay by high-throughput *in vivo* analysis. *Genes & Development, 28*, 1721–1732.

Hafner, M., Landthaler, M., Burger, L., Khorshid, M., Hausser, J., Berninger, P., et al. (2010). Transcriptome-wide identification of RNA-binding protein and microRNA target sites by PAR-CLIP. *Cell, 141*, 129–141.

Keppler-Ross, S., Noffz, C., & Dean, N. (2008). A new purple fluorescent color marker for genetic studies in *Saccharomyces cerevisiae* and *Candida albicans*. *Genetics, 179*, 705–710.

Kuroha, K., Akamatsu, M., Dimitrova, L., Ito, T., Kato, Y., Shirahige, K., et al. (2010). Receptor for activated C kinase 1 stimulates nascent polypeptide-dependent translation arrest. *EMBO Reports, 11*, 956–961.

Letzring, D. P., Dean, K. M., & Grayhack, E. J. (2010). Control of translation efficiency in yeast by codon-anticodon interactions. *RNA, 16*, 2516–2528.

Letzring, D. P., Wolf, A. S., Brule, C. E., & Grayhack, E. J. (2013). Translation of CGA codon repeats in yeast involves quality control components and ribosomal protein L1. *RNA, 19*, 1208–1217.

Matia-Gonzalez, A. M., Laing, E. E., & Gerber, A. P. (2015). Conserved mRNA-binding proteomes in eukaryotic organisms. *Nature Structural & Molecular Biology, 22*, 1027–1033.

Pedelacq, J. D., Cabantous, S., Tran, T., Terwilliger, T. C., & Waldo, G. S. (2006). Engineering and characterization of a superfolder green fluorescent protein. *Nature Biotechnology, 24*, 79–88.

Phizicky, E. M., & Hopper, A. K. (2010). tRNA biology charges to the front. *Genes & Development, 24*, 1832–1860.

Plotkin, J. B., & Kudla, G. (2011). Synonymous but not the same: The causes and consequences of codon bias. *Nature Reviews. Genetics, 12*, 32–42.

Quartley, E., Alexandrov, A., Mikucki, M., Buckner, F. S., Hol, W. G., DeTitta, G. T., et al. (2009). Heterologous expression of *L. major* proteins in *S. cerevisiae*: A test of solubility, purity, and gene recoding. *Journal of Structural and Functional Genomics, 10*, 233–247.

Quivey, R. G., Jr., Grayhack, E. J., Faustoferri, R. C., Hubbard, C. J., Baldeck, J. D., Wolf, A. S., et al. (2015). Functional profiling in *Streptococcus mutans*: Construction and examination of a genomic collection of gene deletion mutants. *Molecular Oral Microbiology, 30*, 474–495.

Raser, J. M., & O'Shea, E. K. (2004). Control of stochasticity in eukaryotic gene expression. *Science, 304*, 1811–1814.

Salas-Marco, J., & Bedwell, D. M. (2005). Discrimination between defects in elongation fidelity and termination efficiency provides mechanistic insights into translational readthrough. *Journal of Molecular Biology, 348*, 801–815.

Sherman, F., Fink, G., & Hicks, J. B. (1986). *Methods in yeast genetics* (pp. 145–149). New York: Cold Spring Harbor Laboratory Press.

Whipple, J. M., Lane, E. A., Chernyakov, I., D'Silva, S., & Phizicky, E. M. (2011). The yeast rapid tRNA decay pathway primarily monitors the structural integrity of the acceptor and T-stems of mature tRNA. *Genes & Development, 25*, 1173–1184.

Wolf, A. S., & Grayhack, E. J. (2015). Asc1, homolog of human RACK1, prevents frameshifting in yeast by ribosomes stalled at CGA codon repeats. *RNA, 21*, 935–945.

Xu, Z., Wei, W., Gagneur, J., Perocchi, F., Clauder-Munster, S., Camblong, J., et al. (2009). Bidirectional promoters generate pervasive transcription in yeast. *Nature, 457*, 1033–1037.

CHAPTER ELEVEN

Fluorescent Protein-Based Quantification of Alternative Splicing of a Target Cassette Exon in Mammalian Cells

N.G. Gurskaya[*,†], **D.B. Staroverov**[*], **K.A. Lukyanov**[*,†,1]
[*]Shemyakin–Ovchinnikov Institute of Bioorganic Chemistry, Moscow, Russia
[†]Nizhny Novgorod State Medical Academy, Nizhny Novgorod, Russia
[1]Corresponding author: e-mail address: kluk@ibch.ru

Contents

1. Introduction	256
2. Description of the Method	257
2.1 Overall Scheme	257
2.2 Reporter for Alternative Splicing of PIG3	259
3. Materials	260
4. Procedure	261
4.1 Preparation of Genetic Constructs	261
4.2 Cell Culture and Transfection	262
4.3 Fluorescence Microscopy	263
4.4 Flow Cytometry	265
4.5 RNA Isolation and Quantitative RT-PCR	266
5. Concluding Remarks	266
Acknowledgments	267
References	267

Abstract

Alternative splicing is an important mechanism of regulation of gene expression and expansion of proteome complexity. Recently we developed a new fluorescence reporter for quantitative analysis of alternative splicing of a target cassette exon in live cells (Gurskaya et al., 2012). It consists of a specially designed minigene encoding red and green fluorescent proteins (Katushka and TagGFP2) and a fragment of the target gene between them. Skipping or inclusion of the alternative exon induces a frameshift; ie, alternative exon length must not be a multiple of 3. Finally, red and green fluorescence intensities of cells expressing this reporter are used to estimate the percentage of alternative (exon-skipped) and normal (exon-retained) transcripts. Here, we provide a detailed description of design and application of the fluorescence reporter of a target alternative exon splicing in mammalian cell lines.

1. INTRODUCTION

Pre-mRNA splicing is a fundamental process required for the expression of plurality of metazoan genes. Most human genes contain multiple exons and express more than one splice variant (Stamm et al., 2000). Protein isoforms, produced by alternative splicing, can vary in different ways, including protein domain composition, subcellular localization, signaling activity, and protein half-life (Stamm et al., 2005). Alternative splicing often brings about nonsense-mediated mRNA decay (NMD) due to introduction of frameshifts and premature terminal codons. Such transcripts are degraded rapidly that result in a very low level of corresponding proteins (Lewis, Green, & Brenner, 2003; Maquat, 2004). The exon skipping is the most numerous alternative splice events (Stamm et al., 2000). It often results in strong effects for the protein functioning, for example, by deleting its transmembrane helix or domain responsible for specific protein–protein interactions (Hiller, Huse, Platzer, & Backofen, 2005; Kriventseva et al., 2003; Resch et al., 2004).

Alternative splicing can be assessed by a number of methods. Classical approaches such as northern blot analysis and reverse transcriptase–polymerase chain reaction (RT-PCR) are commonly used to identify and quantify the presence of alternatively spliced variants of a target gene in an RNA sample. High-throughput sequencing of whole transcriptomes enables detailed global analysis of alternative splicing (Barash et al., 2010; Blencowe, 2006; Shalek et al., 2013). However, these methods usually require a large number of cells and also do not allow working with live cells.

To overcome these limitations, fluorescent protein-based methods for alternative splicing analysis were developed (Kuroyanagi, 2013; Kuroyanagi, Kobayashi, Mitani, & Hagiwara, 2006; Newman et al., 2006; Ohno, Hagiwara, & Kuroyanagi, 2008; Orengo, Bundman, & Cooper, 2006; Somarelli et al., 2013; Stoilov, Lin, Damoiseaux, Nikolic, & Black, 2008; Wang et al., 2004). These methods use specially designed minigenes encoding one or two fluorescent proteins and a fragment of the gene of interest. Alternative splicing of such transcripts results in fluorescence turn-on, turn-off, or color switching that can be detected by fluorescence microscopy or flow cytometry.

Recently, we developed a reporter for quantitative analysis of alternatively spliced exons using two fluorescent proteins (Gurskaya et al., 2012). This chapter describes applications of this method in mammalian cell lines.

2. DESCRIPTION OF THE METHOD
2.1 Overall Scheme

The described method can be used to assess splicing of alternative (cassette) exons whose length is not a multiple of 3. In other words, exon skipping leads to an open reading frameshift. In the case of a nonframeshifting alternative exon, one can artificially add one to two nucleotides to induce the required frameshift.

Fig. 1 illustrates the proposed scheme. An alternative exon of interest together with two adjacent introns and constitutive exons (or fragments of these exons) is cloned between open reading frames encoding red and green fluorescent proteins (RFP and GFP) (Fig. 1A left). The cloning is performed to create a single joint open reading frame from RFP through all three target exons to GFP. Thus, normally spliced full-length (exon-retained) transcript produces a fusion protein that fluoresces both red and green. In contrast, alternatively spliced (exon-skipped) transcript possesses a frameshift and a stop codon before GFP; it produces a protein that fluoresces only red.

A control vector encodes intronless variant of the normally spliced transcript (Fig. 1A right). This vector is required to measure a reference green-to-red ratio characteristic for the particular experimental model (Fig. 1B right). Therefore, potential difference in expression levels of intron-containing and intronless transcripts does not hinder successful implementation of the method as it does not affect green-to-red ratio. Potentially, known mutant of the target gene producing only the exon-retained transcript can be used as a control. However, such mutants are available not for all genes. Thus, intronless construct (Fig. 1A right) appears to be a simple and universal way to determine control green-to-red ratio.

Cells expressing the splicing reporter can be analyzed by fluorescence microscopy and flow cytometry. The latter provides opportunity to assess a large number of cells and should be used whenever possible. On the standard flow cytometry green–red dot plot, the control cells form a diagonal distribution (Fig. 1B right). This sample provides a reference red-to-green ratio characteristic for a particular fusion protein and fluorescence detection parameters. From this reference ratio, percentage of the normally and alternatively spliced transcripts in each splicing reporter-expressing cell can be easily calculated (Fig. 1B left and equation at the bottom).

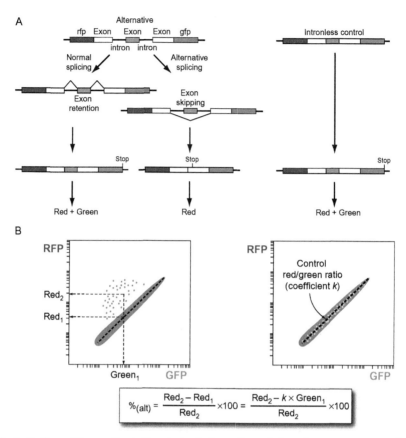

Fig. 1 Outline of the method of analysis of alternative splicing of a cassette exon using two fluorescent proteins. (A) Scheme of genetic constructs. *Left*: A fragment of a gene of interest (consisting of the alternative exon and two adjacent introns and exons) is cloned between rfp- and gfp-coding regions in such a way that normal splicing (exon retention) results in a full-length protein (*red* and *green fluorescence*), whereas alternative splicing (exon skipping) leads to the frameshift and appearance of a stop codon before GFP (only *red fluorescence*). *Right*: A control intronless construct provides reference *red-to-green ratio* characteristic for the 100% normally spliced transcript. (B) Schematics of flow cytometry analysis of the splicing reporter in *green* and *red* channels. Cells expressing the control plasmid (*right plot*) form a diagonal distribution (*gray*) and make it possible to determine a reference *red*-to-*green ratio* (k) characteristic for a particular model and detection parameters. Cell expressing the target transcript undergoing alternative splicing (*left plot*) can be found to the left from the control diagonal. Percentage of the alternative transcript can be estimated for each cell from its fluorescence intensities in *green* and *red* channels (Green$_1$ and Red$_1$) using the equation shown below the plots. (See the color plate.)

Potentially, alternative splicing reporter can be designed using an "opposite" color scheme, in which exon skipping leads to the synthesis of RFP and GFP, whereas exon retention results in a premature stop codon and the synthesis of RFP only. However, in this case the latter transcript can undergo efficient degradation by NMD machinery due to a stop codon before the last exon–exon junction site (Chang, Imam, & Wilkinson, 2007; Shyu, Wilkinson, & van Hoof, 2008). To avoid NMD, stop codon should be situated no more than 40 bases upstream of the end of alternative exon.

The proposed scheme possesses two main advantages over other analogous fluorescent protein-based methods of alternative splicing analysis. First, it allows quantitative analysis of alternative splicing in individual cells due to the presence of RFP-encoding part in both normal and alternative transcripts. Second, a long unaltered target gene fragment is used that ensures a close to native state of main splicing regulatory elements.

2.2 Reporter for Alternative Splicing of PIG3

The scheme described earlier (see Section 2.1) was implemented to assess alternative splicing of human PIG3 (p53-inducible gene 3) gene. This gene has five exons; exon 4 is an alternative exon of 197-bp length, which can be either present or absent in the PIG3 transcripts (Nicholls & Beattie, 2008). Exon 4 skipping results in a frameshift in exon 5.

Splicing reporter was constructed on the base of pTurboFP635-C vector (Evrogen). PIG3 fragment consisting of exons 3, 4, and 5 and introns 3 and 4 was cloned between open-reading frames of far-red FP Katushka and green FP TagGFP2 (plasmid pSplPIGv2; Fig. 2 left). Katushka-encoding region with synonymous mutations eliminating a strong donor splice site was

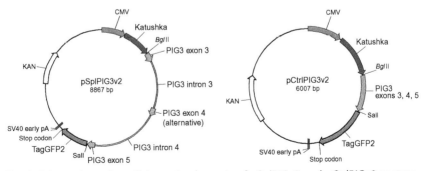

Fig. 2 Schematic outline of the main elements of pSplPIGv2 and pCtrlPIGv2 vectors.

used (Gurskaya, Staroverov, Fradkov, & Lukyanov, 2011). TagGFP2 was used instead of TagGFP in the previously published variant of this reporter (Gurskaya et al., 2012) that ensured brighter and more reliable readout (Pereverzev et al., 2015). A control plasmid pCtrlPIGv2 encoded the normally spliced transcript Katushka–exon3–exon4–exon5–TagGFP2 (Fig. 2, right).

3. MATERIALS

- Reagents
- Vectors pTurboRFP635-N (Evrogen, cat. no. FP721), pTagGFP2-N (Evrogen, cat. no. FP191)
- Vectors SplPIGv2 or pCtrlPIGv2 (GenBank accession numbers KT831770 and KT831769, respectively) for quantitative assessment of alternative splicing (available upon request)
- Plasmid Midi Kit (Qiagen, cat. no. 12143)
- Plasmid Mini Kit (Qiagen, cat. no.12123)
- Mammalian cell line(s) of interest
- DMEM (Invitrogen, cat. no. 41965-039)
- OPTI-MEM (Invitrogen, cat. no. 31985-047)
- Bovine serum (Invitrogen, cat. no. 16170-078)
- Penicillin/streptomycin (Invitrogen, cat. no. 15070-063)
- L-Glutamine (Invitrogen, cat. no. 25030-024)
- BSA (Sigma-Aldrich, cat. no. 9048-46-8)
- Trypsin/EDTA solution (Invitrogen, cat. no. R-001-100)
- PBS (0.5% BSA)
- Versene Solution (EDTA in PBS) 0.02% solution (ThermoFisher Scientific, cat. no. 15040-066)
- FuGene6 transfection reagent (Roche Diagnostics, cat. no. 11988387001)
- Plastic dishes for tissue culture (Sigma-Aldrich, cat. no. CLS 3294)
- Cell Strainer Polypropylene frame 70 μm (SPL Lifesciences, cat. no. 93070)
- cDNA Synthesis Kit Mint-2 (Evrogen, cat. no. SK005)
- Encyclo PCR Kit (Evrogen, cat. no. PK001)
- Phusion Hot Start Flex DNA Polymerase (New England BioLabs Inc., cat. no. M0535S)
- RQ1 RNase-Free DNase (Promega, cat. no. M6101)
- RNeasy Mini Kit (Qiagen, cat. no. 74104)

Equipment
- Standard equipment for eukaryotic cell culturing
- Nucleofector™ 2b Device (Lonza, AAB-1001) (optional, for difficult to transfect cell lines)
- Fluorescence microscope
- Flow cytometer or fluorescence-activated cell sorter
- Standard equipment and reagents for quantitative PCR

Software
- ImageJ software (National Institutes of Health)
- Software for flow cytometry data analysis (eg, Weasel Software: http://en.bio-soft.net/other/WEASEL.html; Flowing Software: http://www.flowingsoftware.com/)
- Spreadsheet processing software (eg, Microsoft Excel or Microcal Origin)

4. PROCEDURE
4.1 Preparation of Genetic Constructs

To create fluorescence reporter for splicing of an alternative exon of interest, vectors analogous to pSplPIG3v2 and pCtrlPIGv2 should be constructed. Careful design of the insert encoding a fragment of the gene of interest and fluorescent proteins is crucial for further successful action of the splicing reporter.

Regulation of alternative splicing is a complex and not fully understood process. It involves many short *cis*-acting sequence elements termed splicing enhancers and splicing silencers distributed over the alternative exon and nearby introns, as well as adjacent constitutive exons (Barash et al., 2010; Blencowe, 2006). In addition, general structural features of pre-mRNA, such as secondary structure and lengths of the exons and introns, can affect alternative splicing. Thus, to minimally perturb native splicing regulation of a gene of interest, a full-length unmodified gene fragment should be taken for the fluorescence reporter construction.

1. Choose a cassette alternative exon of interest. Ensure that its length is not a multiple of 3. If its length is a multiple of 3, addition of one to two nucleotides somewhere in the middle of this exon will be required.
2. Analyze sequence of introns and exons upstream and downstream of the alternative exon of interest. Ideally, the full-length unmodified gene fragment consisting of the upstream exon and intron, alternative exon, and downstream intron and exon should be used to construct splicing reporter. This would ensure including of virtually all splicing regulatory elements and thus a proper regulation of alternative splicing of the

minigene. If this fragment is too long for PCR amplification and cloning, the flanking introns and exons can be shortened. We recommend retaining of at least 300 bp adjacent to the splice sites of each intron and exon.
3. Design PCR primers for amplification of the gene fragment of interest and its further cloning into pCtrlPIG3v2 using restriction sites BglII and SalI (in lieu of PIG3 fragment). Using sequence of pCtrlPIG3v2 (GenBank KT831769), ensure that after cloning Katushka and TagGFP2 open reading frames coincide with that of the first and last exons of the target gene fragment.
4. Amplify the selected gene fragment from an available template, eg, genomic DNA sample or a vector carrying the gene of interest. Use a high-fidelity polymerase and hot start for specific and error-free amplification.
5. Using standard cloning protocols, insert the amplified gene fragment into pCtrlPIG3v2 digested with BglII and SalI. Select recombinant clones and verify correctness of insertion by sequencing. As a result, splicing reporter "pSpl-GOI" (gene of interest) is obtained.
6. Create corresponding control vector "pCtrl-GOI," which is identical to pSpl-GOI except absence of the introns. For this, use PCR amplification of cDNA copy of the target gene fragment with gene-specific primers on a template of cellular total cDNA. Alternatively, the intronless gene fragment can be generated by joining exon sequences by PCR (Matsumoto & Itoh, 2011). Vector pCtrl-GOI should ensure splicing-independent production of the full-length protein including both Katushka and TagGFP2.

4.2 Cell Culture and Transfection

Splicing reporter vectors (pSplPIG3v2 and pCtrlPIGv2, or newly constructed vectors pSpl-GOI and pCtrl-GOI, see Section 4.1) should be used to transfect mammalian cell lines, eg, HEK293 or HeLa. Cell culturing and transfection can be done using standard protocols. Each experiment should include at least two samples: cells transfected with a splicing-dependent vector (eg, pSplPIG3v2) and cells transfected with a corresponding splicing-independent control vector (eg, pCtrlPIGv2). See detailed description of cell culturing and transfection in Section 4.1 of the chapter "Analysis of Nonsense-Mediated mRNA Decay at the Single Cell Level Using Two Fluorescent Proteins" by Gurskaya et al.

4.3 Fluorescence Microscopy

Cells transfected with pSplPIG3v2 and pCtrlPIGv2 as well as analogous reporters for other target genes (pSpl-GOI and pCtrl-GOI) produce far-RFP Katushka and GFP TagGFP2. For the purpose of a preliminary analysis, these cells can be imaged using a wide-field fluorescence or a confocal microscope. Details on fluorescence microscopy of cell expressing these fluorescent proteins are described in Section 4.2 of the chapter "Analysis of Nonsense-Mediated mRNA Decay at the Single Cell Level Using Two Fluorescent Proteins" by Gurskaya et al. Fig. 3 shows an example of fluorescence microscopy of pSplPIG3v2- or pCtrlPIGv2-expressing HEK293 cells.

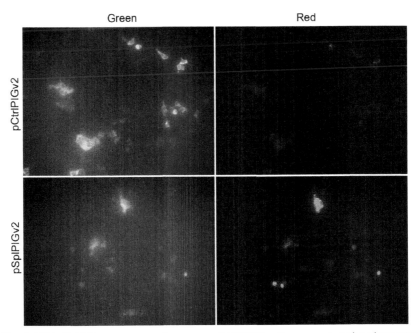

Fig. 3 Wide-field fluorescence microscopy of HEK293 cells expressing the alternative splicing reporter. *Top images*: pCtrlPIGv2-expressing cells. *Bottom images*: pSplPIGv2-expressing cells. Cells were imaged with BZ9000 fluorescence microscope (Keyence) using 20× objective and standard filter sets for *green* (excitation at 450–490, emission at 510–560 nm; *left panels*) and *red* (excitation at 540–580, emission at 600–660 nm; *right panels*) channels. The same settings were applied for the two cell samples. (See the color plate.)

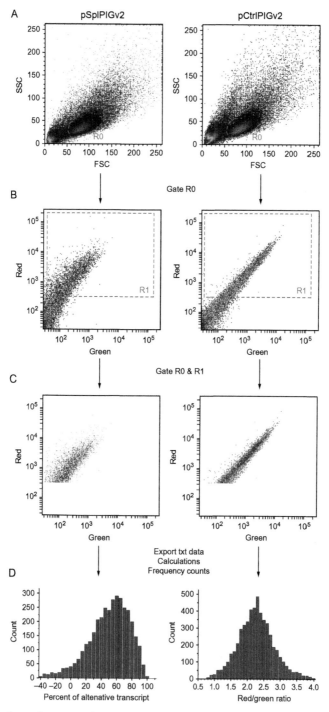

Fig. 4 See legend on opposite page.

4.4 Flow Cytometry

4.4.1 Flow Cytometry of Cell Samples

Procedures of preparation of live cell samples and their analysis by a flow cytometer of fluorescence-activated cell sorter are described in detail in Sections 4.3.1 and 4.3.2 of the chapter "Analysis of Nonsense-Mediated mRNA Decay at the Single Cell Level Using Two Fluorescent Proteins" by Gurskaya et al.

Cells expressing TagGFP2 or Katushka alone should be used to select detection parameters with no cross talk between TagGFP2 and Katushka fluorescent signals.

4.4.2 Analysis of Flow Cytometry Results

Typical results of PIG3 splicing reporter analysis with FACSAria III are shown in Fig. 4. This example is used below to show main steps of quantitative analysis of splicing reporter performance.

1. Use forward and side light-scattering values (FSC vs SSC plots) to select cell populations of appropriate size and structure (Fig. 4A, region R0).
2. Create bivariate green–red dot plots to observe cell fluorescence. Gate cells from R0. pCtrlPIGv2-expressing cells should form a concise diagonal corresponding to the presence of both red Katushka and green TagGFP2 proteins (Fig. 4B right). Be sure that this diagonal is not due to cross talk of the channels by analyzing the control cells expressing TagGFP2 or Katushka alone. pSplPIG3v2-expressing cells should form a diagonal distribution shifted to the left relatively to pCtrlPIGv2 control cells (Fig. 4B left).
3. Select a region R1 to select cells with medium and high level of fluorescence (Fig. 4B) and gate these cells into new dot plots (Fig. 4C).
4. Using appropriate software (eg, Weasel) export digital data in text format from these plots.

Fig. 4 Flow cytometry analysis of HEK293 cells expressing pSplPIGv2 (*left*) or pCtrlPIGv2 (*right*). Analysis was done by FACSAria III (Becton Dickinson) using 488 nm laser and 530/30 nm detector for TagGFP2, and 561 nm laser and 670/14 nm detector for Katushka. (A) *Dot plots* in forward scattering (FSC) and side scattering (SSC) channels. Region R0 was used to select live cells. (B) *Dot plots* in *green* and *red channels*. Region R1 was used to select well-expressing cells. (C) *Dot plots* of cells from R1 region. Digital data from these plots were exported in the text format to perform statistical analysis shown in (D). Cells showing negative percentage of alternative transcript arise from the area below the *middle line* of the control pCtrlPIGv2 diagonal. (See the color plate.)

5. Using a spreadsheet processing software (eg, Microsoft Excel or Microcal Origin) perform the following calculations on fluorescence intensities in the green and red channels for each cell. First, calculate red-to-green ratios in cells expressing pCtrlPIGv2; use mean value k of the ratio on the next step. Second, calculate a percentage of alternatively spliced transcript in each cell using the equation shown in Fig. 1. Estimate frequency counts for both data sets (Fig. 4D). The calculated distribution of alternative transcript abundance is usually quite broad (Fig. 4D left) that shows significant heterogeneity of cells in respect to alternative splicing regulation (Gurskaya et al., 2012; Shalek et al., 2013).

4.5 RNA Isolation and Quantitative RT-PCR

Optionally, reverse transcription and quantitative PCR analysis can be used to check the results obtained with splicing reporter. See Section 4.4 of the chapter "Analysis of Nonsense-Mediated mRNA Decay at the Single Cell Level Using Two Fluorescent Proteins" by Gurskaya et al. for more details on these steps.

In the case of pSplPIGv2, the total amount of the transcript (including both normally and alternatively spliced forms) can be assessed using primers specific to TagGFP2 (5′-agcggggcgaggagctgtt and 5′-cctgtacagctcgtccatgcc), whereas the amount of the normally spliced transcript can be measured using primers that are specific to the alternative exon 4 (5′-cgaggaagtctgatcacc) and reverse TagGFP2-specific primer (5′-cgctgaacttgtggccgt). Comparison of these two qPCR reactions allows estimating percentage of normal and alternative transcripts. For example, in experiment shown in Fig. 4, the average threshold cycles in the qPCR runs of TagGFP2 and exon 4 fragments were 34.2 and 36.3 cycles, respectively (2.1 cycles or 4.29-fold difference). Thus, exon 4-containing normally spliced transcript occupies 23% (100/4.29), whereas the alternative transcript occupies 67%.

5. CONCLUDING REMARKS

Fluorescence-based quantitative assessment of alternative splicing of an exon of interest opens new perspectives for studying regulation of gene expression in live models. Fluorescence reporter is useful to find out exonic and intronic splicing enhancers and splicing silencers, which regulate alternative splicing of the particular exon. Also, high-throughput screening of chemicals affecting alternative splicing can potentially be performed using green-to-red ratio of the reporter as a convenient readout. For heterogeneous cell samples, fluorescence-activated cell sorter can be applied to separate subpopulations with distinct regulation of alternative splicing and study these cells in detail. Finally, for complex multicellular models, such as tissue slices or small transparent embryos and animals, two-channel fluorescence microscopy can reveal spatial patterns of alternative splicing regulation.

ACKNOWLEDGMENTS
This work was supported by the Russian Science Foundation (Project 14-25-00129). The work was partially carried out using equipment provided by the IBCH Core Facility (CKP IBCH).

REFERENCES
Barash, Y., Calarco, J. A., Gao, W., Pan, Q., Wang, X., Shai, O., et al. (2010). Deciphering the splicing code. *Nature, 465,* 53–59.
Blencowe, B. J. (2006). Alternative splicing: New insights from global analyses. *Cell, 126,* 37–47.
Chang, Y. F., Imam, J. S., & Wilkinson, M. F. (2007). The nonsense-mediated decay RNA surveillance pathway. *Annual Review of Biochemistry, 76,* 51–74.
Gurskaya, N. G., Staroverov, D. B., Fradkov, A. F., & Lukyanov, K. A. (2011). Coding region of far-red fluorescent protein Katushka contains a strong donor splice site. *Russian Journal of Bioorganic Chemistry, 37,* 380–382.
Gurskaya, N. G., Staroverov, D. B., Zhang, L., Fradkov, A. F., Markina, N. M., Pereverzev, A. P., et al. (2012). Analysis of alternative splicing of cassette exons at single-cell level using two fluorescent proteins. *Nucleic Acids Research, 40,* e57.
Hiller, M., Huse, K., Platzer, M., & Backofen, R. (2005). Creation and disruption of protein features by alternative splicing—A novel mechanism to modulate function. *Genome Biology, 6,* R58.
Kriventseva, E. V., Koch, I., Apweiler, R., Vingron, M., Bork, P., Gelfand, M. S., et al. (2003). Increase of functional diversity by alternative splicing. *Trends in Genetics, 19,* 124–128.
Kuroyanagi, H. (2013). Switch-like regulation of tissue-specific alternative pre-mRNA processing patterns revealed by customized fluorescence reporters. *Worm, 2,* e23834.
Kuroyanagi, H., Kobayashi, T., Mitani, S., & Hagiwara, M. (2006). Transgenic alternative-splicing reporters reveal tissue-specific expression profiles and regulation mechanisms in vivo. *Nature Methods, 3,* 909–915.
Lewis, B. P., Green, R. E., & Brenner, S. E. (2003). Evidence for the widespread coupling of alternative splicing and nonsense-mediated mRNA decay in humans. *Proceedings of the National Academy of Sciences of the United States of America, 100,* 189–192.
Maquat, L. E. (2004). Nonsense mediated mRNA decay: Splicing, translation and mRNP dynamics. *Nature Reviews. Molecular Cell Biology, 5,* 89–99.
Matsumoto, A., & Itoh, T. Q. (2011). Self-assembly cloning: A rapid construction method for recombinant molecules from multiple fragments. *Biotechniques, 51,* 55–56.
Newman, E. A., Muh, S. J., Hovhannisyan, R. H., Warzecha, C. C., Jones, R. B., McKeehan, W. L., et al. (2006). Identification of RNA-binding proteins that regulate FGFR2 splicing through the use of sensitive and specific dual color fluorescence minigene assays. *RNA, 12,* 1129–1141.

Nicholls, C. D., & Beattie, T. L. (2008). Multiple factors influence the normal and UV-inducible alternative splicing of PIG3. *Biochimica et Biophysica Acta, 1779*, 838–849.

Ohno, G., Hagiwara, M., & Kuroyanagi, H. (2008). STAR family RNA-binding protein ASD-2 regulates developmental switching of mutually exclusive alternative splicing in vivo. *Genes & Development, 22*, 360–374.

Orengo, J. P., Bundman, D., & Cooper, T. A. (2006). A bichromatic fluorescent reporter for cell based screens of alternative splicing. *Nucleic Acids Research, 34*, e148.

Pereverzev, A. P., Gurskaya, N. G., Ermakova, G. V., Kudryavtseva, E. I., Markina, N. M., Kotlobay, A. A., et al. (2015). Method for quantitative analysis of nonsense-mediated mRNA decay at the single cell level. *Scientific Reports, 5*, 7729.

Resch, A., Xing, Y., Modrek, B., Gorlick, M., Riley, R., & Lee, C. (2004). Assessing the impact of alternative splicing on domain interactions in the human proteome. *Journal of Proteome Research, 3*, 76–83.

Shalek, A. K., Satija, R., Adiconis, X., Gertner, R. S., Gaublomme, J. T., Raychowdhury, R., et al. (2013). Single-cell transcriptomics reveals bimodality in expression and splicing in immune cells. *Nature, 498*, 236–240.

Shyu, A. B., Wilkinson, M. F., & van Hoof, A. (2008). Messenger RNA regulation: To translate or to degrade. *EMBO Journal, 27*, 471–481.

Somarelli, J. A., Schaeffer, D., Bosma, R., Bonano, V. I., Sohn, J. W., Kemeny, G., et al. (2013). Fluorescence-based alternative splicing reporters for the study of epithelial plasticity in vivo. *RNA, 19*, 116–127.

Stamm, S., Ben-Ari, S., Rafalska, I., Tang, Y., Zhang, Z., Toiber, D., et al. (2005). Function of alternative splicing. *Gene, 344*, 1–20.

Stamm, S., Zhu, J., Nakai, K., Stoilov, P., Stoss, O., & Zhang, M. Q. (2000). An alternative-exon database and its statistical analysis. *DNA & Cell Biology, 19*, 739–756.

Stoilov, P., Lin, C. H., Damoiseaux, R., Nikolic, J., & Black, D. L. (2008). A high-throughput screening strategy identifies cardiotonic steroids as alternative splicing modulators. *Proceedings of the National Academy of Sciences of the United States of America, 105*, 11218–11223.

Wang, Z., Rolish, M. E., Yeo, G., Tung, V., Mawson, M., & Burge, C. B. (2004). Systematic identification and analysis of exonic splicing silencers. *Cell, 119*, 831–845.

CHAPTER TWELVE

IRAS: High-Throughput Identification of Novel Alternative Splicing Regulators

S. Zheng[1]
University of California, Riverside, CA, United States
[1]Corresponding author: e-mail address: sika.zheng@ucr.edu

Contents

1. Introduction	270
2. Method Design	272
3. Construction of Dual-Fluorescence Minigene Reporters	276
3.1 Materials and Equipment	277
3.2 Protocol	277
3.3 Considerations	278
4. Generation of Stable Cell Clones	278
4.1 Materials and Equipment	279
4.2 Protocol	279
4.3 Considerations	279
5. Library and Array Construction	280
5.1 Materials and Equipment	280
5.2 Protocol	281
6. Cell-Based High-Throughput Screens	281
6.1 Materials and Equipment	281
6.2 Protocol	281
7. Data Acquisition	282
7.1 Imaging Resolution and Speed	282
7.2 The Uniformity of Fluorescence Signals Across Wells and Plates	283
7.3 Materials and Equipment	284
7.4 Protocol	284
8. Background Fluorescence Correction and Normalization	285
9. Data Analysis	285
10. Validation	287
11. Limitations and Perspectives	287
References	288

Abstract

Alternative splicing is a fundamental regulatory process of gene expression. Defects in alternative splicing can lead to various diseases, and modification of disease-causing splicing events presents great therapeutic promise. Splicing outcome is commonly affected by extracellular stimuli and signaling cascades that converge on RNA-binding splicing regulators. These *trans*-acting factors recognize *cis*-elements in pre-mRNA transcripts to affect spliceosome assembly and splice site choices. Identification of these splicing regulators and/or upstream modulators has been difficult and traditionally done by piecemeal. High-throughput screening strategies to find multiple regulators of exon splicing have great potential to accelerate the discovery process, but typically confront low sensitivity and low specificity of screening assays. Here we describe a unique screening strategy, IRAS (identifying regulators of alternative splicing), using a pair of dual-output minigene reporters to allow for sensitive detection of exon splicing changes. Each dual-output reporter produces green fluorescent protein (GFP) and red fluorescent protein (RFP) fluorescent signals to assay the two spliced isoforms exclusively. The two complementary minigene reporters alter GFP/RFP output ratios in the opposite direction in response to splicing change. Applying IRAS in cell-based high-throughput screens allows sensitive and specific identification of splicing regulators and modulators for any alternative exons of interest. In comparison to previous high-throughput screening methods, IRAS substantially enhances the specificity of the screening assay. This strategy significantly eliminates false positives without sacrificing sensitive identification of true regulators of splicing.

1. INTRODUCTION

Most mammalian multiexon genes produce multiple mRNA isoforms because of pre-mRNA alternative splicing, which dramatically increases transcriptome complexity and proteome diversity. Many pre-mRNA splicing patterns are highly regulated to produce functionally distinct gene products during development or in response to extracellular stimuli (Chen & Manley, 2009; Zheng & Black, 2013). The selection of alternative splice sites is typically determined by the interactions between *cis*-regulatory elements in the pre-mRNA and *trans*-acting protein factors that can affect spliceosome assembly (Black, 2003; Fu & Ares, 2014). A single alternative exon can contain multiple *cis*-elements in itself or in its surrounding introns, and thus can be regulated by multiple *trans*-factors. Meanwhile each *trans*-factor can control a multitude of target exons. Only tens of RNA-binding proteins have been described to modulate alternative splicing in eukaryotic cells. Given the myriad genes subject to alternative splicing, many splicing factors probably have not been identified and await discovery. Factors controlling the activity of known splicing regulators are also mostly unclear.

Genome-wide sequencing-based methods allow for identifying many targets of individual splicing factors (Chen & Zheng, 2009; Hartmann & Valcárcel, 2009; Witten & Ule, 2011); however, identifying individual positive and negative regulators of an alternative exon has been challenging and often involved laborious intensive workflow and needed to be done individually. Previous strategies using various cell-based screening techniques were able to identify few new factors that target exons, however, were overall insensitive for identifying a larger sets of regulators of splicing (Kar, Havlioglu, Tarn, & Wu, 2006; Oberdoerffer et al., 2008; Topp, Jackson, Melton, & Lynch, 2008; Wu, Kar, Kuo, Yu, & Havlioglu, 2006). Most of these strategies designed a single output of splicing, such that splicing changes were measured by changes in expression of one of the two isoforms which were then converted into either a fluorescence-based or a luminescence-based readout. These single-output reporters, however, would also directly measure the overall expression, in addition to splicing, of the reporter transcripts. Additionally, the alternative isoform generating the readout must be produced at a very low basal level for any changes in splicing to be measurable, which has occluded the application of reporters on exons spliced at intermediate levels. Because the single-output reporter has to be spliced at either very low or very high level for the screening strategy to work, individual screens can only identify activators or repressors of splicing, but not both. The use of a pair of single-output reporters that encoded two different fluorescent proteins could overcome some of these limitations (Kuroyanagi, Ohno, Sakane, Maruoka, & Hagiwara, 2010). However, the requirement for integrating two minigene reporters into genomes as well as the possibilities of multiple insertion loci and different copy numbers can impose difficulty in analyzing results.

A dual-output reporter, in which both splicing isoforms are assayed, allows for screening of changes in isoform ratio and can detect both increases and decreases in exon inclusion with higher sensitivity. This screening strategy allows targeted exons spliced at intermediate levels and can also reduce the amount of false positives that alter overall reporter expression independently of splicing. Because the expression of the two alternative isoforms is negatively correlated, using the output ratio of the two isoforms to measure splicing level squares the dynamic range given by individual output. One example of dual-output reporters contained open-reading frame (ORF) of green fluorescent protein (GFP) and red fluorescent protein (RFP) in each of the mutually exclusive exons (Kuroyanagi, Kobayashi, Mitani, & Hagiwara, 2006). Because the large GFP and RFP ORFs were in place of the alternative exons, this reporter was tailored to insert an intronic *cis*-element of interest and identify its cognate *trans*-factors and was not designed to capture most regulatory elements within and flanking the exon of interest.

Another dual-fluorescence splicing reporter was successfully used to identify chemicals modulating alternative splicing of MAPT exon 10 (Stoilov, Lin, Damoiseaux, Nikolic, & Black, 2008). This reporter (pflareA) contains tandem ORFs of GFP and RFP. The GFP ORF is split between two constitutive exons by the alternative cassette exon of interest with its flanking introns. The GFP ORF is initiated for translation when the alternative exon is skipped, and the RFP translation will not be initiated. When the alternative exon is included, the GFP ORF loses its start codon, and the downstream RFP is then translated. Some alternative exons may have in-frame start codon for the GFP reading frame. In that case, ATG start codons within the alternative exon are mutated. The RFP/GFP ratio can then be used as a proxy of splicing level and a change in RFP/GFP ratio indicates alteration of splicing. Because the GFP ORF is closer to the $5'$ end of the mRNA than the RFP ORF, modulators of translation would likely have an impact on the RFP/GFP ratio. As seen for many other screening strategies mentioned earlier, variables affecting the readout (ie, RFP/GFP ratio for this reporter) without an impact on pre-mRNA splicing constitute a majority of (often about 90%) hits.

In summary, genome-wide screens for splicing modulators of any single alternative exon confront a multitude of challenges including low sensitivity, narrow dynamic range, substantial false positives, low throughput, and limited applicability. Previous screening strategies using different types of splicing reporters have each addressed a few of these challenges to some degree. To provide a universal screening method broadly applicable to any cassette exon with high sensitivity and specificity, we have designed a novel strategy named IRAS, identifying regulators of alternative splicing, which significantly improves global identification of true positives (Fig. 1).

2. METHOD DESIGN

IRAS uses two complementary dual-output minigene reporters exhibiting opposite changes of the readouts in response to splicing alternation (Fig. 1A). These outputs can be fluorescence based or luminescence based. For the simplicity of description, we herein use fluorescence-based

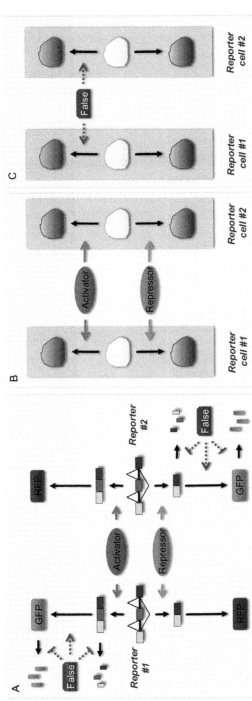

Fig. 1 IRAS strategy illustrated by dual-fluorescence reporters. (A) Two complementary reporters differ in several nucleotides. Reporter #1 translates GFP when the alternative exon is included and RFP when the alternative exon is skipped. Reporter #2 translates RFP from the included isoform and GFP from the skipped isoform. Therefore, the ratios of GFP/RFP or RFP/GFP represent the inclusion ratio of the alternative exon in reporters #1 and #2, respectively. Factors that alter these ratios without affecting the inclusion ratio are false positives in screens. Possible actions of false positives that increase GFP/RFP ratio are labeled in purple (dark gray in the print version). False positives that increase RFP/GFP ratio are not drawn. (B and C) Expected phenotypes induced by true splicing activators and repressors and false positives in screens using cells stably express either reporters #1 or #2. A true splicing regulator changes GFP/RFP ratios in the two reporter cells in the opposite direction, whereas a false positive more likely affects the GFP/RFP ratios in the same direction. False positives that increase RFP/GFP ratio are not drawn.

reporters to explain the concept of IRAS, as we have successfully screened with these reporters. For IRAS to work, one reporter produces RFP when an exon is included and GFP from the mRNA lacking the exon, whereas a second reporter produces GFP and RFP to represent exon inclusion and exclusion respectively. The first reporter is based on the pflareA minigene for which the RFP/GFP ratio represents the exon inclusion ratio (Fig. 2A). Any ATG start codons present within the alternative exon of interest and in frame with the GFP ORF need to be mutated for pflareA minigene to translate RFP when the exon is included. The second minigene (pflareG) reporter is constructed by having the start codon of GFP in the alternative exon such that when the alternative exon is included GFP is translated (Fig. 2B) (Zheng, Damoiseaux, Chen, & Black, 2013). This can be easily done if the alternative exon has an in-frame start codon. Otherwise, an ATG start codon for GFP within the alternative exon can be engineered by site-directed mutagenesis or insertion. When the alternative exon is skipped, GFP does not have an initiation codon and the downstream RFP ORF is used. Therefore, GFP/RFP ratio reflects the exon inclusion ratio in the pflareG reporter.

The idea behind IRAS is the following. Screening with individual reporters can identify hundreds and thousands of hits, many of which are false positives affecting the GFP/RFP ratio without altering the splicing

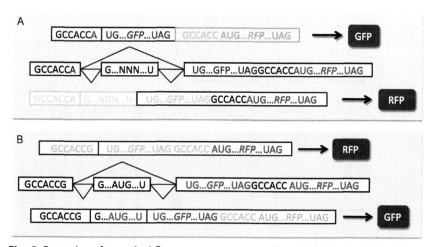

Fig. 2 Examples of two dual-fluorescence reporters that produce opposite splicing-dependent readouts. (A) Reporter translates GFP when the alternative exon is included and RFP when the alternative exon is skipped. (B) Reporter translates RFP from the included isoform and GFP from the skipped isoform.

of the exon. Some false positives may have preferentially stabilized one of the fluorescent proteins. Others may differentially affect ORF translation efficiency between GFP and RFP (Fig. 1A). A true splicing modulator should change the GFP/RFP ratios of the two reporters in the opposite direction (Fig. 1B). Because pflareA and pflareG minigenes differ in only a few nucleotides, a false-positive hit that alters GFP/RFP ratio of the first minigene without affecting splicing is very unlikely to change GFP/RFP ratio of the second minigene in the opposite direction. Instead, this false positive, if affecting only GFP ORF or RFP ORF, could change GFP/RFP ratios of the two reporters in the same direction (Fig. 1C). Using these two dual-fluorescence minigene reporters, we minimize systematic variations associated with fluorescent screening to enable more sensitive and accurate detection of moderate splicing changes. IRAS can simultaneously identify multiple activators and repressors of an alternative exon. Combining the screening results from these two reporters significantly eliminates false positives and enriches for identification of true splicing regulators.

Multiple technologies allow fluorescence-based selection of positive hits. Fluorescence-activated cell sorting (FACS) enriches cell population that deviates from control cells in fluorescence gating. The FACS procedure is relatively fast but requires subsequent profiling of the sorted cells, thus is most suitable for genetic screens and not for chemical screens. Genetic alterations (eg, cDNA overexpression or shRNA knockdown) within the sorted cells can then be amplified by PCR using library-specific primers and identified by next-generation sequencing. Another technique to select positive hits is fluorescence imaging. The versatility of fluorescence imaging has been aided by the development of cataloged libraries and high-throughput robotic liquid-handling systems which has enabled screening at greater depth and the identification of multiple regulators from a single screen. These libraries (eg, cDNA, shRNA, siRNA) are typically prearrayed in multiwell plates in a format of one well one known candidate; therefore, the identities of positive hits are known right after screening.

Here in detail we describe the application of IRAS using a gain-of-function genetic screen as an example (Fig. 3). In addition to cDNAs, IRAS can be used to screen libraries of siRNAs or shRNAs, and of small molecules, and we will discuss issues for different libraries. Overall the method described here greatly reduces false positives while maintaining high sensitivity in detecting regulators. This system allows for genome-wide screening for factors regulating splicing of any exon of interest.

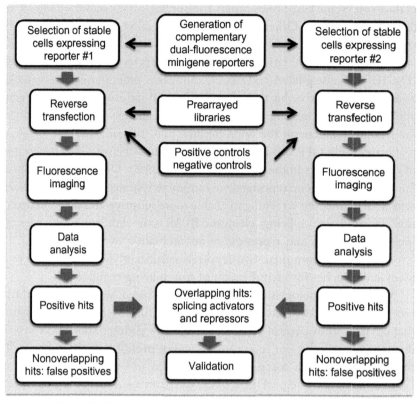

Fig. 3 Workflow of a fluorescent cell-based high-throughput IRAS genetic screen.

3. CONSTRUCTION OF DUAL-FLUORESCENCE MINIGENE REPORTERS

To construct the minigene reporters, the alternative exon and its flanking intronic sequences will need to be inserted into the EcoRI and BamHI sites of either the pflareA or pflareG vectors. The selection of intronic sequence can be optimized with prior knowledge. Conserved intronic sequence is usually indicative of regulation and is recommended to be included (Chen & Zheng, 2008). In most cases, different length of intronic sequences is chosen to generate multiple versions of minigenes. The one with a mid-range inclusion ratio is recommended because this allows screening for both positive and negative regulators simultaneously.

3.1 Materials and Equipment

- Pflare A reporter and pIfareG reporter
- Phusion High-Fidelity DNA polymerase (NEB)
- Purified genomic DNA containing the gene and exon of interest
- QuikChange II XL Site-Directed Mutagenesis Kit (Catalog# 200521, Agilent)
- QIAquick Gel Extraction Kit (Qiagen)
- NanoDrop (ThermoFisher Scientific)
- Alkaline phosphatase (NEB)
- T4 DNA Ligase (NEB)
- TOP10 Competent Cells (ThermoFisher Scientific)
- *Eco*RI (NEB)
- *Bam*HI (NEB)
- LB-kanamycin agar plate
- Plasmid Miniprep Kit (Qiagen)
- Agarose gel electrophoresis setup

3.2 Protocol

1. Design forward and reverse primers for cloning the alternative exon of interest and its flanking introns.
2. Synthesize the forward primer with an *Eco*RI recognition site at the 5′ end. Synthesize the reverse primer with a *Bam*HI recognition site at the 5′ end.
3. PCR amplification using the above forward and reverse primers, Phusion High-Fidelity DNA polymerase, and genomic DNA as the template.
4. Confirm the PCR amplicon by agarose gel electrophoresis.
5. Excise the band of the PCR amplicon and purify the PCR product with QIAquick Gel Extraction Kit.
6. Measure the concentrations of the recovered PCR products with NanoDrop.
7. Restriction enzyme double digestion of the PCR product with *Eco*RI and *Bam*HI for 1 h at 37°C. Meanwhile linearize the pflareA and pflareG vectors with *Eco*RI and *Bam*HI for 1 h at 37°C. Dephosphorylate the linearized vectors with alkaline phosphatase for 0.5 h at 37°C.
8. Confirm the linearized backbone vectors by agarose gel electrophoresis.

9. Gel-purify the digested PCR products and vectors using QIAquick Gel Extraction Kit.
10. Ligate the PCR product to the linearized vector with T4 DNA ligase at 16°C overnight.
11. Transform the ligated products into TOP10 competent cells.
12. Plate the transformed competent cells on LB-kanamycin agar plates and incubate at 37°C overnight.
13. Select several colonies for amplification and purification of plfareA-exon and plfareG-exon reporters using Plasmid Miniprep Kit.
14. For the plfareA-exon vector, mutate any ATG codons present within the alternative exon of interest to ATA using the QuikChange II XL Site-Directed Mutagenesis Kit.
15. For the pflareG-exon reporter, an ATG start codon within the alternative exon in frame with the GFP ORF is required for expression of GFP when the exon is included. If missing, a start codon needs to be created using the QuikChange II XL Site-Directed Mutagenesis Kit. The Kozak consensus sequence gccRccAUGG is also recommended to promote translation initiation for the GFP ORF.

3.3 Considerations

After construction the reporters should be transfected in the cells of choice to confirm their applicability for the screen. The choice of cells will be discussed later. The two reporters should exhibit similar splicing levels of the inserted alternative exons. They should also express both GFP and RFP but with opposite preference. For example, if one expresses a high level of GFP and a low level of RFP, the other one should express a high level of RFP and a low level of GFP. Any known positive or negative modulator of the inserted alternative exon should also be tested on the reporters for its ability to change the GFP/RFP ratio, and more importantly in the opposite direction for the two reporters. Such tests not only examine the performance of the reporters but also confirm a known modulator suitable as a spike-in control in the final screen.

4. GENERATION OF STABLE CELL CLONES

Two criteria need to be met for the parental cell line. First, the cell line ideally needs to contain characteristics of or be derived from the primary cells of interest, so that the identified hits are more likely physiologically relevant. This is particularly important for loss-of-function screens (eg, with shRNA or siRNA libraries) where the most interesting cell-type-specific splicing factors to be targeted are sometimes expressed only in a highly relevant cell line. Second, for genetic screens, the cell line needs to be highly transfectable to increase the signal-to-noise ratio of the fluorescence readouts, because both transfected and untransfected cells within a well contribute to the final fluorescence signals.

4.1 Materials and Equipment
- A parental cell line of choice
- QIAquick Gel Extraction Kit (Qiagen)
- DraIII (NEB)
- Lipofectamine 2000 (ThermoFisher Scientific)
- G418 or Geneticin (ThermoFisher Scientific)
- Phenol red-free media (ThermoFisher Scientific)
- FACS machine

4.2 Protocol
1. Linearize the pflareA-exon vector by DraIII digestion and gel purify the vector with QIAquick Gel Extraction Kit.
2. Transfect the linearized pflareA-exon vector into the target cell line of choice using Lipofectamine 2000.
3. Maintain the transfected cells in G418-containing media for 2 weeks to enrich stable cell clones.
4. Sort double fluorescent (ie, GFP+RFP+) single cells to individual wells of multiwell plates with a FACS machine and allow cell clones to grow and expand. Some cell clones may lose GFP and RFP expression during the selection process and can be discarded. From now on, the cells need to be maintained in phenol red-free media for enhanced fluorescence visualization and imaging.
5. Repeat steps 1–4 to generate stable cell clones expressing pflareG-exon reporter.

4.3 Considerations
The stable cell clones need to be tested for their responsiveness to splicing changes before high-throughput screens. An appropriate clone should not change GFP/RFP ratio when transfected with an empty control vector and should do so when transfected with a known regulator of the exon. An RT-PCR assay to measure the splicing changes at the RNA level is

highly recommended. This step will narrow down the number of appropriate cell clones.

We tested the "transfectability" of each selected stable cell clone using a GFP expression plasmid. A highly transfectable cell clone would show very high GFP signals without affecting the RFP signals. We also compared different transfection reagents and optimized transfection reagent to GFP plasmid ratios as well as cell densities of selected clones to finalize the transfection conditions for the screen. Once an optimal stable cell clone was identified, we expanded them in a large scale.

5. LIBRARY AND ARRAY CONSTRUCTION

Although IRAS is broadly applicable to screen any library, considerations should be taken in light of the library attributes. For example, autofluorescent chemicals interfere with the readouts and are technically not "screenable," even although these false-positive ones can be filtered by IRAS. Loss-of-function screens depend on the specificity of shRNA and siRNA. Multiple shRNA or siRNA hits targeting the same gene are often required to declare a true target. Many splicing regulators have paralogs that cross-regulate each other, such that RNAi depletion of one paralog induces expression of the other paralogs. Paralogous splicing regulators often have similar effects on a target exon, and thus need to be all depleted for a significant splicing change of the target exon. As a result, these regulatory paralogs may not be easily identified in loss-of-function screens. Loss-of-function screens may also miss true regulators that are not expressed in the screened cells. Gain-of-function screen has less concern on off-target effects, paralogous compensation, or missing expression, but overexpression may induce artificial splicing changes. Therefore, the physiological relevance of the hits will require loss-of-function validation. In this report, we use gain-of-function screens to demonstrate the preparation of the arrayed library.

5.1 Materials and Equipment

- The mammalian gene collection (MGC) library: developed by NIH and representing the most extensive collection of mammalian cDNA clones available. All cDNA clones are in the pCMVsport6.0 or pCMVsport6.1 backbone vectors prearrayed in 96-well plates (Life Technology).
- Genetix Qbot (Molecular Devices)
- Biomek FX robot (Beckman Coulter)

5.2 Protocol

1. Prepare an "assay ready" cDNA library by duplicating the MGC collection bacteria plates into 384-well plates using a Genetix Qbot (Molecular Devices). A library of about 16 thousand clones would occupy 45 384-well plates.
2. Prepare the plasmid DNA of the library using plasmid prep kit (Macherey-Nagel) on a Biomek FX robot (Beckman Coulter).
3. Normalize plasmid concentration and spot them into assay plates. We typically spot 40 ng cDNA per well (except the wells of A-H23 and A-H24) in the 384-well plate format. We also spot 40 ng negative control (pCMV sport6.0) plasmids to wells of C23, C24, D23, and D24; 40 ng plasmids of a positive regulator (if available) to wells of E23, E24, F23, and F24; as well as a negative regulator (if available) to wells of G23, G24, H23, and H24 (16).
4. Two identical sets of "assay ready" library plates are prepared for pflareA and pflareG screens, respectively. The assay plates can take days to prepare and should be kept in $-80°C$ until use.

6. CELL-BASED HIGH-THROUGHPUT SCREENS

The number of cells and the transfection conditions need to be optimized. We have been successful in reverse-transfecting 8000 N2a cells per 384-well with 40 ng plasmid DNA and 0.12 µL Lipofectamine 2000. Typically, about 150 million cells and 2 mL Lipofectamine 2000 are needed to screen about 16 thousand cDNA candidates once.

6.1 Materials and Equipment

- Trypsin (ThermoFisher Scientific)
- Report cell lines
- Arrayed library plates
- Lipofectamine 2000 (ThermoFisher Scientific)
- Opti-MEM (ThermoFisher Scientific)
- Multidrop 384 (Thermo LabSystems)

6.2 Protocol

1. Scale up the reporter cell cultures to 10 15-cm culture dishes by the day of performing the screens.

2. Thaw the cDNA plates while the reporter cells and transfection reagents are being prepared.
3. Dissociate the cells by trypsin.
4. Resuspend and dilute the cells in phenol red-free feeding media to be 400 thousand cells per mL.
5. Mix 2 mL Lipofectamine 2000 with 166 mL Opti-MEM.
6. Dispense 10 μL Opti-MEM containing 0.12 μL Lipofectamine 2000 into each well with Multidrop 384 to mix with the plasmid DNA in the plates.
7. Dispense 8000 reporter cells in 20 μL feeding media to every well except A23, A24, B23, and B24 wells using Multidrop 384.
8. Dispense cell-free feeding media into A23, A24, B23, and B24 of all plates. These four wells will be used to derive a scale factor for interplate normalization (see later).
9. Dispense cell-free media to one (or multiple) new 384-well plate as a background plate to measure background GFP and RFP fluorescence intensity of the cell media.

7. DATA ACQUISITION

Multiple fluorescence imagers can be used to obtain the fluorescence signals. We used a Typhoon imager rather than a conventional fluorescence microplate reader. The confocal optics of the Typhoon imager detects light from a limited depth at the bottom of the culture well. This yields an improved signal-to-noise ratio by reducing background signals from media above the cells. Each data point gained from Typhoon imaging represents overall fluorescence intensity of the whole population of cells within individual wells. By contrast, high-content fluorescence imager produces richer data including fluorescence intensity, cell shape, size, and cell count at a single-cell resolution. Despite the versatility of a high-content imager, its imaging time and data processing speed dramatically decrease the throughout and much of its data are probably irrelevant in evaluating splicing changes.

7.1 Imaging Resolution and Speed

Typhoon imagers use laser scanning technology. The imaging resolution and speed are intrinsically inversely correlated. Imaging at a higher resolution generates better signal-to-noise ratio, but slow down scanning and increase the lagging time of image acquisition between the first and the last plate. We have optimized the scanning speed and resolution. The Typhoon 9410 can scan at pixel sizes of 10, 25, 50, 100, 200, 500, and 1000 μm. The corresponding pixel numbers per well and total scanning times for forty-five 384-well plates containing the MGC cDNA library are listed in Table 1. To examine the effect of pixel size, we measured the correlation of signal intensities acquired at different pixel sizes. We have found that the dynamic range of fluorescence changes for our reporter cells can be fully captured at 200 μm, probably because the signal-to-noise ratio starts to plateau at the resolution of 200 μm pixels (or 256 pixels per well). Scanning at larger pixel sizes, eg, 500 μm (or 49 pixels per well) significantly increased data variance at low fluorescence intensities. At a pixel size of 200 μm, the whole MGC library can be scanned within 100 min, which is a reasonable lagging time between the first and last plate. Therefore, 200 μm is optimal for 384-well plates. To meet the similar data quality, 100 μm pixel size should be needed to screen 1536-well plates.

7.2 The Uniformity of Fluorescence Signals Across Wells and Plates

Because all candidates are compared to control wells, systematic noises across wells and plates should be kept at minimum relative to the signals, ie, changes in fluorescence intensity induced by splicing alternation. The sources of systematic noises include plates, wells, the scanner, media, and cell variability. To measure the systematic variance of the fluorescence signals, we added the same amount of media to multiple 384-well plates and measured fluorescence intensities of each well at different position of the scanner,

Table 1 Scanning Pixel Size of Typhoon Imager Determines Signal Intensity and Scanning Speed

Pixel Size (μm)	Number of Pixels Per Individual 384-Well	Total Scanning Time for the MGC Plates (min)
10	102,400	2000
25	16,384	1600
50	4096	400
100	1024	210
200	256	100
500	49	50
1000	16	30

which allowed us to quantify noises derived from plates, wells and the scanner. Satisfyingly, each individual wells varied in fluorescence signals by only 4–7% at different positions on the scanner. Therefore, scanning positions and the scanner only contribute minor variance to the system.

Surprisingly the largest contributor to systemic noises is the well-to-well variance, as the fluorescence intensities fluctuate significantly between wells. In particular, wells on the edge of a plate had (up to 30%) higher signals than inner wells for both GFP and RFP. Such an edge effect was likely due to faster media evaporation along the edges, because the fluorescence difference amplified over time. However, the edge effect was very reproducible across plates. Indeed, wells of the same coordinates varied in fluorescence intensity by only about 5% between plates. In summary, when we compared uniform plates containing the same cell-free media, the fluorescence signal from a specific well coordinate was much similar to the same coordinate in a different plate than to a different coordinate in the same plate.

We also compared well intensities between cell-free media and media-containing parental nonfluorescent N2a cells. The cellular autofluorescence of the parental N2a cells was about on average 1.9% and 6.6% the level of the media fluorescence in the GFP channel and the RFP channel, respectively. Therefore, the native cellular autofluorescence is negligible.

7.3 Materials and Equipment

- Typhoon laser scanner equipped with lasers and filters for GFP and RFP (GE Healthcare Life Sciences, PA, USA)
- ImageQuant TL Software (GE Healthcare Life Sciences, PA, USA)

7.4 Protocol

1. Fourty-eight hours after transfection, arrange and scan the plates on the Typhoon scanner. GFP signals are acquired using the blue laser (488 nm) and 520 nm band pass emission filter. RFP signals are obtained using the green laser (532 nm) and 610 nm band pass emission filter. Keep the scanning parameters (including the PMT voltages) consistent across all plates.
2. Use ImageQuant TL software to analyze the scanned images (.gel files) and obtain raw GFP and RFP fluorescence intensity of every well. Export the GFP/RFP signal values into Data/Text/Excel files.

8. BACKGROUND FLUORESCENCE CORRECTION AND NORMALIZATION

Because the largest source of background comes from the coordinates of the wells rather than the order of the plates or the scanning position, we perform interplate well-by-well background correction using a background plate that contains uniform cell-free media. Nevertheless, the variance originated from plates and scanning positions averaged around 7% and should be taken into consideration. Therefore, we normalized all plates before background correction.

To normalize the plates, we purposefully dispense cell-free media into wells A23, A24, B23, and B24 of each plate to derive a scaling factor for that plate. The GFP and RFP signal intensity of each of the four media-alone wells (A23, A24, B23, and B24) are acquired similarly as any other wells in all cell plates and the background plate. For example, XFP_{ijk} is the raw GFP or RFP intensities of a well at row "i" column "j" in plate k. We calculate the scale factor S_k for plate k based on the A23, A24, B23, and B24 wells by the following equation:

$$S_k = \frac{1}{4} \sum_{\substack{p=A,B \\ q=23,24}} \frac{\overline{\mathrm{XFP}}_{pq}}{\mathrm{XFP}_{pqk}}$$

where $\overline{\mathrm{XFP}}_{pq}$ is the average values across all the plates. The normalized and background-corrected GFP and RFP expression values (XFP_x) of the stable cells after transfection with cDNA x at row "i" column "j" in plate c are calculated with the following equation:

$$\mathrm{XFP}_x = \mathrm{XFP}_{ijc} S_c - \frac{1}{n_b} \sum_{b=1}^{n_b} \mathrm{XFP}_{ijb} S_b$$

where S_c and S_b are the scale factors for the cell plates and the background plates, respectively, and n_b is the total number of background plates used.

9. DATA ANALYSIS

To estimate the splicing ratio of the reporter upon expression of cDNA x, we use $\frac{\mathrm{GFP}x}{\mathrm{RFP}x}$ for the pflareG minigene reporter and $\frac{\mathrm{RFP}x}{\mathrm{GFP}x}$ for the pflareA minigene reporter. To calculate the basal splicing level of the reporter, we used the four wells (ie, C23, C24, D23, and D24) that were transfected with empty control vector in every plate and derive the mean $\frac{\text{GFP}_{\text{ctrl}}}{\text{RFP}_{\text{ctrl}}}$ in the pflareG-exon cells and the mean $\frac{\text{RFP}_{\text{ctrl}}}{\text{GFP}_{\text{ctrl}}}$ in the pflareA-exon cells. To examine the effect of cDNA x on the splicing of the pflareG reporter, we used the following formula to approximate the change in the splicing ratio (M_x):

$$M_x = \log_2\left[\frac{\text{GFP}_x}{\text{RFP}_x}\right] - \log_2\left[\frac{\text{GFP}_{\text{ctrl}}}{\text{RFP}_{\text{ctrl}}}\right]$$

where $\frac{\text{GFP}_{\text{ctrl}}}{\text{RFP}_{\text{ctrl}}}$ was derived from the same plate as cDNA$_x$. Similarly, to examine the action of cDNA x on the splicing of the pflareA reporter, we calculated the change in the splicing ratio (M_x) as followed:

$$M_x = \log_2\left[\frac{\text{RFP}_x}{\text{GFP}_x}\right] - \log_2\left[\frac{\text{RFP}_{\text{ctrl}}}{\text{GFP}_{\text{ctrl}}}\right]$$

where $\frac{\text{RFP}_{\text{ctrl}}}{\text{GFP}_{\text{ctrl}}}$ was derived from the same plate as cDNA$_x$. A value of $M_x > 0$ indicates a possible increase in splicing by cDNA$_x$. A value of $M_x > 0$ indicates a possible decrease in splicing. To assess the overall fluorescence expression of cells after transfection, or the "A" value, we use the following equation:

$$A_x = \frac{1}{2}(\log_2(\text{GFP}_x) + \log_2(\text{RFP}_x))$$

Note that multiple variables, including live cell numbers, transcriptional activity, and translational activity of the reporter, can have an impact on A values. The M values, indicative of splicing changes, can be scatter plotted against the A values to visualize the performance of the whole library.

With the R package locfdr, we calculated z score and local false discovery rate (FDR) for individual cDNA by comparing their M values to the population distribution of M values. The cutoff of the M values in calling a positive hit, as usual, could be arbitrary. We have found that a typical FDR threshold of 0.05 has yielded relatively high sensitivity (80–88%) for our reporter screens.

The two parallel screens using pflareA and pflareG reporters produce two different sets of possible splicing regulators. A simple overlap of the two target sets will eliminate a large proportion of false positives. A gene identified by both screens is much more likely to be the real regulators. In our past experience, overlapping candidates constitute about 5–20% of total hits and the sensitivity of the combined screens is roughly equal to the product of the sensitivity of the individual screens.

10. VALIDATION

The overlapping hits can be subjected to various kinds of secondary validation and gene enrichment analysis. RNA splicing factors and RNA-binding proteins are expected to be enriched in comparison to the library input. More importantly, the candidates need to be examined for their actions on the splicing of the endogenous target exon, because minigene reporters do not necessarily recapitulate the regulation of endogenous exons. For example, the minigene may not contain all of the relevant *cis*-regulatory elements, or the heterologous genomic context may alter its splicing pattern. Cells used for screening may have different transcriptomes from the cells in the organism. Thus, molecules that affect the splicing of the minigene reporter will need to be confirmed on endogenous transcripts in the correct cell type in subsequent analyses. This can be achieved by either overexpressing the cDNA candidate or RNAi depletion of the candidate gene followed by RT-PCR assay of the endogenous spliced variants. Note that gain-of-function and loss-of-function validations have various limitations as mentioned previously.

11. LIMITATIONS AND PERSPECTIVES

Fluorescence-based screening assays have both advantages and disadvantages over luminescence-based screening assays. Chemiluminescence assays typically have a much lower background and allow for a larger dynamic range of screening but require additional steps of cell lysis and enzymatic reactions before data acquisition. Fluorescence-based screening by contrast is noninvasive and allows temporal profiling of the readouts. Although we have used a fluorescence-based genetic screen as an illustrating example, luminescence-based minigenes can be designed in certain circumstances.

Several features of this screen require careful attention. First, reporter minigene expression can vary substantially among stable cell clones. Cell clones of higher expression are not necessarily better, because they may not be as efficiently transfectable or responsive to known regulators, as we have seen. Furthermore, some clones can change GFP/RFP ratios even when transfected with an empty control vector. Therefore, multiple clonal lines should be tested to find the optimal one. Second, the screening parameters and the kinetics of reporter expression need to be examined for screening different libraries or in different cell types. For example, we have seen that cDNA overexpression plateaus at 2 days posttransfection but RNAi depletion typically requires 3 days or longer. To avoid overconfluency within a well, the number of cells initially plated needs to be optimized; this can affect the transfection condition.

The IRAS system can accommodate nearly any cassette exon and its flanking introns. However, mutagenesis in one or both reporters is often required and must be assessed carefully, as it may generate or alter *cis*-acting splicing regulatory elements. Because a mutation is introduced into only one of the two vectors, a gain of false positives and/or a loss of true positives will manifest in only one of the two screens. Since the candidate regulators are selected from the overlap from both screens, the mutagenesis, if causing any effect, will most likely lead to a loss of true positives rather than the gain of false positives. Finally, altering the amount of flanking intron sequence included with the test exon can affect the basal splicing level of the reporter. A minigene exhibiting an intermediate level of exon inclusion (eg, 15–85%) will allow more effective simultaneous screening for both activators and repressors.

REFERENCES

Black, D. L. (2003). Mechanisms of alternative pre-messenger RNA splicing. *Annual Review of Biochemistry*, 72, 291–336. http://dx.doi.org/10.1146/annurev.biochem.72.121801.161720.

Chen, M., & Manley, J. L. (2009). Mechanisms of alternative splicing regulation: Insights from molecular and genomics approaches. *Nature Reviews. Molecular Cell Biology*, 10(11), 741–754. http://dx.doi.org/10.1038/nrm2777.

Chen, L., & Zheng, S. (2008). Identify alternative splicing events based on position-specific evolutionary conservation. *PLoS One*, 3(7), e2806. http://dx.doi.org/10.1371/journal.pone.0002806.

Chen, L., & Zheng, S. (2009). Studying alternative splicing regulatory networks through partial correlation analysis. *Genome Biology*, 10(1), R3. http://dx.doi.org/10.1186/gb-2009-10-1-r3.

Fu, X., & Ares, M., Jr. (2014). Context-dependent control of alternative splicing by RNA-binding proteins. *Nature Reviews. Genetics*, 15(10), 689–701. http://dx.doi.org/10.1038/nrg3778.

Hartmann, B., & Valcárcel, J. (2009). Decrypting the genome's alternative messages. *Current Opinion in Cell Biology*, 21(3), 377–386. http://dx.doi.org/10.1016/j.ceb.2009.02.006.

Kar, A., Havlioglu, N., Tarn, W., & Wu, J. Y. (2006). RBM4 interacts with an intronic element and stimulates tau exon 10 inclusion. *The Journal of Biological Chemistry*, 281(34), 24479–24488. http://dx.doi.org/10.1074/jbc.M603971200.

Kuroyanagi, H., Kobayashi, T., Mitani, S., & Hagiwara, M. (2006). Transgenic alternative-splicing reporters reveal tissue-specific expression profiles and regulation mechanisms in vivo. *Nature Methods*, *3*(11), 909–915. http://dx.doi.org/10.1038/nmeth944.

Kuroyanagi, H., Ohno, G., Sakane, H., Maruoka, H., & Hagiwara, M. (2010). Visualization and genetic analysis of alternative splicing regulation in vivo using fluorescence reporters in transgenic *Caenorhabditis elegans*. *Nature Protocols*, *5*(9), 1495–1517. http://dx.doi.org/10.1038/nprot.2010.107.

Oberdoerffer, S., Moita, L. F., Neems, D., Freitas, R. P., Hacohen, N., & Rao, A. (2008). Regulation of CD45 alternative splicing by heterogeneous ribonucleoprotein, hnRNPLL. *Science (New York, NY)*, *321*(5889), 686–691. http://dx.doi.org/10.1126/science.1157610.

Stoilov, P., Lin, C., Damoiseaux, R., Nikolic, J., & Black, D. L. (2008). A high-throughput screening strategy identifies cardiotonic steroids as alternative splicing modulators. *Proceedings of the National Academy of Sciences of the United States of America*, *105*(32), 11218–11223. http://dx.doi.org/10.1073/pnas.0801661105.

Topp, J. D., Jackson, J., Melton, A. A., & Lynch, K. W. (2008). A cell-based screen for splicing regulators identifies hnRNP LL as a distinct signal-induced repressor of CD45 variable exon 4. *RNA (New York, NY)*, *14*(10), 2038–2049. http://dx.doi.org/10.1261/rna.1212008.

Witten, J. T., & Ule, J. (2011). Understanding splicing regulation through RNA splicing maps. *Trends in Genetics: TIG*, *27*(3), 89–97. http://dx.doi.org/10.1016/j.tig.2010.12.001.

Wu, J. Y., Kar, A., Kuo, D., Yu, B., & Havlioglu, N. (2006). SRp54 (SFRS11), a regulator for tau exon 10 alternative splicing identified by an expression cloning strategy. *Molecular and Cellular Biology*, *26*(18), 6739–6747. http://dx.doi.org/10.1128/MCB.00739-06.

Zheng, S., & Black, D. L. (2013). Alternative pre-mRNA splicing in neurons: Growing up and extending its reach. *Trends in Genetics: TIG*, *29*(8), 442–448. http://dx.doi.org/10.1016/j.tig.2013.04.003.

Zheng, S., Damoiseaux, R., Chen, L., & Black, D. L. (2013). A broadly applicable high-throughput screening strategy identifies new regulators of Dlg4 (Psd-95) alternative splicing. *Genome Research*, *23*(6), 998–1007. http://dx.doi.org/10.1101/gr.147546.112.

CHAPTER THIRTEEN

Analysis of Nonsense-Mediated mRNA Decay at the Single-Cell Level Using Two Fluorescent Proteins

N.G. Gurskaya*[,][†], A.P. Pereverzev*, D.B. Staroverov*, N.M. Markina*, K.A. Lukyanov*[,][†][,1]

*Shemyakin-Ovchinnikov Institute of Bioorganic Chemistry, Moscow, Russia
[†]Nizhny Novgorod State Medical Academy, Nizhny Novgorod, Russia
[1]Corresponding author: e-mail address: kluk@ibch.ru

Contents

1. Introduction	292
2. Description of the Method	294
3. Materials	296
3.1 Reagents	296
3.2 Equipment	297
3.3 Software	297
4. Procedure	297
4.1 Cell Culture and Transfection	297
4.2 Fluorescence Microscopy	300
4.3 Flow Cytometry	301
4.4 RNA Isolation and Quantitative RT-PCR	305
4.5 Inhibitors of NMD Activity	308
5. Further Possible Modifications of Genetic Constructs	309
5.1 Changes of the Vector Backbone	309
5.2 Replacement of Fluorescent Indicators	310
5.3 Introduction of a new NMD-Inducing Fragment	311
6. Concluding Remarks	311
Acknowledgments	312
References	312

Abstract

Nonsense-mediated mRNA decay (NMD) is an evolutionarily conserved mechanism of specific degradation of transcripts with a premature stop codon. NMD eliminates aberrant mRNAs arising from mutations, alternative splicing, and other events in cells. In addition, many normal transcripts undergo NMD. Recent studies demonstrated that

NMD activity is specifically regulated and that NMD can play a role of global regulator of gene expression. Recently, we developed dual-color fluorescent protein-based reporters for quantification of NMD activity using fluorescence microscopy and flow cytometry (Pereverzev, Gurskaya, et al., 2015). Due to ratiometric fluorescence response, these reporters make it possible to assess NMD activity in live cells at the single-cell level and to reveal otherwise hidden heterogeneity of cells in respect of NMD activity. Here we provide a detailed description of applications of the NMD reporters in mammalian cell lines.

1. INTRODUCTION

Nonsense-mediated mRNA decay (NMD) is a ubiquitous mechanism of degradation of transcripts with a premature termination codon (PTC) (Behm-Ansmant et al., 2007; Bhuvanagiri, Schlitter, Hentze, & Kulozik, 2010; Chang, Imam, & Wilkinson, 2007). NMD eliminates aberrant mRNAs derived from gene mutations, alternative splicing, DNA rearrangements, and other sources (Huang & Wilkinson, 2012; Lewis, Green, & Brenner, 2003). NMD-targeted transcripts potentially can lead to harmful C-truncated proteins with altered biochemical properties if not degraded (Bhuvanagiri et al., 2010).

In addition, recent data demonstrate that NMD activity undergoes specific regulation, and thus this cascade functions as an important mechanism of global gene expression regulation (Lykke-Andersen & Jensen, 2015; Mendell, Sharifi, Meyers, Martinez-Murillo, & Dietz, 2004; Rehwinkel, Letunic, Raes, Bork, & Izaurralde, 2005). The regulated physiological transcripts include mRNAs involved in basic biological processes such as embryonic development, cell differentiation, carcinogenesis, and stress survival (Bruno et al., 2011; Colak, Ji, Porse, & Jaffrey, 2013; Gardner, 2008, 2010; Huang et al., 2011; Hwang & Maquat, 2011; Schoenberg & Maquat, 2012; Wang et al., 2011). Known regulation mechanisms of NMD activity include intracellular calcium level and specific microRNA (Bruno et al., 2011; Nickless et al., 2014). In particular, the recent study has elucidated the role of specific NMD downregulation in neurogenesis and brain development (Bruno et al., 2011). The emerging concept of the regulatory importance of NMD calls for the development of new methods for the noninvasive quantitative monitoring of NMD activity.

NMD functions in all eukaryotes, but the mechanism of PTC recognition and target mRNA degradation varies (Nicholson & Mühlemann, 2010). According to the unified model of "faux" 3′UTRs, NMD is triggered by an abnormal composition of mRNA–protein complexes downstream of the terminating ribosome (Rebbapragada & Lykke-Andersen, 2009; Schweingruber, Rufener, Zünd, Yamashita, & Mühlemann, 2013). In yeasts and invertebrates, NMD targets are marked by a long 3′UTR that is not bound by a required set of proteins for proper translation termination and can recruit an NMD-triggering protein UPF1 (ie, the PTC recognition is splicing independent). In addition to this basic mechanism, vertebrates possess a splicing-dependent pathway of PTC recognition, where stop codons >50 nucleotides upstream of the last exon–exon junction are recognized as PTCs (Schweingruber et al., 2013). The key trigger of this mechanism is an interaction between the ribosome and the multiprotein exon junction complex (EJC) that is deposited on mRNA near exon–exon junctions after splicing event (Shyu, Wilkinson, & van Hoof, 2008). A stop codon normally is present in the last exon after all the exon–exon junctions and the first round of translation results in the removal of all EJCs from the mRNA by the ribosome. In contrast, when the ribosome terminates on PTCs, it prevents the ribosome from displacement of all EJCs from the mRNA. The remaining RNA-bound EJCs act as a signal for activation of the NMD machinery and mRNA degradation.

NMD activity can be evaluated by classical RNA quantification methods—northern blotting or quantitative PCR. Further, luciferase- and GFP-based reporter systems have been designed to assess NMD activity. The bioluminescence reporters (Boelz, Neu-Yilik, Gehring, Hentze, & Kulozik, 2006; Nickless et al., 2014) utilized two luciferases with different spectral properties and substrate specificity for the quantitative ratiometric readout of NMD activity, but were inapplicable at the single-cell level. Therefore, spatial patterns of NMD activity in tissues and organisms and NMD heterogeneity in cell cultures cannot be addressed using bioluminescence reporters. In contrast, the single-color GFP fluorescence reporter (Paillusson, Hirschi, Vallan, Azzalin, & Mühlemann, 2005) allowed for the analysis of single cells but lacked an internal expression control. This drawback does not allow for correcting for inherent noise in the reporter expression level.

Recently, we developed new genetically encoded reporters of NMD activity (Pereverzev, Gurskaya, et al., 2015). These reporters are based on two fluorescent proteins with distinct colors (green and red) that make it possible to quantify NMD activity in single live cells using fluorescence microscopy or flow cytometry. In this chapter we describe the experimental protocols for application of the NMD reporters in mammalian cell cultures.

2. DESCRIPTION OF THE METHOD

The method for quantitative ratiometric assessment of NMD activity is based on a pair of related genetic constructs each bearing two independent expression cassettes encoding two fluorescent proteins in a single vector (Fig. 1). In the NMD-sensitive construct (NMD+) one fluorescent protein

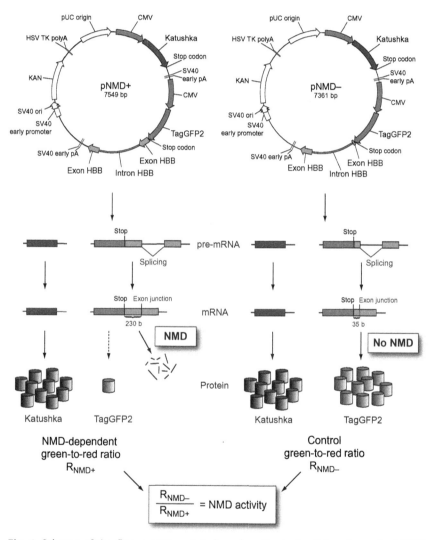

Fig. 1 Scheme of the fluorescent protein-based method of splicing-dependent NMD analysis.

is encoded by mRNA that is targeted by NMD, and the other is used as an expression control. In the control construct (NMD−) both transcripts are not targeted to NMD to establish a reference ratio of fluorescence intensities in the two channels.

We used two fluorescent proteins with well-distinguishable excitation and emission spectra: green TagGFP2 and far-red Katushka (TurboFP635). Vectors pNMD+ and pNMD− were constructed based on the pTurboFP635-N and pTagGFP2-N vectors (Evrogen). Both, pNMD+ and pNMD−, carried two fluorescent protein-encoding genes, each including the CMV promoter (PCMV IE) and transcription terminator (two SV40 early mRNA polyadenylation signals) (Fig. 1). The first gene encoded the far-red fluorescent protein Katushka (TurboFP635) with one synonymous mutation, which eliminates strong splice donor site (Gurskaya et al., 2011; Shcherbo et al., 2007). Katushka was used as a control of the overall expression level of the reporter, which can vary widely from cell to cell. The second gene encoded the green fluorescent protein TagGFP2 and a fragment of the human β-globin gene downstream of the TagGFP2 stop codon. In pNMD+, β-globin gene fragment consisted of exons 2, intron 2, and exon 3. This mRNA undergoes splicing-dependent NMD due to the β-globin exon–exon junction situated 230 nucleotides downstream of the TagGFP2. Vector pNMD− was identical to pNMD+ except for a shortened fragment of β-globin exon 2 (Fig. 1). In this case, the exon–exon junction becomes too close to the stop codon (35 nucleotides), making this transcript unrecognizable by the NMD machinery.

Fluorescence of Katushka and TagGFP2 in cells can be easily quantified by fluorescence microscopy and flow cytometry. Cells expressing pNMD− provide a reference ratio of the green and red signals. A change of this green-to-red ratio measured under the same conditions for pNMD+-expressing cells corresponds to NMD-dependent degradation of TagGFP2 transcript. Therefore, we used a ratio of green-to-red ratios (ie, $(green/red)_{pNMD-}/(green/red)_{pNMD+}$) as a measure of NMD activity (Fig. 1).

In addition, we constructed vectors for analysis of splicing-independent NMD (see Fig. 5A). For this, long 3′UTR of human SMG5 gene was used to trigger NMD, whereas human TRAM1 3′UTR, which is similarly long but insensitive to NMD, was used as a control (Hogg & Goff, 2010; Singh, Rebbapragada, & Lykke-Andersen, 2008). SMG5 and TRAM1 3′UTRs were cloned into pNMD+ in lieu of the β-globin gene fragment resulting in pSMG5-NMD+ and pTRAM1-NMD− vectors. These vectors can be used similarly to pNMD+ and pNMD− to access activity of the NMD pathway, which recognizes a long 3′UTR.

3. MATERIALS

3.1 Reagents

- Vectors pTurboRFP635-N (Evrogen, cat. no. FP721), pTagGFP2-N (Evrogen, cat. no. FP191)
- Vectors pNMD+ and pNMD− for quantitative ratiometric assessment of splicing-dependent NMD activity (available upon request)
- Vectors pSMG5-NMD+ and pTRAM1-NMD− for quantitative ratiometric assessment of splicing-independent NMD activity (available upon request)
- Plasmid Midi Kit (Qiagen, cat. no. 12143)
- Plasmid Mini Kit (Qiagen, cat. no. 12123)
- Mammalian cell line(s) of interest
- DMEM (Invitrogen, cat. no. 41965-039)
- DMEM/F12 (Thermo Fisher Scientific, cat. no. 21331-046)
- RPMI-1640 (Thermo Fisher Scientific, cat. no. 11875-085)
- OPTI-MEM (Invitrogen, cat. no. 31985-047)
- Bovine serum (Invitrogen, cat. no. 16170-078)
- Penicillin/streptomycin (Invitrogen, cat. no. 15070-063)
- L-Glutamine (Invitrogen, cat. no. 25030-024)
- BSA (Sigma-Aldrich, cat. no. 9048-46-8)
- Trypsin/EDTA solution (Invitrogen, cat. no. R-001-100)
- PBS (0.5% BSA)
- Versene Solution (EDTA in PBS) 0.02% solution (Thermo Fisher Scientific, cat. no. 15040-066)
- FuGene6 transfection reagent (Roche Diagnostics, cat. no. 11988387001)
- Plastic dishes for tissue culture (Sigma-Aldrich, cat. no. CLS 3294)
- Nucleofector™ Kit (V) (Lonza) (optional, for difficult to transfect cell lines)
- Cell Strainer Polypropylene frame 70 μm (SPL Lifesciences, cat. no. 93070)
- RNeasy Mini Kit (Qiagen, cat. no. 74104)
- MMLV RT kit (Evrogen)
- RQ1 RNase-Free DNase (Promega, cat. no. M6101)
- cDNA synthesis kit Mint-2 (Evrogen, cat. no.SK005)
- Encyclo PCR kit (Evrogen, cat. no. PK001)
- pGreenPuro™ shRNA Cloning and Expression Lentivector (System Biosciences, cat. no. SI505A-1)

3.2 Equipment
- Standard equipment for eukaryotic cell culturing
- Nucleofector™ 2b Device (Lonza, AAB-1001) (optional, for difficult to transfect cell lines)
- Fluorescence microscope
- Flow cytometer or fluorescence-activated cell sorter
- Standard equipment and reagents for quantitative PCR

3.3 Software
- ImageJ software (National Institutes of Health)
- Software for flow cytometry data analysis (eg, Weasel Software: http://en.bio-soft.net/other/WEASEL.html; Flowing Software: http://www.flowingsoftware.com)
- Spreadsheet processing software (eg, Microsoft Excel or Microcal Origin)

4. PROCEDURE
4.1 Cell Culture and Transfection

For cell lines that are easy to transfect (such as HEK, HeLa, and CT26), we recommend the use of liposome transfection. We obtain the best transfection efficiency with FuGene6 transfection reagent (Jacobsen, Calvin, Colvin, & Wright, 2004). For hard-to-transfect cell types (LLC, Jurkat, and HaCaT), it is preferable to use Nucleofector Kit (Lonza) and electroporate with a Nucleofector Device (Lonza). Cell counting in suspensions can be done either manually with standard Neubauer Chamber or with an automated cell counter such as Countess II (Thermo Fisher Scientific) or TC20 (Bio-Rad). For the tips on generation of stably transfected cell lines see Section 5.1.

4.1.1 Cell Culture for Liposome-Based Transfection
1. HEK293T, HeLa Kyoto, CT26 cell lines should be cultured in complete growth medium DMEM at 37°C, 5% CO_2 in a humidified incubator. The cells should be replated every 2–3 days to avoid reaching too high confluence (>60%).
2. Plate the cells on plastic or glass-bottom (for fluorescence microscopy with high numerical aperture objectives) dishes in complete growth medium at density of approximately 5×10^4 to 8×10^4 cells per 35-mm dish. Incubate for at least 12 h after plating.

3. Transfect the cells with the 1 μg of reporter plasmids using FuGene6 transfection reagent following the manufacturer's protocol or by any other appropriate method. Useful recommendations for liposome-based transfection are highlighted in the manufacturer's protocols (Dalby et al., 2004; Jacobsen et al., 2004). We recommend using Opti-MEM for DNA and transfection reagent mixing step. The next day the medium should be changed to fresh prewarmed liposome-free complete culture medium.
4. Culture the cells for 24–48 h at 37°C, 5% CO_2 in humidified incubator. Change culture medium every day after transfection.

Critical. Avoid high confluence of cells (>60%) in the dish as this can change NMD activity. Replate the transfected cells the next day if necessary.

4.1.2 Cell Culture for Nucleofection (for Difficult-to-Transfect Cells)
4.1.2.1 Human T-lymphocyte Jurkat Cell Culture
1. Maintain Jurkat cells in RPMI with L-glutamine supplemented with 10% fetal bovine serum at 37°C, 5% CO_2 in a humidified incubator. Grow cells to no more than 2×10^5 cells/mL and passage cells 2–3 times a week.
2. For transient transfection of Jurkat E6-1 cells, we recommend the Lonza Nucleofector Device and the Cell Line Nucleofector Kit V using cell type-specific protocol. Before nucleofection seed out 2×10^5 cells/mL and culture for 2 days.
3. Centrifuge about 1×10^6 cells at $300 \times g$ for 5 min at room temperature.
4. Remove supernatant completely and resuspend the cell pellet carefully in 100 μL room-temperature Nucleofector Solution per sample.
5. Combine 100 μL of cell suspension with 5 μg DNA. Transfer cell/DNA suspension into certified cuvette. Select the appropriate Nucleofector Program (we recommend the program number X-001).
6. Once the program is finished add about 500 μL of the preequilibrated culture medium to the cuvette and gently transfer the sample into the plastic dish. We recommend growing cells in RPMI with 20% fetal bovine serum after nucleofection.
7. The next day the medium should be changed to fresh prewarmed complete culture medium (RPMI with 10% FBS). Useful tips are described in the manufacturer's protocols: http://bio.lonza.com.

4.1.2.2 Spontaneously Transformed Human Keratinocyte Cell Culture (HaCaT)
1. Maintain HaCaT cells in RPMI with L-glutamine supplemented with 10% fetal bovine serum at 37°C, 5% CO_2 in a humidified incubator. Passage cells at no more than 80% confluence.

2. Seed out 2.5×10^5 cells per 25-cm^2 flask and culture for 2–3 d before nucleofection. We recommend the Lonza Nucleofector Device and the Cell Line Nucleofector Kit V using cell type-specific protocol.
3. For harvesting, incubate the cells 8–10 min at 37°C with 1 mL of trypsinization reagent. *Optional*: Treat cells with 0.2 mg/mL EDTA in PBS before trypsinization.
4. Centrifuge about 1×10^6 cells at $300 \times g$ for 5 min at room temperature.
5. Remove supernatant completely. Resuspend the cell pellet carefully in 100 μL room-temperature Nucleofector Solution (Kit V) per sample.
6. We recommend using program U-020 of Lonza Nucleofector and transfect about 1×10^6 cells with 5 μg of a plasmid.
7. After nucleofection grow cells in RPMI with 20% fetal bovine serum.

4.1.2.3 Lewis Lung Carcinoma Cell Culture

1. Maintain Lewis lung carcinoma (LLC) in DMEM/F12 with L-glutamine supplemented with 10% fetal bovine serum at 37°C, 5% CO_2 in a humidified incubator. Passage cells at no more than 80% confluence.
2. Seed out 2.5×10^5 cells per 25-cm^2 flask and culture for 2–3 days before nucleofection.
3. Collect medium containing nonadherent cells into a sterile centrifuge tube.
4. Wash any remaining attached cells with Versene solution and collect the washings.
5. Pipette trypsin/EDTA onto the washed cell monolayer using 1 mL per 25 cm^2 of the surface area. After trypsinization transfer the cells into the centrifuge tube containing the collected medium and cells.
6. Centrifuge the entire cell suspension at $300 \times g$ for 5 min.
7. Remove the supernatant and resuspend the cell pellet carefully in 100 μL room-temperature Nucleofector Solution per sample.
8. We recommend using the Cell Line Nucleofector Kit V and program T-019* of Lonza Nucleofector Device. Transfect about 1×10^6 cells with 5 μg of a plasmid.
9. After nucleofection grow cells in DMEM/F12 with 20% fetal bovine serum.

Critical. After Jurkat and HaCaT cells transfection, 96 h may be required to detect fluorescence.

Critical. Use of high-quality plasmid DNA greatly increases transfection efficiency. We recommend the QIAGEN Plasmid Midi Kit.

4.2 Fluorescence Microscopy

Fluorescence microscopy measurement is preferable for cells, which are difficult to detach from the culture dish and suspend, such as neurons and tissue samples. In all other cases, the flow cytometry measurements are preferred. However, fluorescent microscopy can be used as an auxiliary readout before the cytometry analysis.

Adherent cells transfected with NMD reporter plasmids can be analyzed by any research wide-field or confocal fluorescence microscope. TagGFP2 possesses excitation and emission maxima at 483 and 506 nm, respectively; it can be imaged using standard GFP filter set or 488 nm laser. Katushka possesses excitation and emission maxima at 588 and 635 nm, respectively; standard Texas Red filter set or green-yellow lasers (eg, 543 or 561 nm) are suitable for this protein. Optimal filter sets and lasers for detection of TagGFP2 and Katushka (TubroFP635) can be selected using interactive Spectra Viewer at the Evrogen website (http://evrogen.com/spectra-viewer/viewer.shtml). In our work we imaged cells with BZ9000 motorized fluorescence microscope (Keyence) using 20× objective and standard filter sets for green (excitation at 450–490, emission at 510–560 nm) and red (excitation at 540–580, emission at 600–660 nm) channels (see Fig. 2).

1. Place a dish with cells transfected with pNMD− on the microscope stage and focus on the cell layer. Take images in green (for TagGFP2) and red (for Katushka) channels for several fields of view.

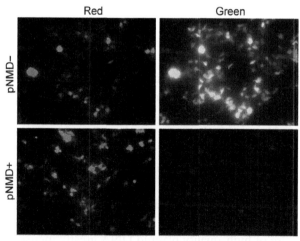

Fig. 2 Fluorescence microscopy of HEK293 cells transiently transfected with pNMD− or pNMD+ vectors. Pictures were taken in *red* (*left*) and *green* (*right*) *channels* using the same settings for the two cell samples. (See the color plate.)

Critical. Standard plastic dishes with thick bottom are suitable for low numerical aperture (NA) objectives only. For high-resolution microscopy, glass-bottom dishes should be used.

Critical. For further quantitative comparison, fluorescence signals should be bright enough but not saturated. Adjust excitation light power and exposure time correspondingly.

2. Use exactly the same microscope settings to image pNMD+-transfected cells. Fluorescence of TagGFP2 in the green channel is expected to be much dimmer compared to that in the pNMD− cells. If the green signal is too low for reliable detection, increase exposure time to get sufficient brightness. Do not increase excitation light power as this change is more difficult to quantify.
3. Calculate green-to-red ratio for 20–40 cells in pNMD− and pNMD+ cells using an appropriate image analysis program (eg, ImageJ or Adobe Photoshop). Compare mean ratio for pNMD+ sample with that for pNMD− sample; difference between these values corresponds to NMD activity in the cells under investigation.

4.3 Flow Cytometry

4.3.1 Sample Preparation for Flow Cytometry of Live Cells

1. Wash the cells twice with 1 mL of Versene solution and trypsinize in 0.5–1 mL of Trypsin/EDTA solution for 5 min at 37°C.
2. Centrifuge the required number of cells (10^4–10^6 cells per sample) at $300 \times g$ for 5 min at room temperature. Remove supernatant completely. Resuspend the cell pellet carefully in 0.5–1 mL of PBS/0.5% BSA.
3. Optionally, take the cell suspension aliquot for RNA isolation (see Section 4.4).
4. To prepare single-cell suspension, filter samples immediately before sorting the cells through a 70 μm Cell Strainer and transfer the filtered suspension to 5-mL round bottom polystyrene test tube for flow cytometry.

Critical. Avoid keeping cells at an unnecessary high concentration and do not pellet them dry at any point. Keep the samples on ice, unless cells are susceptible to cold. Bring some extra culture medium with you to dilute your sample, if necessary.

4.3.2 Analysis of the Cells by Flow Cytometry

Various flow cytometers or fluorescence-activated cell sorters (FACS) can be used for analysis of cells expressing the NMD reporters. Ideally, it should be equipped with blue (488 nm) and green-yellow (eg, 561 nm) lasers for separate excitation of TagGFP2 and Katushka, respectively. Suitable lasers and detection filters for TagGFP2 and Katushka (TubroFP635) can be selected using interactive Spectra Viewer at the Evrogen website (http://evrogen.com/spectra-viewer/viewer.shtml). Use equipment available to you according to the standard practice of flow cytometry analysis of live mammalian cells expressing green and red fluorescent proteins. Below we provide recommendations for FACSAria III (Becton Dickinson) and Cytomics FC500 (Beckman Coulter), which were used in our work.

For FACSAria III, use 488 nm laser and 530/30 nm detector for TagGFP2, and 561 nm laser and 670/14 nm detector for Katushka. For Cytomics FC500, both TagGFP2 and Katushka should be excited with 488 nm laser line and distinguished from one another using two different filter sets for TagGFP2 (510–540 nm) and for Katushka (660–690 nm).

Critical. To avoid the cross talk between TagGFP2 and Katushka fluorescent signals the acquisition protocol should include the predetermined PMT voltage and compensation parameters. The detection parameters should be selected using the preliminary experiment with cells of the same type expressing TagGFP2 or Katushka alone, such that the cells expressing one fluorescent protein are not detected in the second channel (Fig. 3). The cells should be transfected with pTurboFP635-N or pTagGFP2-N vectors

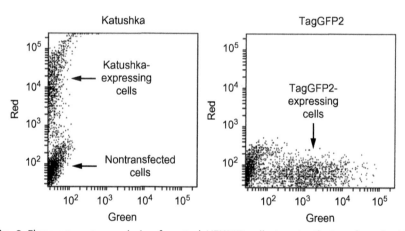

Fig. 3 Flow cytometry analysis of control HEK293 cells transiently transfected with pTurboFP635-N (*left*) or pTagGFP2-N (*right*) vectors using FACSAria III. Settings of the flow cytometer were adjusted to avoid cross talks of the green and red signals (Katushka fluorescence is detected only in the red channel, whereas TagGFP2 fluorescence is detected only in the green channel). These settings were used in all further experiments with NMD reporters.

and analyzed 24–48 h after transfection. The nontransfected cells can be analyzed as well.

Critical. Use forward and side light scattering (FSC vs SSC plots) for gating the cell populations of appropriate size and structure (Fig. 4A).

Typical results of NMD reporter analyses with FACSAria III are shown in Figs. 3–5. Examples of flow cytometry with Cytomics FC500 are shown in Fig. 6 and can also be found in Pereverzev, Gurskaya, et al. (2015). It is convenient to represent data as dot plots (TagGFP2 vs Katushka). Cells expressing the control NMD-insensitive vector should form a diagonal distribution with bright signals in both green and red channels (Figs. 4B right; 5B right; 6 right). Cells expressing NMD-sensitive transcripts (pNMD+ or pSMG5-NMD+) should form a diagonal distribution shifted to the left compared to the control cells (Figs. 4B left; 5B left; 6 left). In some cell lines (Jurkat, HaCaT) or under some conditions (eg, overgrown HEK293) a strong heterogeneity of cells in respect of NMD activity can be observed (left panels in Figs. 4B and 6A and B).

4.3.3 Analysis of Flow Cytometry Results

Different ways can be used to calculate NMD activity on the base of obtained flow cytometry data as described later.

1. Use the Quadrants tool as a simple and fast way to compare NMD+ and NMD− samples. Place the cross manually at the point of highest density (in the middle of the diagonal) of pNMD+-expressing cells (Figs. 4C and 5B left). Export the value in green channel. At the same level of red signal, place the cross at the point of highest density (in the middle of the diagonal) of pNMD−-expressing cells (Figs. 4C and 5B right). Export the value in green channel. Ratio of the determined green signal values pNMD−/pNMD+ presents NMD activity. In the case of heterogeneity of pNMD+-expressing cells, determine NMD activity for each cell subpopulation (Fig. 4C left). Repeat the calculation for areas with different levels of the red signal.

2. Export digital data for detailed characterization of the samples. To remove nontransfected and low-fluorescent cells from analysis, apply gates for regions corresponding to well-expressing cells (R1 and R2 in Figs. 4B,C and 5B). Export text data using a program capable to do this, such as Weasel or Flowing Software. For example, in Weasel right-click on dot plot and select "Export–Text" from the drop-down menu. Then import the text data into spreadsheet processing software (eg, Microsoft Excel or Microcal Origin) and calculate green/red ratio for each cell. Perform regular statistical description of the obtained

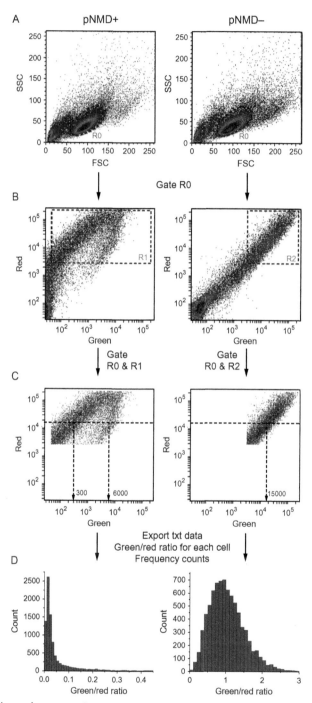

Fig. 4 See legend on opposite page.

datasets, eg, mean, standard deviation, and frequency counts (Figs. 4D and 5C).
3. For homogeneous cell populations (eg, see Fig. 5B), calculate mean values for well-expressing cells in the green and red channels separately using histogram tool in the flow cytometry software. Ratio of green/red ratios for the NMD− and NMD+ samples presents NMD activity. In this way of calculation, any heterogeneity within the analyzed cell populations becomes hidden. However, it is a useful way to compare results of fluorescence analysis with classical methods such as qPCR.

4.4 RNA Isolation and Quantitative RT-PCR

Optionally, performance of the fluorescence NMD reporter can be verified at the RNA level by reverse-transcription and quantitative PCR analysis.
1. Isolate total RNA from the cell samples following one of the standard protocols (Chomczynski & Sacchi, 1987; Gehring, Hentze, & Kulozik, 2008) or by RNeasy Mini Kit.
2. To eliminate the plasmid DNA, treat the RNA samples by RQ DNAase (1 unit per sample) according to manufacturer's recommendation.
3. Use RNAeasy column or phenol–chloroform extraction to inactivate RQ DNAase.
4. Use this RNA samples (approximately 3 μg of total RNA) for reverse-transcription reaction by the MMLV RT kit (Evrogen) (or analogous kit) with an oligo(dT)$_{20}$ primer, according to the manufacturer's protocol.

   Critical. If the amount of total RNA is low (<0.2–0.3 μg), we do not recommend using RQ DNAse treatment and subsequent purification. Instead use total cDNA amplification by the cDNA synthesis kit Mint-2 and Encyclo PCR kit (Evrogen).

Fig. 4 Flow cytometry analysis of HEK293 cells expressing pNMD+ (*left*) or pNMD− (*right*) using FACSAria III. (A) *Dot plots* in forward scattering (FSC) and side scattering (SSC) channels. Region R0 was used to select live cells. (B) *Dot plots* in *green* and *red channels*. Regions R1 and R2 were used to select well-expressing cells. (C) *Dot plots* of cells from R1 and R2 regions. *Dashed lines* illustrate simplified ("by eye") assessment of NMD activity by comparison of *green signals* in pNMD+ and pNMD− samples at the equal level of the *red signal*. Note strong cell heterogeneity resulting in a major subpopulation with a high NMD activity (15,000/300 = 50-fold) and a minor subpopulation with a low NMD activity (15,000/6000 = 2.5-fold). Full data from these plots were exported in the text format to perform statistical analysis shown in (D). (See the color plate.)

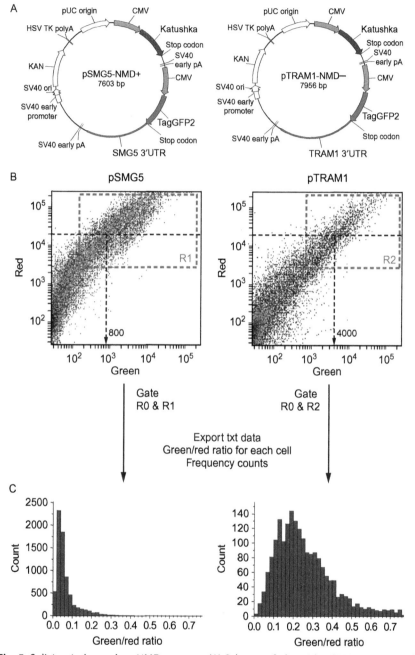

Fig. 5 Splicing-independent NMD reporter. (A) Scheme of plasmids pSMG5-NMD+ and pTRAM1-NMD−. (B) Flow cytometry analysis of HEK293 cells expressing pSMG5-NMD+ (*left*) or pTRAM1-NMD− (*right*) using FACSAria III. Dashed lines show simple way to estimate NMD activity by comparison of green signals in pNMD+ and pNMD− samples at the equal level of the red signal. (C) Statistical analysis of data for cells from regions R1 and R2. See Fig. 4 legend for details. (See the color plate.)

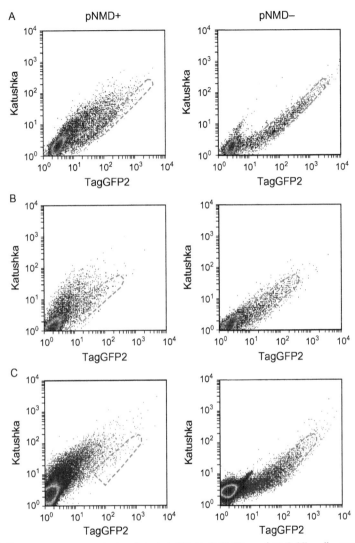

Fig. 6 Flow cytometry analysis of Jurkat (A), HaCaT (B), and LLC (C) cells expressing pNMD+ (*left*) or pNMD− (*right*) using Cytomics FC500. To simplify comparison, diagonal areas of the control pNMD− cells are marked with *cyan dashed lines*. (See the color plate.)

5. Perform quantitative real-time PCR analysis for quantification of the NMD reporter transcripts. Use standard qPCR protocols, eg, described in Jozefczuk and Adjaye (2011). The abundance of the TagGFP2- and Katushka-encoding transcripts can be measured with the primers 5′-cgccatcagcaaggaccg and 5′-actggtggggtcaattctttgc for TagGFP2, and 5′-tgagagcggattgacaggccat and 5′-gagtccggattgatcccccagtttgct for Katushka.

To estimate NMD activity, compare abundances of TagGFP2-encoding transcript in pNMD+ and pNMD− samples (in each sample, the level of TagGFP2-encoding transcript should be normalized to the level of Katushka-encoding transcript in the same sample). The primers 5′-cgctaccggactccag and 5′-caagttaacaacaacaattgc should be used to control for the presence of nonspliced or alternatively spliced forms of the TagGFP2-encoding transcript.

4.5 Inhibitors of NMD Activity

NMD inhibition by wortmannin and caffeine can be performed to examine effects of NMD regulation. These chemicals were shown to block phosphorylation of the key NMD factor UPF1 (Noensie & Dietz, 2001; Usuki et al., 2004). If the inhibition of NMD cascade occurs, a severalfold increase of pNMD+ green fluorescence would be observed, whereas green-to-red fluorescence ratio in pNMD− samples would not change significantly (Pereverzev, Gurskaya, et al., 2015).

4.5.1 NMD Inhibition by Wortmannin

To test the effects of wortmannin, treat transfected cells with 20 µM wortmannin 14 h after transfection. Treat control sample of transfected cells with DMSO as the vehicle control. After 8 h of incubation subject cells to flow cytometry analysis. To estimate the value of NMD inhibition, compare the ratio of green/red signal values for the treated and untreated NMD reporter samples.

4.5.2 NMD Inhibition by Caffeine

To test the effects of caffeine, perform treatment of transfected cells with 10 mM caffeine 14 h after transfection. Change cell medium to fresh prewarmed complete culture medium with 10 mM caffeine every 12 h. Perform flow cytometry analysis of the cells 48 h after transfection. Treat control sample of transfected cells with ethanol as a vehicle control.

4.5.3 NMD Inhibition by RNAi

For specific NMD downregulation we recommend using RNAi experiments. For this purpose vector encoding short hairpin RNA (shRNA) against the key NMD factor UPF1 should be constructed. The design of shUPF1 oligonucleotides and its application were described earlier (Boelz et al., 2006; Paillusson et al., 2005; Pereverzev, Gurskaya, et al., 2015). The UPF1 target sequence for shRNA can be the following: 5′-gagaatcgcctacttcact (pshUPF1 plasmid). A plasmid without anti-UPF1 shRNA sequence (shControl) should be used as a control. The shControl vector could be created by cutting off a copGFP sequence using the *Xba*I and *Pst*I sites of pGreenPuro (System Biosciences). shRNA-coding sequences can be inserted as double-stranded oligonucleotides into the shControl vector between the *Bam*HI and *Eco*RI sites, according to the supplier's recommendations.

Perform the following contransfections of the cells: (1) pNMD+ and pshUPF1, (2) pNMD+ and shControl, (3) pNMD− and pshUPF1, and (4) pNMD− and shControl. Take pshUPF1 and shControl plasmids in twofold excess over pNMD+/−. Culture the transfected cells for 5–7 days to ensure efficient decrease of UFP1 protein level. During this time, replate the cells regularly to avoid dense monolayer formation. Excess cells are used for flow cytometry analysis at sequential time points. A gradual decrease of NMD activity is expected to be observed in these experiments. Note that UFP1 is important for cell survival; thus, significant cell death occurs during prolonged expression of pshUPF1.

5. FURTHER POSSIBLE MODIFICATIONS OF GENETIC CONSTRUCTS

The proposed analysis method can potentially be tailored for particular experimental goals by modifications of the reporter constructs using modular approach. Each of the reporter functional modules (vector backbone, promoters, FPs, and NMD-inducing gene fragment) can be replaced for a required functional equivalent to address specific experimental needs.

5.1 Changes of the Vector Backbone

The appropriate vector backbone for modified NMD reporters should contain two separated independent expression cassettes with independent promoters and transcription terminators. It is convenient to use pNMD− vector backbone for construction of modified NMD reporters.

Critical. IRES elements or 2A peptides cannot be used instead.

Cell- or tissue-specific or inducible promoters can be cloned in place of the CMV promoters using unique *Ase*I + *Age*I and *Mlu*I + *Bam*HI cleavage sites in pNMD− and pNMD+ vectors. The selection of appropriate promoter can be based on relative promoter strength, promoter methylation, cell type, and other considerations.

Critical. The promoter strength is important for NMD reporters because the overload of cells with chimeric NMD target can saturate the NMD machinery in particular experimental system and lead to artifact NMD inhibition. To test the saturation of NMD pathway it is useful to compare the measured NMD activity for cells with different levels of reporter expression using flow cytometry data. The NMD activity for the brightest transfected cells should not be lower than NMD activity in low-fluorescent cells.

Critical. The promoters should be identical for both expression cassettes to avoid effects of differential changes in promoter activity.

The two-promoter fragment of pNMD− and pNMD+ vectors can be transferred to a different auxiliary vector backbone, eg, a transposon vector to obtain stably transfected cell line. Due to presence of multiple restriction sites it is preferable to PCR-amplify the two-promoter fragment using long-distant high-fidelity PCR with primers introducing new unique restriction sites compatible with the new vector backbone.

Critical. It is not possible to use lentiviral vectors to generate stably transfected cell lines with the NMD reporters. The reporters contain internal pA sites after expression cassettes and a spliceable intron in the case of splicing-dependent reporter that make impossible generation of lentiviruses. Transposon-based technologies could be used instead (Koga et al., 2003). However, the identical promoters of two expression cassettes represent direct repeats that can lead to recombination and decrease the efficiency of stable transfection.

5.2 Replacement of Fluorescent Indicators

In the developed NMD reporters we used green fluorescent protein TagGFP2 and far-red fluorescent protein Katushka (TurboFP635), that give bright and well-distinguishable fluorescent signals in mammalian cells. These fluorescent proteins can potentially be replaced with different fluorescent proteins using *Age*I + *Bam*HI and *Bgl*II + *Hin*dIII restriction sites.

Critical. It is necessary to ascertain that new pair of FPs has non-overlapping emission spectra and the equipment for fluorescence measuring is compatible with fluorescent probes (ie, is able to excite the FPs and has appropriate filters for emission).

Critical. In case of replacement of fluorescent proteins it is required to inspect the FP's cDNA for cryptic donor splice sites that can interfere with normal splicing of beta-globin gene. Numerous online tools for splice site prediction are available, eg, http://www.fruitfly.org/seq_tools/splice.html and http://www.umd.be/HSF/#.

5.3 Introduction of a new NMD-Inducing Fragment

Gene fragments other than β-globin and SMG5 that target distinct routes of the NMD pathway or provide greater dynamic range could be used.

For the splicing-dependent NMD reporter the stop codon of TagGFP2 in pNMD+ construct should be placed at a distance of more than 50 nucleotides upstream of the exon–exon junction to make the TagGFP2 coding transcript recognizable by the NMD machinery.

The NMD-inducing gene fragment for splicing-dependent NMD reporter should contain at least one constitutive spliceable intron. The presence of alternative donor and acceptor splice sites can be assessed as described earlier.

Critical. The absence of alternatively spliced products and unspliced transcript should be confirmed after the transfection using comprehensive RT-PCR method or northern blot hybridization.

Potentially two types of control pNMD− constructs for splicing-dependent NMD reporter can be designed. One approach is to shorten the distance between FP's stop codon and the exon junction to less than 50 bp (Pereverzev, Gurskaya, et al., 2015). An alternative way is to replace the NMD-inducing gene fragment with its intronless splice variant. We strongly recommend using the first approach because splicing itself can greatly influence the mRNA fate and protein expression level (Pereverzev et al., 2014).

For development of new splicing-independent NMD reporters, 3′UTRs from genes with known NMD-sensitive and NMD-insensitive long 3′UTRs could be used. Since the structural basis of long 3′UTR-mediated NMD is not completely understood, it is preferable to clone complete 3′UTR from the stop codon to native polyadenylation site. The main requirement for a long 3′UTR control construct is approximately the same length of the NMD-insensitive 3′UTR fragment.

6. CONCLUDING REMARKS

The dual-color NMD reporters enable quantitative ratiometric assessment of NMD activity at the single-cell level. This is especially important for complex models with potentially heterogeneous spatiotemporal NMD activity patterns. For example, possible stage- and tissue-specific differences in NMD activity can be visualized using fluorescence microscopy of live developing embryos (eg, in the well-developed *Caenorhabditis*, *Drosophila*,

Danio, or *Xenopus* models easily accessible to light microscopy). For nontransparent models, such as mice, fluorescence microscopy of tissue slices can be used to detect specific cell types with altered NMD activity. Suspension samples such as cell cultures or blood cells are well suited for analysis by flow cytometry. Notably, dual-color fluorescence reporter allowed to reveal strong heterogeneity of NMD activity in some mammalian cell lines that cannot be detected neither by classical methods of RNA isolation and quantification nor by luminescence or single-color GFP fluorescence reporters (Pereverzev, Gurskaya, et al., 2015; Pereverzev, Matlashov, Staroverov, Lukyanov, & Gurskaya, 2015). Potentially, fluorescence-activated cell sorter enables to separate cell subpopulations with different NMD activity for further detailed analysis of their transcriptomes and biochemical features. Thus, fluorescence NMD reporters represent a useful tool to study mechanisms of NMD regulation and influence of NMD on cell physiology in various experimental models.

ACKNOWLEDGMENTS

This work was supported by the Russian Science Foundation (Project 14-25-00129). The work was partially carried out using equipment provided by the IBCH Core Facility (CKP IBCH).

REFERENCES

Behm-Ansmant, I., Kashima, I., Rehwinkel, J., Saulière, J., Wittkopp, N., & Izaurralde, E. (2007). mRNA quality control: An ancient machinery recognizes and degrades mRNAs with nonsense codons. *FEBS Letters*, *581*, 2845–2853.
Bhuvanagiri, M., Schlitter, A. M., Hentze, M. W., & Kulozik, A. E. (2010). NMD: RNA biology meets human genetic medicine. *Biochemical Journal*, *430*, 365–377.
Boelz, S., Neu-Yilik, G., Gehring, N. H., Hentze, M. W., & Kulozik, A. E. (2006). A chemiluminescence-based reporter system to monitor nonsense-mediated mRNA decay. *Biochemical & Biophysical Research Communications*, *349*, 186–191.
Bruno, I. G., Karam, R., Huang, L., Bhardwaj, A., Lou, C. H., Shum, E. Y., et al. (2011). Identification of a microRNA that activates gene expression by repressing nonsense-mediated RNA decay. *Molecular Cell*, *42*, 500–510.
Chang, Y. F., Imam, J. S., & Wilkinson, M. F. (2007). The nonsense-mediated decay RNA surveillance pathway. *Annual Review Biochemistry*, *76*, 51–74.
Chomczynski, P., & Sacchi, N. (1987). Single-step method of RNA isolation by acid guanidinium thiocyanate-phenol-chloroform extraction. *Analytical Biochemistry*, *162*, 156–159.
Colak, D., Ji, S. J., Porse, B. T., & Jaffrey, S. R. (2013). Regulation of axon guidance by compartmentalized nonsense-mediated mRNA decay. *Cell*, *153*, 1252–1265.
Dalby, B., Cates, S., Harris, A., Ohki, E. C., Tilkins, M. L., Price, P. J., et al. (2004). Advanced transfection with Lipofectamine 2000 reagent: Primary neurons, siRNA, and high-throughput applications. *Methods*, *33*, 95–103.

Gardner, L. B. (2008). Hypoxic inhibition of nonsense-mediated RNA decay regulates gene expression and the integrated stress response. *Molecular and Cellular Biology, 28*, 3729–3741.

Gardner, L. B. (2010). Nonsense-mediated RNA decay regulation by cellular stress: Implications for tumorigenesis. *Molecular Cancer Research, 8*, 295–308.

Gehring, N. H., Hentze, M. W., & Kulozik, A. E. (2008). Tethering assays to investigate nonsense-mediated mRNA decay activating proteins. *Methods in Enzymology, 448*, 467–482.

Gurskaya, N. G., Staroverov, D. B., Fradkov, A. F., & Lukyanov, K. A. (2011). Coding region of far-red fluorescent protein Katushka contains a strong donor splice site. *Russian Journal of Bioorganic Chemistry, 37*, 380–382.

Hogg, J. R., & Goff, S. P. (2010). Upf1 senses 3'UTR length to potentiate mRNA decay. *Cell, 143*, 379–389.

Huang, L., Lou, C., Chan, W., Shum, E., Shao, A., Stone, E., et al. (2011). RNA homeostasis governed by cell type-specific and branched feedback loops acting on NMD. *Molecular Cell, 43*, 950–961.

Huang, L., & Wilkinson, M. F. (2012). Regulation of nonsense-mediated mRNA decay. *Wiley Interdisciplinary Reviews: RNA, 3*, 807–828.

Hwang, J., & Maquat, L. E. (2011). Nonsense-mediated mRNA decay (NMD) in animal embryogenesis: To die or not to die, that is the question. *Current Opinion in Genetics & Development, 21*, 422–430.

Jacobsen, L. B., Calvin, S. A., Colvin, K. E., & Wright, M. (2004). FuGENE 6 transfection reagent: The gentle power. *Methods, 33*, 104–112.

Jozefczuk, J., & Adjaye, J. (2011). Quantitative real-time PCR-based analysis of gene expression. *Methods in Enzymology, 500*, 99–109.

Koga, A., Iida, A., Kamiya, M., Hayashi, R., Hori, H., Ishikawa, Y., et al. (2003). The medaka fish Tol2 transposable element can undergo excision in human and mouse cells. *Journal of Human Genetics, 48*, 231–235.

Lewis, B. P., Green, R. E., & Brenner, S. E. (2003). Evidence for the widespread coupling of alternative splicing and nonsense-mediated mRNA decay in humans. *Proceedings of the National Academy of Sciences of the United States of America, 100*, 189–192.

Lykke-Andersen, S., & Jensen, T. H. (2015). Nonsense-mediated mRNA decay: An intricate machinery that shapes transcriptomes. *Nature Reviews Molecular Cell Biology, 16*, 665–677.

Mendell, J. T., Sharifi, N. A., Meyers, J. L., Martinez-Murillo, F., & Dietz, H. C. (2004). Nonsense surveillance regulates expression of diverse classes of mammalian transcripts and mutes genomic noise. *Nature Genetics, 36*, 1073–1078.

Nicholson, P., & Mühlemann, O. (2010). Cutting the nonsense: The degradation of PTC-containing mRNAs. *Biochemical Society Transactions, 38*, 1615–1620.

Nickless, A., Jackson, E., Marasa, J., Nugent, P., Mercer, R. W., Piwnica-Worms, D., et al. (2014). Intracellular calcium regulates nonsense-mediated mRNA decay. *Nature Medicine, 20*, 961–966.

Noensie, E. N., & Dietz, H. C. (2001). A strategy for disease gene identification through nonsense-mediated mRNA decay inhibition. *Nature Biotechnology, 19*, 434–439.

Paillusson, A., Hirschi, N., Vallan, C., Azzalin, C. M., & Mühlemann, O. (2005). A GFP based reporter system to monitor nonsense-mediated mRNA decay. *Nucleic Acids Research, 33*, e54.

Pereverzev, A. P., Gurskaya, N. G., Ermakova, G. V., Kudryavtseva, E. I., Markina, N. M., Kotlobay, A. A., et al. (2015). Method for quantitative analysis of nonsense-mediated mRNA decay at the single cell level. *Scientific Reports, 5*, 7729.

Pereverzev, A. P., Markina, N. M., Yanushevich, Y. G., Gorodnicheva, T. V., Minasyan, B. E., Lukyanov, K. A., et al. (2014). Intron 2 of human beta-globin in

3'-untranslated region enhances expression of chimeric genes. *Russian Journal of Bioorganic Chemistry*, *40*, 269–271.

Pereverzev, A. P., Matlashov, M. E., Staroverov, D. B., Lukyanov, K. A., & Gurskaya, N. G. (2015). Differences in nonsense-mediated mRNA decay activity in mammalian cell lines revealed by a fluorescence reporter. *Russian Journal of Bioorganic Chemistry*, *41*, 547–550.

Rebbapragada, I., & Lykke-Andersen, J. (2009). Execution of nonsense-mediated mRNA decay: What defines a substrate? *Current Opinion in Cell Biology*, *21*, 394–402.

Rehwinkel, J., Letunic, I., Raes, J., Bork, P., & Izaurralde, E. (2005). Nonsense-mediated mRNA decay factors act in concert to regulate common mRNA targets. *RNA*, *11*, 1530–1544.

Schoenberg, D. R., & Maquat, L. E. (2012). Regulation of cytoplasmic mRNA decay. *Nature Reviews Genetics*, *13*, 246–259.788.

Schweingruber, C., Rufener, S. C., Zünd, D., Yamashita, A., & Mühlemann, O. (2013). Nonsense-mediated mRNA decay—Mechanisms of substrate mRNA recognition and degradation in mammalian cells. *Biochimica et Biophysica Acta*, *1829*, 612–623.

Shcherbo, D., Merzlyak, E. M., Chepurnykh, T. V., Fradkov, A. F., Ermakova, G. V., Solovieva, E. A., et al. (2007). Bright far-red fluorescent protein for whole-body imaging. *Nature Methods*, *4*, 741–746.

Shyu, A. B., Wilkinson, M. F., & van Hoof, A. (2008). Messenger RNA regulation: To translate or to degrade. *The EMBO Journal*, *27*, 471–481.

Singh, G., Rebbapragada, I., & Lykke-Andersen, J. (2008). A competition between stimulators and antagonists of Upf complex recruitment governs human nonsense-mediated mRNA decay. *PLoS Biology*, *6*, e111.

Usuki, F., Yamashita, A., Higuchi, I., Ohnishi, T., Shiraishi, T., Osame, M., et al. (2004). Inhibition of nonsense-mediated mRNA decay rescues the phenotype in Ullrich's disease. *Annals of Neurology*, *55*, 740–744.

Wang, D., Zavadil, J., Martin, L., Parisi, F., Friedman, E., Levy, D., et al. (2011). Inhibition of nonsense-mediated RNA decay by the tumor microenvironment promotes tumorigenesis. *Molecular and Cellular Biology*, *17*, 3670–3680.

CHAPTER FOURTEEN

Developing Fluorogenic Riboswitches for Imaging Metabolite Concentration Dynamics in Bacterial Cells

J.L. Litke[*,†], M. You[†], S.R. Jaffrey[*,†,1]
[*]Tri-Institutional Chemical Biology Program at Weill-Cornell Medical College, Rockefeller University, Memorial Sloan-Kettering Cancer Center, New York, NY, United States
[†]Weill Medical College, Cornell University, New York, NY, United States
[1]Corresponding author: e-mail address: srj2003@med.cornell.edu

Contents

1. Introduction 316
2. Structure-Guided Design of Spinach Riboswitch Sensors 319
   2.1 Structure-Guided Design of Switching Sequences and Transducer Sequences 321
   2.2 Structure-Guided Design of Linker Sequences 322
3. Identifying Spinach Riboswitch Sensors In Vitro 323
   3.1 In vitro Preparation of RNA-Based Sensor Variants 324
   3.2 Quantification of Fluorescence Sensor Characteristics 325
4. Live-Cell Imaging of Metabolites with Spinach Riboswitch Sensors 327
   4.1 Preparation of Sensor-Expressing E. coli 327
   4.2 Imaging Dynamics of Metabolite Concentration in live E. coli (TPP Sensor) 328
   4.3 Data Analysis of Metabolite Concentration Variations (TPP Sensor) 330
5. Summary and Concluding Remarks 331
Acknowledgments 332
References 332

Abstract

Genetically encoded small-molecule sensors are important tools for revealing the dynamics of metabolites and other small molecules in live cells over time. We recently developed RNA-based sensors that exhibit fluorescence in proportion to a small-molecule ligand. One class of these RNA-based sensors are termed Spinach riboswitches. These are RNAs that are based on naturally occurring riboswitches, but have been fused to the Spinach aptamer. The resulting RNA is a fluorogenic riboswitch, producing fluorescence upon binding the cognate small-molecule analyte. Here, we describe how to design and optimize these sensors by adjusting critical sequence elements, guided by structural insights from the Spinach aptamer. We provide a stepwise procedure to characterize sensors in vitro and to express sensors in bacteria for live-cell imaging of metabolites. Spinach riboswitch sensors offer a simple method for fluorescence measurement of a wide range of metabolites for which riboswitches exist, including nucleotides and their derivatives, amino acids, cofactors, cations, and anions.

1. INTRODUCTION

Measuring the flux of cellular metabolites is important because metabolites exert a direct influence on cell physiology (Patti, Yanes, & Siuzdak, 2012). Metabolic pathways are highly linked to both signaling pathways and nutrient availability, and various lines of evidence show that alterations in cellular metabolite dynamics are a mediator of diverse diseases. In addition to imaging endogenously generated metabolites, there is considerable interest in imaging exogenously derived molecules such as antibiotics, drugs, and other xenobiotics.

Traditional methods for quantifying metabolite levels rely on liquid chromatography, NMR spectroscopy, or mass spectrometry. However, these techniques require harvesting cells and thus are not "continuous" assays. They do not enable quantification of metabolite levels in the same cells over time. Additionally, they do not reveal cell-to-cell variation in metabolite levels, which is important for understanding population heterogeneity.

Fluorescence imaging approaches can provide a continuous assay of metabolite concentrations in cells. There are two popular approaches for designing metabolite sensors. One approach is fluorescent protein (FP)-based sensors, which typically depend on the FRET signal generated when two fused FPs become closer together. This is regulated by a ligand-binding domain, such as calmodulin (Tian et al., 2009), that changes structure in response to binding the target ligand. The advantages of this approach are that it can be easily genetically encoded and that the fluorescence signal is ratiometric (Palmer, Qin, Park, & McCombs, 2011); however, the changes in signal are often not very pronounced. Another noteworthy approach is to derivatize organic dyes such that the dye fluorescence is activated by a metabolite-recognizing group, as is the case for Fura-2 and many reactivity-based sensors (Chan, Dodani, & Chang, 2012; Malgaroli, Milani, Meldolesi, & Pozzan, 1987; Valeur, 2000). Despite producing a robust signal, this approach can suffer from difficulty in delivering sensors intracellularly, toxicity problems, and the labor-intensive task of synthesizing novel dye-based sensors.

Recently a new type of sensor technology has been developed based on fluorogenic aptamers (Paige, Nguyen-Duc, Song, & Jaffrey, 2012; You, Litke, & Jaffrey, 2015) (Fig. 1). These fluorescent RNA aptamers have been synthetically evolved to activate the fluorescence of an otherwise nonfluorescent dye that easily crosses membranes, shows minimal nonspecific fluorescence in cells, and is nontoxic. As an example, Spinach (Paige, Wu, & Jaffrey, 2011) binds to 3,5-difluoro-4-hydroxybenzylidene imidazolinone (DFHBI), an analogue of the chromophore formed in green FP. Binding of the target metabolite is linked to the folding of Spinach,

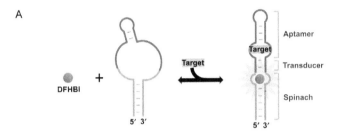

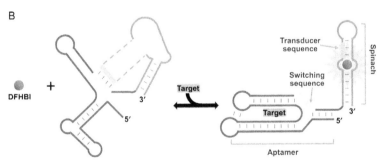

Fig. 1 Two types of metabolite sensors based on fluorogenic aptamers. (A) Spinach-RNA-based sensor using an allosteric mechanism of metabolite (target) detection. Spinach can only bind DFHBI when the target is available to correctly fold Spinach's binding pocket. (B) Structural switching approach to sensor design using Spinach and riboswitch RNA. Transducer sequence of Spinach and switching sequence of riboswitch interact such that Spinach cannot bind DFHBI until target is bound, which precludes this interaction, such that Spinach can bind DFHBI and fluoresce. *Sensor mechanism diagrams originally printed in You, M., Litke, J. L., & Jaffrey, S. R. (2015). Imaging metabolite dynamics in living cells using a Spinach-based riboswitch.* Proceedings of the National Academy of Sciences of the United States of America, 112(21), E2756–E2765. http://doi.org/10.1073/pnas.1504354112. Copyright (2015) National Academy of Sciences, USA.

leading to fluorescence. This approach is genetically encodable since the RNA can be expressed within the cell.

Two types of approaches have been described: allosteric Spinach sensors (Paige et al., 2012) and Spinach riboswitches (You et al., 2015). The structure of allosteric Spinach sensors is shown in Fig. 1A. The metabolite-binding aptamer is linked to a critical stem in Spinach via a short transducer region (Paige et al., 2012). In the absence of the metabolite target, the aptamer is unstructured and prevents Spinach from folding. However, metabolite binding results in hybridization of the transducer domain and folding of Spinach, leading to fluorescence. Optimal performance requires optimizing the transducer sequence. This approach has led to the development of sensors for adenosine diphosphate, cyclic-di-GMP, cyclic-di-AMP, etc. (Kellenberger, Wilson, Sales-Lee, & Hammond, 2013; Song, Strack, & Jaffrey, 2013; Strack, Song, & Jaffrey, 2013). These allosteric Spinach sensors allow rapid and dynamic monitoring of intracellular dynamics of metabolites. However, one key limitation is to identify a metabolite-binding aptamer that has suitable selectivity in a complex cellular environment.

The Spinach riboswitch approach takes advantage of the high specificity of naturally occurring riboswitches. Riboswitches are found in the 5′-UTR of certain mRNAs from many lower organisms and undergo a conformational switch upon binding metabolites that affects mRNA translation or stability (Mandal & Breaker, 2004; Winkler, Nahvi, & Breaker, 2002). The key element of this regulation mechanism is the switching of a single strand, ie, switching sequence, between binding to the riboswitch's aptamer domain and to a semistructured region adjacent that influences mRNA translation or stability, known as the expression platform.

Given the known structure of Spinach and several riboswitches, we modified several known riboswitches so that the switching triggered the formation of the DFHBI-binding pocket of Spinach (Fig. 1B). This strategy takes advantage of the similarity between the structure of Spinach and the expression platform. The strategy also takes advantage of the similarity between sequences in the switching strand and the fluorophore-binding pocket in Spinach. The straightforward process of developing Spinach riboswitch sensors (Fig. 2) allows the development of an array of sensors for many important cellular metabolites recognized by riboswitches. Here, we describe the procedure to design, optimize, and apply Spinach riboswitch sensors for imaging of metabolite dynamics in bacteria cells.

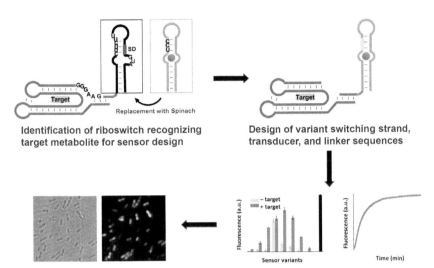

Fig. 2 Scheme for development of RNA-based structural switching sensor. A riboswitch for the target metabolite is identified and the expression platform is replaced with the Spinach sequence. The switching sequence of the riboswitch is identified that can interact with Spinach's transducer sequence or a slightly modified transducer sequence. Variants of putative sensor with changes in the switching strand, transducer, and linker regions are tested in vitro for properties such as signal response, affinity, and kinetics of binding. Cloning of best sensor variants into vector for bacterial T7-induced expression allows imaging of metabolite concentration changes in the presence of DFHBI. (See the color plate.)

2. STRUCTURE-GUIDED DESIGN OF SPINACH RIBOSWITCH SENSORS

For rational design of RNA Spinach riboswitch sensors, structural information is very useful in finding RNA sequences with sensor-like properties and optimizing their behavior, shown in Fig. 3A. The 24 different classes of riboswitches are comprised of sequences that recognize nucleotides, amino acids, sugars, cofactors, and cations/anions (Baker et al., 2012; Breaker, 2011). To date, the Nucleic Acid Database Project lists over 180 different crystal structures of known riboswitches (http://ndbserver.rutgers.edu/). Some riboswitches have structures in multiple conformations, which can help indicate which sequences might make suitable Spinach riboswitch sensor candidates.

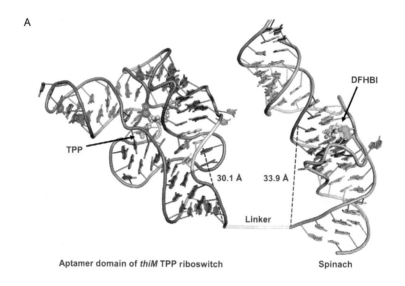
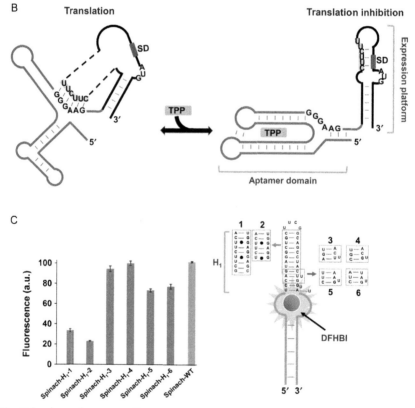

Fig. 3 See legend on opposite page.

2.1 Structure-Guided Design of Switching Sequences and Transducer Sequences

Spinach forms a unique G-quadruplex structure (Warner et al., 2014), which along with a Hoogsteen U•A•U base triple forms a binding pocket for DFHBI that activates the fluorophore's fluorescence. In Spinach, the stem adjacent to the fluorophore-binding pocket (H_1 stem) holds a 5′-UCC-3′ sequence (Figs. 2 and 3A). This sequence, called the transducer sequence, overlaps with the base triple that helps form the DFHBI-binding pocket and is critical for Spinach fluorescence (Paige et al., 2012; You et al., 2015). If this transducer sequence interacts with a complementary sequence from the riboswitch, then we would expect destabilization the U•A•U triplet and reduced DFHBI binding and fluorescence. If this complementary sequence also occupies different conformations in the riboswitch structure when the metabolite is bound or unbound, then we have the basis for a sensor.

Nature has already evolved the necessary complementarity in the switching sequence of many riboswitches. The key element of the riboswitch mechanism is the switching of this sequence between binding to the aptamer domain and to the expression platform, which, for example, contains a Shine–Dalgarno (SD) sequence. This switching is mediated by riboswitch conformation, such that metabolite binding to the riboswitch releases the SD sequence and enables ribosome binding, illustrated in Fig. 3A. Furthermore, the switching sequence can be optimized to allow

Fig. 3 (A) Model of the *thiM*-based Spinach riboswitch sensor for TPP. Riboswitch in *blue* with its switching sequence in *yellow*, while Spinach in *green* with transducer sequence in *gray*. Spinach U•A•U base triple is shown in *cyan* (overlap with most proximal residue of transducer sequence) and G-quadruplex shown in *purple*. TPP and DFHBI are shown in *ball-and-stick* representation. The distance from the switching sequence and transducer sequence to the base of *thiM* and Spinach, respectively, is quite similar, which would not be apparent in a two-dimensional representation. (B) Mechanism of translational control by the *thiM* riboswitch in response to TPP. The transducer sequence shows complementarity to the 5′-AGGAGG-3′ Shine–Dalgarno sequence and switching sequence of the riboswitch's aptamer domain. (C) H_1 stem of Spinach tolerates certain modifications with only modest fluorescence changes. Such mutants expand the modularity of the approach and the range of transducer sequences that can be used for Spinach-based sensor design. *Structural model, sensor mechanism, and Spinach stem diagrams originally printed in You, M., Litke, J. L., & Jaffrey, S. R. (2015). Imaging metabolite dynamics in living cells using a Spinach-based riboswitch. Proceedings of the National Academy of Sciences of the United States of America, 112(21), E2756–E2765. http://doi.org/10.1073/pnas.1504354112. Copyright (2015) National Academy of Sciences, USA.* (See the color plate.)

efficient switching between interaction with Spinach and the riboswitch in response to metabolite concentration.

Knowledge of riboswitch structure is immensely useful in identifying a switching sequence that will interact with Spinach's transducer sequence. PyMOL is an invaluable open-source software tool for making structure-based hypotheses and is available for free from Schrödinger, LLC. A community-supported wiki (http://pymolwiki.org/) along with various online sources contain tutorials for the skills needed to observe and measure aspects of a riboswitch structure as in Fig. 3B. An algorithmic prediction of RNA structure, such as mFold (http://unafold.rna.albany.edu/?q=mfold) can also help guide identification of a good switching sequence. Structural information has also been collected in the apo-states of certain riboswitches (Garst, Héroux, Rambo, & Batey, 2008; Huang, Serganov, & Patel, 2010) and many have been probed using 2′-hydroxyl probing (SHAPE) to identify bases that have altered conformations upon metabolite binding (Steen, Malhotra, & Weeks, 2010). In several riboswitch classes and putative classes, genomic alignment of homologous riboswitches (Barrick & Breaker, 2007; Weinberg et al., 2010) suggests which bases are more conserved and perhaps more essential for control of gene regulation. Without much prior knowledge of the aptamer domain, a stem that holds the 5′ and/or 3′ ends together in a riboswitch may also contain a good candidate for a switching sequence.

Even if a riboswitch does not contain a clear candidate for a switching sequence in the native conformation, there are additional steps to consider that may lead to an effective sensor. Critically, the U bases of the Spinach H_1 stem can base pair with A or G in RNA, and Spinach tolerates some mutations in the H_1 stem, as long of it retains a stem structure, forming A-U, G-C, or G-U base pairs. There is evidence that certain modifications of the 3′ end of this transducer sequence do not impair the function of Spinach-based metabolite sensors, as shown in Fig. 3C. Thus, there is flexibility to change the "native" 5′-UCC-3′ of Spinach into other sequences that can act as the transducing sequence that will interact with the switching sequence. Changing the sequence of the riboswitch to adjust transduction of the sensor is also possible, but may impair sensitivity of the sensor for the metabolite target. These principles allow us to design sensors from a wider variety of riboswitches for detection of many different metabolites.

2.2 Structure-Guided Design of Linker Sequences

The way in which the riboswitch and Spinach are linked can change the behavior of the sensor. Previous sensors have placed the riboswitch on the 5′ end of Spinach, but in principle the order of domains in the 5′-to-3′ sequence is not restricted. Once putative switching sequences in the riboswitch and transducer sequences in Spinach have been determined, the linker sequence between the riboswitch and Spinach should be considered, which also benefits greatly from a structural approach.

The purpose of adjusting the linker sequence is to optimize positioning of the transducer sequence relative to the switching sequence. The crystal structure of Spinach demonstrates that the distance between its 5′ end and the CUU within the transducer sequence and the base triplet is roughly 3.4 nm (Warner et al., 2014). The distance between the riboswitch's 3′ end and switching sequence normally cannot be changed (Serganov, Polonskaia, Phan, Breaker, & Patel, 2006), but the linker length can be controlled to fine tune a sensor, as illustrated in Fig. 3A. The optimal distance from the linker to the switching sequence should be around this 3.4 nm distance. For example, as shown in the structural model of the optimized thiamine pyrophosphate (TPP) Spinach riboswitch sensor (Fig. 3A), the predicted distance from the linker to the switching sequence (3.0 nm) is nearly the same as the distance to the corresponding U base of the transducer sequence (3.4 nm).

Optimized length and flexibility of the linker can also enhance sensor function. Long lengths of single-stranded linkers allow flexibility that lets the sensor explore different orientations between the domains. On the other hand, a long linker may allow "too much flexibility" if the domains of the sensor are already in an ideal position. Less flexible double-stranded linkers that extend from the stem of the riboswitch or Spinach can also be tested. One consideration is that double-stranded linkers will both extend and rotate the Spinach or riboswitch helixes, thereby inducing steric effects that may facilitate or disturb the interaction between the transducer and switching sequences. Once again, structural information is a critical guide for optimizing linker types and the behavior of RNA-based sensors.

3. IDENTIFYING SPINACH RIBOSWITCH SENSORS IN VITRO

Once a series of putative sensors have been conceived, the next step toward imaging cellular metabolites is to characterize these RNA sequences in vitro. As discussed earlier, there are several structure-guided design approaches that may lead to discovery of an optimized Spinach riboswitch sensor. To identify which designs of switching sequence, transducer sequence, and linker sequence are most successful, one must enzymatically synthesize each RNA and then observe fluorescence changes with different levels of metabolite concentration. It is also important to measure other sensor characteristics, such as kinetic response and specificity of recognition. All of these in vitro tests are critical to identifying potential Spinach riboswitch sensors for intracellular imaging.

3.1 In vitro Preparation of RNA-Based Sensor Variants

3.1.1 Materials and Equipment

PCR thermal cycler

PCR clean-up kit (eg, QIAquick PCR purification kit, Qiagen, Hilden, Germany, Cat. No. 28104)

Ampliscribe T7-Flash Transcription Kit (Epicentre, Madison, WI, Cat. No. ASF3507)

Incubator, 37°C

RNA purification kit (eg, RNA Clean & Concentrator, Zymo Research, Irvine, CA, Cat. No. R1018; Bio-Spin Columns, BioRad, Hercules, CA, Cat. No. 732-6251)

Nanodrop or spectrophotometer for UV–Vis absorption

3.1.2 Procedures

1. Once the riboswitch is selected, and the transducer and switching sequences are identified, double-stranded DNA (dsDNA) templates for variants of these sequences should be generated. The T7 promoter sequence (5′-TAATACGACTCACTATAGGG-3′) should be added to the 5′ end of the template for successful in vitro transcription.
2. The protocol for standard PCR can be easily found elsewhere. Here, we briefly introduce the procedure for overlap-extension PCR, a method for generating a long (>100 nt) dsDNA template that is difficult to synthesize chemically and can be expensive when purchased. During overlap-extension PCR, two equimolar template fragments are assembled together to generate the full dsDNA template. These two template fragments contain an overlap region with melting temperature typically around 60°C. A typical thermocycler program can be as follows: heat samples to 98°C for 20 s, followed by 15 repeated cycles of: denaturing primers at 98°C for 5 s, annealing at 58°C for 20 s, and extending template primers at 72°C for 20 s. After the repeated cycles, a final extension at 72°C is held for 5 min, followed by cooling to hold samples stable at 4°C.
3. After overlap-extension PCR, standard primers complementary to the ends of the product are added to the template mix. Amplification of

the template follows the same thermocycler program but with 30 cycles rather than 15 cycles.
4. PCR products should be verified for size by mobility on a 2% agarose gel. If the samples are sufficiently pure, then they can be purified using a PCR clean-up kit (eg, QIAquick PCR purification kit, Qiagen, Cat. No. 28104); otherwise they must be separated from impurities by agarose gel electrophoresis, followed by excision and purification from gel.
5. The reaction for generating RNA from the dsDNA templates consists of using a T7 RNA polymerase recognizing the T7 promoter in the 5′ end of the DNA template. The T7 polymerase transcribes through to the end of the template. Following the manufacturer's protocol, an in vitro transcription reaction can be carried out using the Ampliscribe T7-Flash Transcription Kit (Epicentre, Cat. No. ASF3507) in a 20 µL scale. Allow the reaction to incubate at 37°C for at least 1 h. If the maximum RNA yield is desired, the reaction can proceed overnight. Afterward, treat the reaction with DNase at 37°C for 30 min.
6. Before proceeding, RNA must be purified from the transcription reaction components. Purification of transcribed RNA can be carried out by column purification or an ethanol precipitation method. Column purification can be achieved using one of several kits (eg, RNA Clean & Concentrator, Zymo Research, Cat. No. R1018; Bio-Spin Columns, BioRad, Cat. No. 732-6251). The elution must be made with RNase-free water or TE buffer to prevent base hydrolysis and should be concentrated enough that quantification of concentration by UV absorbance (eg, with a Nanodrop Instrument) is appropriate (0.1–5 µg/µL). It is best to store RNA at −80°C to reduce degradation by incidental RNases.

3.2 Quantification of Fluorescence Sensor Characteristics
3.2.1 Materials and Equipment
Fluorimeter
Incubator, 37°C
DFHBI or DFHBI-1T fluorophore (Lucerna, New York, NY, SKU: 400-1 mg or SKU: 410-1 mg, respectively)

3.2.2 Procedures
Testing the fluorescence characteristics of each sensor begins with measurements of fluorescence response to physiological amounts of metabolite.

Sensors that exhibit the highest fold increase in fluorescence after adding metabolite are tested further for dose response and kinetics. Fluorescent aptamers like Spinach, which activate dye fluorescence, should be measured with fluorophore in excess because of the low background of fluorophore, when free in solution. In this way, the fluorescence observed will be determined exclusively by the percent of the RNA sensor that adopts a folded conformation. If the RNA is in excess and the fluorophore is limiting, any unfolded sensor will not be detectable since the maximal fluorescence level would be determined by the concentration of the fluorophore. The suitable fluorophore for these sensors is DFHBI or its trifluoroethyl derivative, DFHBI-1T (Lucerna). Other measurement conditions may vary from sensor to sensor; however, when using Spinach as a fluorescing domain, sufficient magnesium concentration is necessary for proper folding and fluorescence. Previous Spinach-based sensors are most effective at 2–5 mM magnesium concentration in vitro, which resembles bacterial levels, with fluorophore concentrations roughly 100-fold higher than RNA concentrations.

A starting sample buffer for testing Spinach-based RNA sensors may contain 40 mM HEPES pH 7.4, 100 mM KCl, and 5 mM MgCl$_2$ with 10 µM DFHBI-1T and 0.1 µM of RNA. Assessment of sensors begins with measurement of fluorescence before and after incubation with physiological concentration of metabolite (eg, 100 µM TPP) at 37°C for 30 min. Any fluorescence that is detected after incubation should also be compared to fluorescence of Spinach without the target-binding aptamer domain (Fig. 3C), and any modifications to the Spinach transducer region must be reflected in this control. Measurement of fluorescence with a fluorimeter should be made with 470 nm excitation and 503 nm emission.

The best candidate sensors normally have the highest fold change upon adding metabolite, and it is critical to see which candidates have a suitable affinity for the metabolite. A suitable affinity (EC$_{50}$) should fall well within the physiological concentration range of the target and preferably be slightly higher than the physiological concentration range. In this case, a doubling of the metabolite concentration should result in approximately a doubling in the binding of the metabolite to the sensor and a corresponding increase in fluorescence. Thus, the sensor will produce a fluorescence in a nearly linear manner compared to the metabolite concentration. Titration curves are measured as the fold-change measurement for separate samples with a range of metabolite concentrations or for a single sample with progressive addition of metabolite. These fluorescence measurements form a curve that can be fit by a nonlinear regression to a model for substrate binding.

Kinetic measurements of high-affinity sensors should be made as well to help pick the best sensor and to get a sense of the time course for sensor function in bacteria. Continuous or time-lapsed measurements of fluorescence should be conducted to identify sequences with the highest rate of response to metabolite. An acceptable sensor will have rapid metabolite-induced fluorescence signal change relative to the timescale of physiological dynamics of metabolite. For example, we developed a TPP-targeting Spinach riboswitch sensor that reaches 90% of its maximal fluorescence signal within 10 min. This time frame was suitable for experiments in which we monitored TPP biosynthesis kinetics that occurred over several hours in cells (You et al., 2015).

Furthermore, the specificity of RNA sensors can be measured by determining the fluorescence response to metabolites that are known to occur in cells and which are molecularly similar to the metabolite, eg, thiamine and thiamine monophosphate for TPP sensors. For sensors designed from naturally occurring riboswitches, this step is still important, even though riboswitches tend to have high specificity for their intended metabolites (Mandal & Breaker, 2004).

4. LIVE-CELL IMAGING OF METABOLITES WITH SPINACH RIBOSWITCH SENSORS

After in vitro characterization of RNA-based sensors, it is critical to validate these sensors in live cells. RNA-based metabolite sensors are still difficult to use in mammalian cells, mainly due to the limited stability and abundance of these short and structured RNAs. Despite this, these RNAs typically maintain a steady state level in cells due to a balance of new transcription and degradation. RNA-based sensors with fast kinetic response and readily detectable fluorescence signal are becoming popular for metabolite analysis in bacterial cells (Kellenberger et al., 2013; You et al., 2015). Shown below is a protocol using Spinach riboswitch sensor for TPP imaging in live *Escherichia coli* cells.

4.1 Preparation of Sensor-Expressing *E. coli*
4.1.1 Materials and Equipment
PCR thermal cycler
Water bath or heat block, 42°C

Shaking incubator, 37°C
LB agar plate: dissolve 10 g of tryptone, 5 g of yeast extract, 20 g of agar, and 10 g of NaCl in a final volume of 1 L with deionized water. Autoclave for 15 min at 121°C. After cooling, add 50 µg/mL kanamycin and then pour into 10-cm Petri dishes.

4.1.2 Procedures
1. Clone Spinach riboswitch sensor into bacterial expression vector. Traditional restriction site cloning can be used to generate a bacterial expression plasmid containing the in vitro characterized sensor. For example, sensor templates can be PCR amplified with primers containing either *Bgl*II or *Xho*I restriction sites on the 5′ and 3′ ends for insertion into the T7 expression cassette of pET28c. These restriction sites place the sensor upstream of a T7 terminator in pET28c. A pETDuet vector, which expresses the sensor and a far-red FP from the same promoter as an expression control can also help to gauge sensor function within *E. coli*. All the following described steps should be performed in a sterile environment.
2. Take three tubes of chemically competent *E. coli* Rosetta2 (DE3) cells thawed on ice. Add 40 ng of pET28c-TPP Spinach riboswitch sensor, pET28c-Spinach2 (positive control), and pET28c (negative control), respectively, to each tube. Mix gently by flicking the tubes and incubate on ice for 5 min.
3. Heat shock in a 42°C water bath for 45 s, then immediately place back on ice to chill for 2 min. After chilling, add 250 µL of SOC medium and incubate at 37°C for 45 min with shaking.
4. Plate 50 µL of incubated cells onto an LB agar plate supplemented with 50 µg/mL kanamycin. Incubate at 37°C overnight. Pick several colonies to isolate plasmids. Sequence or use restriction digestion to confirm correct transformation.

4.2 Imaging Dynamics of Metabolite Concentration in live *E. coli* (TPP Sensor)
4.2.1 Materials and Equipment
Shaking incubator, 37°C
Nanodrop or spectrophotometer for UV–Vis absorption
Inverted wide-field fluorescent microscope
Glass-bottom 24-well imaging plates
Poly-D-lysine or poly-L-lysine

LB medium: dissolve 10 g of tryptone, 5 g of yeast extract, and 10 g of NaCl in a final volume of 1 L with deionized water. Autoclave for 15 min at 121°C. The medium can be stored at room temperature.
M9 medium: prepare 5× M9 salts solution by dissolving 64 g of Na_2HPO_4, 15 g of KH_2PO_4, 5 g of NH_4Cl, and 2.5 g of NaCl in a final volume of 1 L with deionized water. Autoclave for 15 min at 121°C. Final M9 medium contain 1× M9 salts solution, 2 mM $MgSO_4$, 100μM $CaCl_2$, and 2 mg/mL glucose. Sterilize the medium by vacuum filtration through a 0.22-μm filter.

4.2.2 Procedures

1. Inoculate a single colony inside a sterile culture tube with 5 mL of LB medium and 50 μg/mL kanamycin. Grow the cultures at 37°C with shaking overnight.
2. Measure the optical density of an overnight culture at 600 nm (OD_{600}). Dilute the culture to a 0.05–0.1 OD_{600} units/mL with LB medium with 50 μg/mL kanamycin, grow at 37°C with shaking till 0.4 OD_{600} units/mL. Add 1 mM isopropyl β-D-1-thiogalactopyranoside (IPTG), and continue to grow at 37°C with shaking for 2 h. During this final induction, 300 μg/mL adenosine, an inhibitor of de novo TPP biosynthesis, may be added together with IPTG.
3. Prepare a glass-bottomed imaging plate, eg, 24-well MatTek plate, by adding 300 μL of poly-D-lysine solution to each well and incubating at 37°C for 3 h. Add 500 μL of sterile water twice to rinse each well and remove excess poly-D-lysine.
4. Spin down 100 μL of an IPTG-induced culture at 5000 g for 2 min at room temperature to pellet the culture. Resuspend in 1 mL of M9 medium. Take a 200 μL aliquot of resuspended culture to plate on the poly-D-lysine-coated plate. Incubate at 37°C for 45 min to allow cell adherence.
5. Wash adherent cells twice with 500 μL of M9 medium to remove unattached cells. Then, add 200 μL of M9 medium supplemented with 200 μM DFHBI. Incubate at 25°C for 1 h.
6. Image the plate through a 60× oil objective. An example filter set includes a 470/40 nm excitation filter, a dichroic mirror 495 nm (long pass), and an emission filter 525/50 nm. Start imaging with the cells expressing pET28c-Spinach2 (positive control). While imaging, focus on a field containing 50–100 evenly adhered cells. Proper exposure time

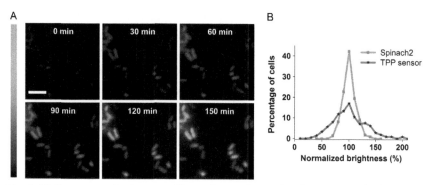

Fig. 4 (A) Representative images of bacteria transformed with TPP Spinach riboswitch. Time course of fluorescence over 3 h demonstrates intracellular sensor function. (B) Comparison of distributions of fluorescence signal from bacteria expressing TPP Spinach riboswitch and the corresponding Spinach sequence by itself. Demonstrates that variations in cell-to-cell fluorescence are due to differences in TPP concentrations rather than differences in expression from T7 promoter. *Fluorescence images adapted from You, M., Litke, J. L., & Jaffrey, S. R. (2015). Imaging metabolite dynamics in living cells using a Spinach-based riboswitch.* Proceedings of the National Academy of Sciences of the United States of America, 112(21), E2756–E2765. http://doi.org/10.1073/pnas. 1504354112. Copyright (2015) National Academy of Sciences, USA. (See the color plate.)

is normally in the range of 100 ms to 2 s, which allows the highest signal without saturated pixels.

7. To image dynamics of intracellular concentration dynamics, for example, of TPP, add 10 μM thiamine to each well. Thiamine can be converted to TPP by intracellular thiamine kinase. Acquire images immediately (time zero) and every 15 min afterward for each well. Between each image acquisition, the shutter should be closed to minimize photobleaching of fluorophore and cell damage. Live fluorescence images were taken for 3 h, as demonstrated in Fig. 4A. Proper exposure time is normally in the range of 100 ms to 2 s. Time intervals and the required duration may vary based on the metabolite and the biological process being studied.

4.3 Data Analysis of Metabolite Concentration Variations (TPP Sensor)

1. Data analysis can be carried out using image analysis software such as NIS-Elements. To subtract the background fluorescence, which is primarily due to cellular autofluorescence, the image for pET28c-expressing (negative control) cells in media containing DFHBI is used.

2. Choose cells that are fully adhered and ensure that the field is entirely in focus in each frame. Circle these cells as regions of interest (ROIs). Track their fluorescence intensity through all background-subtracted timepoint image frames. Calculate the mean fluorescence intensity per unit volume by dividing the total fluorescence by the calculated volume for each ROI. Volume was estimated using the equation $V = \pi r^2 [(4/3)r + a]$, where V is the cell volume, r is the radius, and a is the side length of the cell. Collect the mean fluorescence intensity change relative with respect to time for 150–300 cells in each well.
3. Normalize the sensor fluorescence values to the mean pET28c-Spinach2 (positive control) signal from the corresponding time point. If a vector simultaneously expressing the sensor and an FP has been used, the sensor fluorescence can be normalized to the fluorescence protein signal in the same cells. These normalized values represent the fluorescence signal change due to dynamics of TPP concentration.
4. The fold increase in fluorescence over time is calculated as the ratio of mean intensity at a desired time point to that at time 0. After this value is calculated, cells are binned according to their percentage of total signal relative to the average signal, which is normalized to 100%.
5. Generate heat maps as a pictorial representation of fold change in fluorescence signal as in Fig. 4A. Use the background-subtracted image with the most intense TPP sensor signal to adjust the upper lookup table (LUT) arrow. Assign the brightest cell the red color. Apply the same LUT setting to all other images.
6. By expressing Spinach alone in the same expression vector, the cell-to-cell variation in RNA expression can be seen. This can help to rule out the possibility that the variations in cellular fluorescence are due to variations in Spinach riboswitch expression level rather than differences in metabolite concentration (Fig. 4B).

5. SUMMARY AND CONCLUDING REMARKS

Versatile RNA-based sensor technologies have been developed to monitor dynamics of metabolites inside live cells. Here we describe protocols for designing, optimizing, and implementing modular RNA-based metabolite sensors. We have illustrated a structure-guided strategy to identify riboswitches that can be fashioned into genetically encoded fluorogenic riboswitch sensors, using both their metabolite sensing ability and their structure-switching capacity. Our example of imaging bacterial TPP dynamics based on the *thiM* TPP riboswitch demonstrates that one can ultimately use a sensor to discern different individual cellular levels of metabolite.

This approach to sensor design could be greatly improved by a number of anticipated advancements in RNA imaging (Filonov, Kam, Song, & Jaffrey, 2015) and new types of metabolite-recognizing riboswitches (Lynch & Gallivan, 2009; Weigand & Suess, 2007). By exploiting the diversity of naturally occurring riboswitches, it should be possible to image a large number of intracellular metabolites. Additionally, improvements in fluorogenic RNA aptamers may lead to enhanced sensitivity and future extension of this technology for dynamic imaging of mammalian metabolites and proteins.

ACKNOWLEDGMENTS

We thank Grigory S. Filonov for helpful suggestions and comments. Support was provided by the Tri-Institutional PhD Program in Chemical Biology (MSKCC, Weill-Cornell Medical College, and Rockefeller University). This work was also supported by NIH Grants R01 NS064516 and R01 EB010249 (to S.R.J.).

REFERENCES

Baker, J. L., Sudarsan, N., Weinberg, Z., Roth, A., Stockbridge, R. B., & Breaker, R. R. (2012). Widespread genetic switches and toxicity resistance proteins for fluoride. *Science (New York, N.Y.)*, *335*(6065), 233–235. http://doi.org/10.1126/science.1215063.

Barrick, J. E., & Breaker, R. R. (2007). The distributions, mechanisms, and structures of metabolite-binding riboswitches. *Genome Biology*, *8*(11), R239. http://doi.org/10.1186/gb-2007-8-11-r239.

Breaker, R. R. (2011). Prospects for riboswitch discovery and analysis. *Molecular Cell*, *43*(6), 867–879. http://doi.org/10.1016/j.molcel.2011.08.024.

Chan, J., Dodani, S. C., & Chang, C. J. (2012). Reaction-based small-molecule fluorescent probes for chemoselective bioimaging. *Nature Chemistry*, *4*(12), 973–984. http://doi.org/10.1038/nchem.1500.

Filonov, G. S. S., Kam, C. W. W., Song, W., & Jaffrey, S. R. R. (2015). In-gel imaging of RNA processing using broccoli reveals optimal aptamer expression strategies. *Chemistry & Biology*, *22*(5), 649–660. http://doi.org/10.1016/j.chembiol.2015.04.018.

Garst, A. D., Héroux, A., Rambo, R. P., & Batey, R. T. (2008). Crystal structure of the lysine riboswitch regulatory mRNA element. *Journal of Biological Chemistry*, *283*(33), 22347–22351. http://doi.org/10.1074/jbc.C800120200.

Huang, L., Serganov, A., & Patel, D. J. (2010). Structural insights into ligand recognition by a sensing domain of the cooperative glycine riboswitch. *Molecular Cell*, *40*(5), 774–786. http://doi.org/10.1016/j.molcel.2010.11.026.

Kellenberger, C. A., Wilson, S. C., Sales-Lee, J., & Hammond, M. C. (2013). RNA-based fluorescent biosensors for live cell imaging of second messengers cyclic di-GMP and cyclic AMP-GMP. *Journal of the American Chemical Society*, *135*, 4906–4909. http://doi.org/10.1021/ja311960g.

Lynch, S. A., & Gallivan, J. P. (2009). A flow cytometry-based screen for synthetic riboswitches. *Nucleic Acids Research*, *37*(1), 184–192. http://doi.org/10.1093/nar/gkn924.

Malgaroli, A., Milani, D., Meldolesi, J., & Pozzan, T. (1987). Fura-2 measurement of cytosolic free Ca^{2+} in monolayers and suspensions of various types of animal cells. *The Journal of Cell Biology, 105*, 2145–2155. http://doi.org/10.1083/jcb.105.5.2145.

Mandal, M., & Breaker, R. R. (2004). Gene regulation by riboswitches. *Nature Reviews. Molecular Cell Biology, 5*(6), 451–463. http://doi.org/10.1038/nrm1403.

Paige, J. S., Nguyen-Duc, T., Song, W., & Jaffrey, S. R. (2012). Fluorescence imaging of cellular metabolites with RNA. *Science (New York, N.Y.), 335*(6073), 1194. http://doi.org/10.1126/science.1218298.

Paige, J. S., Wu, K. Y., & Jaffrey, S. R. (2011). RNA mimics of green fluorescent protein. *Science (New York, N.Y.), 333*(6042), 642–646. http://doi.org/10.1126/science.1207339.

Palmer, A. E., Qin, Y., Park, J. G., & McCombs, J. E. (2011). Design and application of genetically encoded biosensors. *Trends in Biotechnology, 29*(3), 144–152. http://doi.org/10.1016/j.tibtech.2010.12.004.

Patti, G. J., Yanes, O., & Siuzdak, G. (2012). Innovation: Metabolomics: The apogee of the omics trilogy. *Nature Reviews. Molecular Cell Biology, 13*(4), 263–269. http://doi.org/10.1038/nrm3314.

Serganov, A., Polonskaia, A., Phan, A. T., Breaker, R. R., & Patel, D. J. (2006). Structural basis for gene regulation by a thiamine pyrophosphate-sensing riboswitch. *Nature, 441*(7097), 1167–1171. http://doi.org/10.1038/nature04740.

Song, W., Strack, R. L., & Jaffrey, S. R. (2013). Imaging bacterial protein expression using genetically encoded RNA sensors. *Nature Methods, 10*(9), 873–875. http://doi.org/10.1038/nmeth.2568.

Steen, K. A., Malhotra, A., & Weeks, K. M. (2010). Selective $2'$-hydroxyl acylation analyzed by protection from exoribonuclease. *Journal of the American Chemical Society, 132*(29), 9940–9943. http://doi.org/10.1021/ja103781u.

Strack, R. L., Song, W., & Jaffrey, S. R. (2013). Using spinach-based sensors for fluorescence imaging of intracellular metabolites and proteins in living bacteria. *Nature Protocols, 9*(1), 146–155. http://doi.org/10.1038/nprot.2014.001.

Tian, L., Hires, S. A., Mao, T., Huber, D., Chiappe, M. E., Chalasani, S. H., ... Looger, L. L. (2009). Imaging neural activity in worms, flies and mice with improved GCaMP calcium indicators. *Nature Methods, 6*(12), 875–881. http://doi.org/10.1038/nmeth.1398.

Valeur, B. (2000). Design principles of fluorescent molecular sensors for cation recognition. *Coordination Chemistry Reviews, 205*(1), 3–40. http://doi.org/10.1016/S0010-8545(00)00246-0.

Warner, K. D., Chen, M. C., Song, W., Strack, R. L., Thorn, A., Jaffrey, S. R., & Ferré-D'Amaré, A. R. (2014). Structural basis for activity of highly efficient RNA mimics of green fluorescent protein. *Nature Structural & Molecular Biology, 21*(8), 658–663. http://doi.org/10.1038/nsmb.2865.

Weigand, J. E., & Suess, B. (2007). Tetracycline aptamer-controlled regulation of pre-mRNA splicing in yeast. *Nucleic Acids Research, 35*(12), 4179–4185. http://doi.org/10.1093/nar/gkm425.

Weinberg, Z., Wang, J. X., Bogue, J., Yang, J., Corbino, K., Moy, R. H., & Breaker, R. R. (2010). Comparative genomics reveals 104 candidate structured RNAs from bacteria, archaea, and their metagenomes. *Genome Biology, 11*(3), R31. http://doi.org/10.1186/gb-2010-11-3-r31.

Winkler, W., Nahvi, A., & Breaker, R. R. (2002). Thiamine derivatives bind messenger RNAs directly to regulate bacterial gene expression. *Nature, 419*(6910), 952–956. http://doi.org/10.1038/nature01145.

You, M., Litke, J. L., & Jaffrey, S. R. (2015). Imaging metabolite dynamics in living cells using a Spinach-based riboswitch. *Proceedings of the National Academy of Sciences of the United States of America, 112*(21), E2756–E2765. http://doi.org/10.1073/pnas.1504354112.

AUTHOR INDEX

Note: Page numbers followed by "*f*" indicate figures.

A

Aakalu, G., 53–54
Abrahamsson, S., 142–143
Accornero, N., 109–110, 125–126, 139
Achilefu, S., 90
Adams, S.R., 107–108
Adiconis, X., 256, 266
Adjaye, J., 307
Aguilera, T.A., 90–91
Ahmad, M., 3, 5
Ahonen, I., 90
Aitken, C.E., 139–140
Akamatsu, M., 243
Alexandrov, A., 239–241
Allen, P.B., 136
Alm, C., 106
Almo, S.C., 126, 161
Amari, K., 108–109
Anacker, J., 99*f*
Anderson, C.J., 90
Andreiev, A.I., 107–108
Androsavich, J.R., 136–137, 139
Angenon, G., 110, 114
Aoyama, Y., 107–108
Apostolopoulos, N., 14
Apweiler, R., 256
Arduise, C., 144
Ares, M., 270–271
Ashby, J., 108–109
Ashwal-Fluss, R., 216–218
Aslanidis, C., 239
Awad, A.M., 207
Azzalin, C.M., 293, 308–309

B

Babak, T., 106
Babcock, H., 40
Babendure, J.R., 107–108
Backofen, R., 256
Baker, J.L., 319
Baker, L., 106
Balagopal, V., 3

Baldeck, J.D., 240–241
Banerjee, A.K., 106
Banerjee, J., 207
Banerjee, R., 114–115
Bao, G., 107–108
Barash, Y., 256, 261–262
Barbarese, E., 124–125
Barrick, J.E., 322
Barron, A.E., 209
Bartok, O., 216–218
Barton, G.J., 53–54
Bassell, G.J., 124–125
Bastiaens, P.I., 204
Batey, R.T., 322
Bathe, M., 174–175
Batish, M., 3–4, 68–69
Baulcombe, D.C., 106, 114–115
Beasley, J., 195–198, 202, 209, 211–212
Beattie, T.L., 259
Beaufils, F., 130
Bedwell, D.M., 242–243
Behm-Ansmant, I., 292
Belaya, K., 148
Beliveau, B.J., 14
Bell, K., 82–84, 107–108
Ben-Ari, S., 256
Bencsics, C.E., 107–108, 126
Bender, G., 128–129
Bendickson, L., 194–212
Bentley, D.L., 160–161, 184
Ben-Yishay, R., 139
Berger, S., 129
Bergsten, S.E., 148
Berninger, P., 238
Bertrand, E., 52, 69, 107–109, 112, 125–126, 128, 139, 161, 194–195, 223
Besse, F., 3
Bewersdorf, J., 77
Bhardwaj, A., 292
Bhuvanagiri, M., 292
Black, D.L., 256, 270–274
Blanchard, J.M., 109–110, 125–126, 139

Blencowe, B.J., 256, 261–262
Boelz, S., 293, 308–309
Boettiger, A.N., 3–5, 7–13, 36–37, 43–44
Boevink, P., 114–115
Bogdanov, A., 90–91
Bogue, J., 322
Bolze, F., 107–108
Bonano, V.I., 256
Bork, P., 256, 292
Bosma, R., 256
Bothma, J.P., 174
Bouche-Pillon, S., 106
Bouthcko, R., 204
Boyko, V., 108–109
Braeckmans, K., 143
Bramsen, J.B., 216
Brandman, O., 243
Brandner, K., 108–109, 111
Bratu, D.P., 68–69, 88–100
Breaker, R.R., 318–319, 322–323, 327
Brechbiel, J.L., 149
Bredow, S., 90–91
Bremer, C., 90–91
Brenner, S.E., 256, 292
Brew, K., 91
Brody, Y., 69, 139
Brooks, R., 114–115
Broude, N.E., 69–70
Brouwer, A.E., 19
Brown, P.O., 216
Brule, C.E., 238–252
Bruno, I.G., 292
Bucher, T., 196–197
Buckner, F.S., 239–241
Bujard, H., 128–129
Bullock, S.L., 148
Bundman, D., 256
Burd, C.E., 216
Burge, C.B., 256
Burger, L., 238
Buschmann, H., 114
Buxbaum, A.R., 3, 108–109, 119, 138, 141–143

C

Cabantous, S., 239–240
Cai, L., 3–5
Cai, W., 90

Calarco, J.A., 256, 261–262
Calvin, S.A., 297–298
Camblong, J., 251–252
Campbell, Z.T., 238
Canio, W., 106
Cannon, D., 53–54
Cantor, C.R., 69–70
Carmo-Fonseca, M., 126, 161
Carrington, J.C., 106
Carson, J.H., 124–125
Carvalho, C., 126, 161
Casadio, A., 53–54
Cates, S., 298
Catrina, I.E., 88–100
Cerovina, T., 106
Cha, B.J., 68–69
Chakrabarty, R., 114–115
Chalasani, S.H., 316
Chamberlain, C., 107–108
Chan, C., 88–89
Chan, J., 316
Chan, P.P., 218
Chan, W., 292
Chang, C.J., 316
Chang, Y.F., 259, 292
Chao, J.A., 52–54, 124–153, 161, 194–195
Chartrand, P., 52, 69, 107–109, 112, 128, 161, 194–195
Chatterjee, M., 106
Chen, C.Y., 216–218
Chen, J., 88–89, 107–108, 126, 130
Chen, K.H., 3–5, 7–13, 36–37, 43–44
Chen, L., 271–274, 276
Chen, M., 53–54, 270–271
Chen, M.C., 321, 323
Chen, P.S., 107–108
Chen, R.E., 216
Chen, X., 90
Chenouard, N., 141–143
Cheong, C.G., 70–72, 74, 107–108
Chepurnykh, T.V., 295
Chernyakov, I., 249
Chiappe, M.E., 316
Chilkoti, A., 209
Chimera, J.A., 116
Chiu, D.T., 136
Chmyrov, A., 139–140
Cho, H., 91, 96, 99f

Cho, S., 136
Choi, S.B., 106, 108–110
Choi, S.Y., 128, 131
Chomczynski, P., 305
Chou, T.B., 148–149
Chow, C.C., 126, 160–162, 164, 172–174, 173f, 176–177, 179, 185, 188–189
Christensen, N.M., 82–84, 107–108
Chubb, J.R., 53–54
Clark, A., 148
Clauder-Munster, S., 251–252
Clausen, B.H., 216
Clavarino, G., 124–125
Clift, K.L., 131
Close, T.J., 116
Coffin, J.M., 107–108, 126
Cohen-Fix, O., 160
Cojocaru, R., 107–108
Colak, D., 292
Collins, K., 211–212
Colvin, K.E., 297–298
Commandeur, U., 110
Condeelis, J.S., 3
Conn, S.J., 216
Conn, V.M., 216
Constantin, T.P., 107–108
Cooper, T.A., 256
Cooperman, B.S., 124–125
Corbino, K., 322
Correa, I.R., 130
Corrigan, A., 53–54
Coskun, A.F., 3, 5
Coue, M., 139–140
Coulon, A., 126, 160–190, 173f
Cox, E.C., 53–54
Craig, N., 160
Cranfill, P.J., 135
Curtis, M.D., 111
Custer, T.C., 77
Czaplinski, K., 3, 52–53, 131

D

Da Costa, J.B., 107–108
Dahm, R., 110
Daigle, N., 107–108, 126
Dalby, B., 298
Damoiseaux, R., 256, 272–274
Darnell, R., 66–67

Darty, K., 222
Darzacq, X., 52–53, 66–67, 69, 131, 139
David, A., 124–125
Davidson, M.W., 130, 135
Davis, I., 68, 148
de Chaumont, F., 141–143
de Jong, P.J., 239
De Paepe, A., 52
De Preter, K., 52
De Rosa, D., 240–241
De Rycke, R., 110, 114
De Smedt, S.C., 143
de Turris, V., 69, 126, 161–162, 164, 172–174, 173f, 176–177, 179, 185, 188–189
de Vries, C., 240–241
Dean, K.M., 238–252
Dean, N., 240
Deckert, J., 211–212
DeLuca, J.G., 190
DeLuca, K.F., 190
Demeester, J., 143
Denise, A., 222
Depicker, A., 110
Deschout, H., 143
DeTitta, G.T., 239–241
Dictenberg, J., 69
Dieckmann, T., 107–108
Dieterich, D.C., 124–125
Dietl, J., 99f
Dietz, H.C., 292, 308
Dimitrova, L., 243
Dinakarpandian, D., 91
Ding, B., 106
Disney, M.D., 194–195, 218
Dodani, S.C., 316
Dolan, B.P., 124–125
Dolgosheina, E.V., 107–108
Donnelly, S.K., 131
Doonan, J.H., 114
Dosset, P., 144
Dow, W.C., 90
Drossard, J., 110
D'Silva, S., 249
Duan, L., 88–89
Dunoyer, P., 106
Dupont, A., 143

E

Eggeling, C., 139–140
Elefsinioti, A., 216
Eliceiri, G.L., 91, 96, 99f
Eliscovich, C., 190
Elledge, S.J., 16
Ellenberg, J., 107–108, 126, 140
Ellies, L.G., 90–91
Elowitz, M.B., 52
Emans, N., 110
Encell, L.P., 130
English, B.P., 130, 190
Ephritikhine, G., 108–109
Ephrussi, A., 3, 66–67, 106, 124–153, 182
Erhardt, M., 108–109
Ermakova, G.V., 124–125, 259–260, 293, 295, 303, 308–309, 311–312
Espenel, C., 144
Esposito, E., 174
Etkin, L.D., 106
Etzion, T., 19
Evans, R.M., 128–129

F

Fague, K., 107–108
Fang, X., 90, 135–136
Farina, K.L., 66–67
Faustoferri, R.C., 240–241
Fay, F.S., 2, 188–189
Femino, A.M., 2, 188–189
Feng, X., 88–89
Ferguson, M.L., 126, 161–162, 164, 172–174, 173f, 176–177, 179, 185, 188–189
Ferralli, J., 108–109
Ferré-D'Amaré, A.R., 321, 323
Filipovska, A., 70–72, 74
Filleur, S., 108–109
Filonov, G.S., 107–108, 194–195, 216–235
Filonov, G.S.S., 332
Fingleton, B., 90–91
Fink, G., 249
Finsen, B., 216
Fiorini, G.S., 136
Fischer, R., 110
Fogarty, K., 2, 188–189
Follenzi, A., 53–54, 136–137

Fonseka, C.Y., 14
Forrest, K.M., 148–149
Fradkov, A.F., 124–125, 256, 259–260, 266, 295
Freitas, R.P., 271
Freundlieb, S., 128–129
Frey, J.U., 53–54
Frey, U., 53–54
Friedman, E., 292
Fu, X., 270–271
Fujiwara, T., 143
Fujiwara, T.K., 77
Fusco, D., 66–67, 109–110, 125–126, 139

G

Gagneur, J., 251–252
Galka-Marciniak, P., 107–108, 125–126
Gallivan, J.P., 332
Ganesan, P., 136
Gao, W., 88–89, 256, 261–262
Garcia, H.G., 174
Gardner, L.B., 292
Garst, A.D., 322
Gaspar, I., 148
Gaublomme, J.T., 256, 266
Gautam, S., 90–91
Gautier, A., 130
Gavis, E.R., 66–67, 148–149
Gawad, C., 216
Gehring, N.H., 293, 305, 308–309
Gelfand, M.S., 256
Gendreizig, S., 130
Geraskin, I.M., 195–197, 207
Gerber, A.P., 238
Gerhardt, I., 136
Gerst, J.E., 3–4
Gertner, R.S., 256, 266
Ghaemmaghami, S., 124–125, 128
Ghosh, I., 72
Ghosh, S., 148
Gibson, U.E.M., 52
Gierlinski, M., 53–54
Gilbert, S.G., 135
Gilchrist, M.J., 148
Gillespie, D.T., 189–190
Gillespie, T., 114–115
Goff, S.P., 295
Gold, B., 216

Author Index

Golding, I., 3, 53–54
Gong, Y., 141–143
Good, P.D., 223
Goodhouse, J., 148
Goodin, M.M., 114–115
Gorlich, D., 129
Gorlick, M., 256
Gorodnicheva, T.V., 311
Gossen, M., 128–129
Gotze, B., 110
Gouzer, G., 124–125
Grayhack, E.J., 238–252
Green, P., 114
Green, R., 160
Green, R.E., 256, 292
Gregor, T., 174
Greider, C., 160
Grima, D., 148–149
Grimm, J.B., 130
Gronemeyer, T., 130
Grossniklaus, U., 111
Grunwald, D., 138–139, 142–143
Guo, S.-M., 174–175
Gupta, R., 216–218, 228
Gupta, V., 194–212
Gurha, P., 216–218, 228
Gurskaya, N.G., 256–267, 292–312
Guttman, M., 66–67
Guy, M.P., 238–239, 243, 250

H

Ha, T., 139–140
Hachet, O., 148
Hacohen, N., 271
Hafner, M., 238
Hagiwara, M., 256, 271–272
Hahn, K.M., 107–108
Haimovich, G., 3, 108–109
Halfpenny, K.C., 88–89
Hall, T.M., 70–72, 74, 107–108
Halstead, J.M., 124–153, 182
Hamada, S., 108–110
Hamilton, A.D., 72
Hamilton, T.P., 107–108
Hamm, B., 90
Hammond, M.C., 318, 327
Hannapel, D.J., 106
Hannon, G.J., 16

Hansen, T.B., 216
Harris, A., 298
Hartley, A., 66–67
Hartmann, B., 271
Hartzell, D.D., 130
Hatkevich, T.L., 216–218, 221, 226, 228
Haupt, S., 114–115
Hausser, J., 238
Havlioglu, N., 271
Hayakawa, T., 131
Hayami, M., 107–108
Hayashi, R., 310
Hazelwood, K.L., 130
He, J., 174–175
Heckel, A., 142–143
Hegedus, L., 91, 96, 99f
Heid, C.A., 52
Heinicke, L.A., 77
Heinis, C., 130
Heinlein, M., 106–119
Hellens, R., 115
Hentze, M.W., 292–293, 305, 308–309
Héroux, A., 322
Hickman, H.D., 124–125
Hicks, J.B., 249
Higashida, C., 136
Higuchi, I., 308
Hillen, W., 128–129
Hiller, M., 256
Hilson, P., 110
Himber, C., 106
Hires, S.A., 316
Hirosawa, K.M., 77
Hirschi, N., 293, 308–309
Hocine, S., 126, 194–195
Hodas, J.J., 124–125
Hodgson, L., 131
Hofmann, C., 108–109, 111
Hogg, J.R., 211–212, 295
Hol, W.G., 239–241
Holden, S.J., 40
Holt, C.E., 3, 66–67, 124
Hong, H., 90
Hong, J., 107–108
Hopper, A.K., 243
Hori, H., 310
Hovhannisyan, R.H., 256
Hu, Q., 108–109

Hu, W.S., 126
Huang, L., 292, 322
Huang, S., 88–89
Huang, T.S., 107–108
Huang, Y., 88–89
Hubbard, C.J., 240–241
Huber, D., 316
Hulur, I., 131
Hurwitz, R., 211–212
Huse, K., 256
Huttelmaier, S., 66–67
Hwang, J., 292

I

Ifrim, M.F., 124–125
Iida, A., 310
Iino, R.; 143
Ilgu, M., 194–212
Imam, J.S., 259, 292
Imamoto, N., 77, 134–135
Inaguma, A., 67, 70–74, 81
Ingolia, N.T., 124–125, 128
Isambert, H., 196–197
Ishigaki, Y., 126
Ishii-Watabe, A., 131
Ishikawa, Y., 310
Ishiyama, K., 108–110
Ito, T., 243
Itoh, T.Q., 262
Itzkovitz, S., 3–4
Ivanov, A., 216–218
Iwane, A.H., 77
Izaurralde, E., 292

J

Jackson, D.P., 106
Jackson, E., 292–293
Jackson, J., 271
Jackson, S.R., 88–100
Jacobsen, L.B., 297–298
Jaffrey, S.R.R., 107–108, 194–195, 218, 292, 316–332
Jakt, L.M., 3–5
Jambor, H., 148
Jan, C.H., 124–125
Jang, J., 136
Janicki, S.M., 66–67, 139
Jaramillo, A.M., 148

Jayagopal, A., 88–100
Jeck, W.R., 216
Jeng, S.C., 107–108
Jennings, P.A., 90–91
Jens, M., 216
Jensen, T.H., 292
Jensen, T.I., 216
Jeon, I., 207
Ji, L., 88–89
Ji, S.J., 292
Jiang, C., 53–54
Jiang, T., 90–91
Jiao, X., 148–149
Johnsson, K., 130
Jones, R.B., 256
Joseph, A., 142–143
Joyce, E.F., 14
Jozefczuk, J., 307
Juillerat, A., 130
Jung, M.E., 208

K

Kado, C.I., 116
Kairemo, K., 90
Kajava, A.V., 124–125
Kam, C.W., 218
Kam, C.W.W., 332
Kamagata, Y., 207
Kaminski, T.P., 142–143
Kamiya, M., 310
Kammerer, U., 99f
Kanagawa, T., 207
Kandel, E.R., 53–54
Kang, H., 53–54
Kangasniemi, A., 90
Kapanidis, A.N., 40
Kapila, J., 110, 114
Kapp, M., 99f
Kar, A., 271
Karam, R., 292
Karassina, N., 130
Karimi, M., 110
Kasai, R.S., 77, 143
Kashima, I., 292
Kato, Y., 243
Kaukinen, S., 90
Kawai, N., 108–110
Kelic, S., 66–67

Kellenberger, C.A., 318, 327
Kemeny, G., 256
Kentner, D., 129
Kenworthy, C.A., 131
Keppler, A., 130
Keppler-Ross, S., 240
Kerr, K., 152
Kessler, S., 106
Khodjakov, A., 138
Khorshid, M., 238
Kiebler, M.A., 110
Kiledjian, M., 148–149
Kim, J.H., 128, 131
Kim, M., 106
Kim, S.B., 69–70, 72
Kim, W.J., 208
Kim-Ha, J., 152
Kindermann, M., 130
Kirchhausen, T., 126, 161
Kirillov, S.V., 124–125
Kitagawa, G., 181
Kitamura, K., 77
Klee, H., 115
Kloc, M., 106
Knowlton, J.J., 124–125
Kobayashi, T., 256, 271–272
Koch, I., 256
Koga, A., 310
Kohrmann, A., 99f
Koivunen, E., 90
Kon, Y., 238–239, 243, 250
Konishi, S., 181
Kopperud, K., 114–115
Korza, G., 124–125
Kotlobay, A.A., 259–260, 293, 303, 308–309, 311–312
Koyama-Honda, I., 143
Kramer, F.R., 68–69
Kramer, M.C., 216
Kraus, G.A., 194–212
Krause, H.M., 3
Kremers, G.J., 135
Krikos, A.J., 223
Kriventseva, E.V., 256
Krueger, J., 216
Krzyzosiak, W.J., 107–108, 125–126
Kubitscheck, U., 142–143
Kudla, G., 241–242

Kudryavtseva, E.I., 259–260, 293, 303, 308–309, 311–312
Kulozik, A.E., 292–293, 305, 308–309
Kumano, G., 124
Kuo, D., 271
Kuo, J.S., 136
Kurane, R., 207
Kurata, S., 207
Kuroha, K., 243
Kuroyanagi, H., 256, 271–272
Kurtsiefer, C., 136
Kusumi, A., 77, 143
Kuyper, C.L., 136

L

Lacayo, N., 216
Lacomme, C., 82–84, 107–108
LaCount, D.J., 240–241
Lahn, B.T., 131
Laing, E.E., 238
Laliberte, J.F., 107–108
Lamas-Linares, A., 136
Lamb, D.C., 143
Lamm, M.H., 196–197
Landthaler, M., 238
Lane, E.A., 249
Lao, K., 124–125
Larson, A., 243
Larson, D.R., 3, 52–54, 126, 138–139, 160–190, 173f
Lavis, L.D., 130
Lavoie, B., 109–110, 125–126, 139
Le Grimellec, C., 144
Lécuyer, E., 3, 106
Lee, C., 256
Lee, H., 136
Lee, J.J., 66–67
Lee, K.Y., 128, 131
Lee, N.S., 223
Lee, S.R., 128, 131
Lehmann, R., 148–149
Lenstra, T.L., 161–162, 174, 185, 188–189
Lepage, M., 90
Letunic, I., 292
Letzring, D.P., 240–243
Levandoski, M., 68–69
Levesque, M.J., 3–5, 8–10
Levin, M., 124–125

Levine, A.J., 52
Levine, M., 174
Levsky, J.M., 3–5
Levy, D., 292
Lewis, B.P., 256, 292
Li, G.W., 243
Li, L.H., 128, 131
Li, N., 135–136
Li, S.X.L., 223
Li, W.P., 90
Li, X., 126
Liang, D., 216
Liang, K., 90
Liao, G., 124
Libri, D., 184
Lim, H., 53–54, 136–137
Lin, A.C., 66–67
Lin, C., 272
Lin, C.H., 256
Lin, M.D., 148–149
Lin, M.Z., 130
Lin, T., 106
Lindemann, D., 143
Linnik, O., 82–84, 107–108
Lionnet, T., 52–53, 124–153, 182
Lippincott-Schwartz, J., 107–108
Litke, J.L., 316–332
Little, S.C., 66–67
Liu, G., 124
Liu, Y., 216
Liu, Z., 136–137, 139
Lloyd, C.W., 114
Lofdahl, P.A., 139–140
Long, R.M., 52, 69, 107–109, 112, 128, 161, 194–195
Looger, L.L., 316
Lopez-Jones, M., 53–54, 136–137
Los, G.V., 130
Lou, C.H., 292
Low, R., 129
Lowe, T.M., 218
Lu, X., 88–89
Lu, Z., 216–218, 221, 226, 228
Lubeck, E., 3–5
Lucas, W.J., 106
Lukyanov, K.A., 256–267, 292–312
Lunsford, L., 148
Luo, J., 88–89

Lux, J., 107–108
Lykke-Andersen, J., 292–293, 295
Lykke-Andersen, S., 292
Lynch, C.C., 90–91
Lynch, K.W., 271
Lynch, S.A., 332
Lyon, K., 190
Lyu, J., 88–89

M
Ma, H., 88–89
Ma, J., 136–137, 139
Macchi, P., 110
Macdonald, P.M., 152
Magidson, V., 138
Magliery, T.J., 72
Mahmood, U., 90–91
Mai, L., 136
Malgaroli, A., 316
Malhotra, A., 322
Mandal, M., 318, 327
Manley, J.L., 270–271
Mao, T., 316
Maquat, L.E., 126, 256, 292
Marasa, J., 292–293
March, Z.M., 216
Marchand, V., 148
Marechal-Drouard, L., 108–109
Margeat, E., 144
Markina, N.M., 256, 259–260, 266, 292–312
Marras, S.A., 68–69, 91–92, 96
Marshall, R.A., 139–140
Martens, T., 143
Martin, K.C., 3, 53–54, 66–67, 106, 114–115
Martin, L., 292
Martin, R.M., 126, 161
Martinez-Murillo, F., 292
Maruoka, H., 271
Maska, M., 141–143
Matera, A.G., 216–235
Mathur, J., 114–115
Matia-Gonzalez, A.M., 238
Matlashov, M.E., 311–312
Matrisian, L.M., 90–91
Matsumoto, A., 262
Mawson, M., 256

McCole, R.B., 14
McCombs, J.E., 316
McCullough, R.M., 69–70
McDougall, M.G., 130
McIntosh, J.R., 139–140
McIntyre, J.O., 90–91
McKeehan, W.L., 256
McKeown, M.R., 130
McKinney, S.A., 139–140
McLachlan, J., 70–72
McNeil, J., 107–108
Medina, O.P., 90
Mehanovic, S., 208
Meijering, E., 141–143
Melchior, M., 90
Meldolesi, J., 316
Melnyk, C.W., 106
Melton, A.A., 271
Memczak, S., 216
Mendell, J.T., 292
Mercer, R.W., 292–293
Merzlyak, E.M., 295
Mestre, P., 114
Meyer, D., 106
Meyer, D.E., 209
Meyer, M., 216–218
Meyers, J.L., 292
Mhlanga, M.M., 68–69
Michaud, M., 108–109
Michelotti, N., 136–137, 139
Mikucki, M., 239–241
Milani, D., 316
Milhiet, P.E., 144
Miller, W.A., 106
Minasyan, B.E., 311
Mingle, L., 124
Mishler, D., 72
Miskolci, V., 131
Mitani, S., 256, 271–272
Mitra, D., 204
Mizuguchi, H., 131
Modrek, B., 256
Moffitt, J.R., 2–48
Moita, L.F., 271
Mokhtarian, M., 208
Molnar, A., 106
Monnier, N., 174–175
Monypenny, J., 136

Moon, H.C., 52–63
Moon, J.D., 107–108, 194–195, 218
Moon, T., 7–8
Mor, A., 139
Morisaki, T., 190
Moriwaki, S., 3–5
Morris, R.G.M., 53–54
Moy, R.H., 322
Mueller, S., 148
Muh, S.J., 256
Mühlemann, O., 292–293, 308–309
Muller, G., 128–129
Müller, S., 216–218
Mullineaux, P., 115
Munsky, B., 3
Murakoshi, H., 143
Muramoto, T., 53–54
Musunuru, K., 66–67
Mutterer, J., 108–109, 111
Myers, E., 142–143

N

Nagase, H., 91
Nahvi, A., 318
Nakai, K., 256
Narita, A., 107–108
Narumiya, S., 136
Natori, Y., 70–74, 80, 107–108
Neems, D., 271
Nelson, R.S., 107–108
Neuert, G., 3
Neu-Yilik, G., 293, 308–309
Newbury, S.F., 148–149
Newman, E.A., 256
Newman, J.R., 124–125, 128
Ngo, J.T., 124–125
Nguyen, L., 106
Nguyen, N., 53–54
Nguyen, Q.T., 90–91
Nguyen-Duc, T., 317–318, 321
Nicholls, C.D., 259
Nicholson, P., 292–293
Nickless, A., 292–293
Nicoud, J.F., 107–108
Niehl, A., 106, 108–109
Nikolaitchik, O., 107–108, 126
Nikolic, J., 256, 272
Nilsen-Hamilton, M., 194–212

Nishikawa, S., 3–5
Niu, G., 90
No, D., 128–129
Noensie, E.N., 308
Noffz, C., 240
Nogami, S., 72
Noto, J.J., 216–235
Nugent, P., 292–293
Nwokafor, C., 53–54, 136–137
Nygard, K.K., 70–72, 74

O

Oberdoerffer, S., 271
Ohki, E.C., 298
Ohnishi, T., 308
Ohno, G., 256, 271
Ohya, Y., 72
Okita, T.W., 106
Olejniczak, M., 107–108, 125–126
Oleynikov, Y., 66–67
Olsen, M.N., 216
Olson, E.S., 90–91
Orengo, J.P., 256
Orr-Weaver, T.L., 70–72
Osame, M., 308
O'Shea, E.K., 240
Ozawa, T., 66–84, 107–108

P

Padovan-Merhar, O., 3
Paige, J.S., 107–108, 194–195, 218, 317–318, 321
Paillusson, A., 293, 308–309
Palacios, I.M., 106
Palangat, M., 126, 161–162, 164, 172–174, 173f, 176–177, 179, 185, 188–189
Palmer, A.E., 316
Pamudurti, N.R., 216–218
Pan, D., 124–125
Pan, Q., 256, 261–262
Pan, W., 72, 88–89
Panchapakesan, S.S., 107–108
Pang, Y., 88–89
Parisi, F., 292
Park, H.J., 128, 131
Park, H.Y., 52–63, 119, 136–138, 141–143
Park, J.G., 316

Park, J.H., 128, 131
Park, Y., 136
Parker, R., 3
Parthasarathy, N., 106
Parthasarathy, R., 141–142
Parton, R.M., 68
Parvin, B., 207
Patel, D.J., 322–323
Patskovsky, Y., 126, 161
Patti, G.J., 316
Pattyn, F., 52
Paulsson, J., 53–54
Payea, M.J., 238–239, 243, 250
Pecreaux, J., 148
Pedelacq, J.D., 239–240
Peña, E.J., 106–119
Pereverzev, A.P., 256, 259–260, 266, 292–312
Perez, J.W., 88–89
Perocchi, F., 251–252
Petkovic, S., 216–218
Pezo, R.C., 3–5
Phair, R.D., 69
Phan, A.T., 323
Phizicky, E.M., 243, 249
Pick, H., 130
Pierre, P., 124–125
Pillman, K.A., 216
Piston, D.W., 90–91, 135
Pitchiaya, S., 77, 136–137, 139
Pittet, M.J., 90–91
Piwnica-Worms, D., 292–293
Platzer, M., 256
Plotkin, J.B., 241–242
Polonskaia, A., 323
Ponty, Y., 222
Poppe, B., 52
Porrua, O., 184
Porse, B.T., 292
Pozzan, T., 316
Preibisch, S., 142–143
Price, P.J., 298
Puglisi, J.D., 139–140

Q

Qin, J.Y., 131
Qin, Y., 316
Quarles, C.C., 90

Quartley, E., 239–241
Quinodoz, S., 66–67
Quivey, R.G., 240–241

R

Rabut, G., 140
Rackham, O., 70–72, 74
Raes, J., 292
Rafalska, I., 256
Raj, A., 2–5, 7–10, 14, 27, 68–69
Rambo, R.P., 322
Rao, A., 271
Raser, J.M., 240
Rasnik, I., 139–140
Ray, J., 194–212
Raychowdhury, R., 256, 266
Raymond, P., 126, 194–195
Razif, M.F., 70–72, 74
Rebbapragada, I., 292–293, 295
Rehwinkel, J., 292
Remaut, K., 143
Ren, B.Z., 131
Ren, X., 124–125
Resch, A., 256
Revyakin, A., 130
Reymann, K.G., 53–54
Rhee, W.J., 107–108
Rieder, R., 211–212
Rieger, B., 142–143
Rifkin, S.A., 2, 7, 14, 27, 68–69
Rigler, R., 165–166
Riley, R., 256
Rinne, J.S., 142–143
Rino, J., 126, 161
Ritchie, K., 143
Rivas, S., 114
Roberts, A.G., 114–115
Roberts, I.M., 82–84, 107–108
Robertson, K.L., 107–108
Rodriguez, A.J., 3
Rogowsky, P.M., 116
Rojas-Triana, M., 106
Rolish, M.E., 256
Rose, J.C., 53–54
Roth, A., 319
Rouillard, J.-M., 14
Roy, M., 90–91
Royer, C.A., 144

Rudloff, F., 90
Rufener, S.C., 292–293
Ruiz, M.T., 114–115
Ruiz-Medrano, R., 106
Russell, D.W., 201

S

Sacchi, N., 305
Sack, M., 110
Said, N., 211–212
Saito, K., 77
Sakane, H., 271
Sakata-Sogawa, K., 77, 134–135
Salas-Marco, J., 242–243
Sales-Lee, J., 318, 327
Salgia, S.R., 216–218, 228
Salmanidis, M., 216
Salzman, J., 216
Sambade, A., 108–109, 111, 114
Sambrook, J., 201
Sanchez, A., 3
Sando, S., 107–108
Sanoff, H.K., 216
Sarnow, P., 216–218
Satija, R., 256, 266
Sato, H., 131
Sato, M., 70–74, 80, 107–108
Saulière, J., 292
Sbalzarini, I.F., 141–143
Schaefer, M., 52, 69, 107–109, 112, 128, 161, 194–195
Schaeffer, D., 256
Schatzel, K., 165–166
Schellenbaum, P., 108–109
Schellenberger, E., 90
Scherer, K., 142–143
Schlabach, M.R., 16
Schlissel, G., 174
Schlitter, A.M., 292
Schmidt, C.A., 216–235
Schnell, S.A., 61–62
Schnorr, J., 90
Schoenberg, D.R., 292
Schonig, K., 129
Schott, G., 106
Schudoma, C., 106
Schuman, E.M., 3, 53–54, 124–125
Schupbach, T., 148

Schweingruber, C., 292–293
Seemanpillai, M., 108–109, 111
Segel, I.H., 160–161
Seibenhener, M.L., 57–58
Seidel, C.A., 139–140
Serganov, A., 322–323
Serin, G., 126
Shadrin, I.Y., 124–125
Shaffer, S.M., 8–10
Shah, K., 68–69
Shai, O., 256, 261–262
Shalek, A.K., 256, 266
Shaner, N.C., 130
Shao, A., 292
Sharifi, N.A., 292
Sharma, P., 106
Sharpless, N.E., 216
Shav-Tal, Y., 52–53, 66–67, 69, 131, 139
Shaw, J.J., 116
Shcherbo, D., 295
Shenoy, S.M., 3–5, 52, 66–67, 69, 107–110, 112, 125–126, 128, 139, 161, 194–195
Sherman, F., 249
Shi, J., 88–89
Shin, I., 194–212
Shirahige, K., 243
Shiraishi, T., 308
Shum, E.Y., 292
Shyu, A.B., 259, 292–293
Siebert, P., 124–125
Sigal, Y.M., 40
Siggia, E.D., 52
Silva, G.L., 107–108
Singer, R.H., 2–5, 52–54, 66–67, 69, 107–110, 112, 119, 125–126, 128, 131, 136–139, 141–144, 160–161, 182, 188–190, 194–195
Singh, G., 295
Singh, J., 107–108, 126
Singh, S., 108–110
Singh, S.K., 216–218, 228
Sinha, N., 106
Sinha, S., 68–69
Sinsimer, K.S., 66–67
Siuzdak, G., 316
Slaughter, J.P., 130
Slevin, M.K., 216
Smal, I., 141–143

Smith, C.S., 142–143
Smith, W.B., 53–54
Sohn, J.W., 256
Solovieva, E.A., 295
Somarelli, J.A., 256
Song, C., 136
Song, W., 218, 317–318, 321, 323, 332
Sorrentino, J.A., 216
Spector, D.L., 66–67, 139
Speleman, F., 52
Spille, J.H., 142–143
Sprague, J.E., 90
St Johnston, D., 3, 106, 148
Staines, W.A., 61–62
Stamm, S., 256
Stanton, P.K., 66–67
Staroverov, D.B., 256–267, 292–312
Stecyk, E., 106
Steen, K.A., 322
Steinbach, P.A., 130
Steitz, J.A., 128
Sternlicht, M.D., 89–90
Stewart-Ornstein, J., 243
Stirnnagel, K., 143
Stockbridge, R.B., 319
Stoilov, P., 256, 272
Stone, E., 292
Storz, G., 160
Stoss, O., 256
Strack, R.L., 194–195, 218, 318, 321, 323
Sudarsan, N., 319
Suess, B., 332
Suetsugu, S., 136
Suh, S.G., 106
Suliman, S., 139
Sun, G., 174–175
Svensen, N., 107–108, 194–195, 218
Swain, P.S., 52

T

Takeda, A., 106
Takizawa, P.A., 126
Tan, P., 124–125
Tang, B., 88–89
Tang, Y., 256
Tarn, W., 271
Tatavarty, V., 124–125
Tatomer, D.C., 216

Taupitz, M., 90
Terskikh, A., 124–125
Terwilliger, T.C., 239–240
Thieme, C.J., 106
Thompson, R.E., 138–139
Thorn, A., 321, 323
Tian, F., 88–89
Tian, J., 88–89
Tian, L., 316
Tilkins, M.L., 298
Tilsner, J., 82–84, 107–108
Tiruchinapalli, D.M., 66–67
Toiber, D., 256
Tokunaga, M., 77, 134–135
Toomre, D., 77
Topp, J.D., 271
Torimura, M., 207
Torti, F., 216
Toth, R., 114–115
Toubia, J., 216
Tran, T., 239–240
Travis, A.R., 88–100
Triller, A., 124–125
Tsien, R.Y., 90–91, 107–108, 130
Tsourkas, A., 107–108
Tsuji, T., 136
Tsunoyama, T.A., 77
Tung, C.H., 90–91
Tung, V., 256
Tutucci, E., 131
Tyagi, S., 2–4, 7, 14, 27, 68–69

U

Ubrig, E., 108–109
Uchida, E., 131
Ule, J., 271
Umezawa, Y., 70–74, 80, 107–108
Uphoff, S., 40
Urbanek, M.O., 107–108, 125–126
Urlaub, H., 211–212
Usuki, F., 308

V

Valcárcel, J., 271
Vale, R.D., 126
Valencia-Burton, M., 69–70
Valentine, T., 114–115
Valeur, B., 316

Vallan, C., 293, 308–309
Valtanen, H., 90
Van De Water, L., 124
van den Bogaard, P., 2, 7, 14, 27, 68–69
Van Etten, W.C., 166
van Hoof, A., 259, 292–293
Van Montagu, M., 110, 114
Van Oudenaarden, A., 2–4, 7, 14, 27, 68–69
Van Roy, N., 52
Vandesompele, J., 52
Vaquero-Martin, C., 110
Vardy, L., 70–72
Vargas, D.Y., 68–69
Veerapaneni, R., 136–137, 139
Vercauteren, D., 143
Verveer, P.J., 204
Vigers, G.P., 139–140
Vignali, M., 240–241
Vingron, M., 256
Vogel, H., 130, 165–166
Vogel, J., 211–212
Voinnet, O., 114–115

W

Waggoner, A.S., 107–108
Waldo, G.S., 239–240
Walter, N.G., 77
Wang, C., 108–110
Wang, D., 292
Wang, J.X., 322
Wang, K., 216
Wang, P.L., 216
Wang, S., 3–5, 7–13, 36–37, 43–44
Wang, T., 195–197, 207
Wang, X., 70–72, 88–89, 256, 261–262
Wang, Y., 216–218
Wang, Z., 216–218, 256
Ward, D.C., 52
Warmuth, C., 90
Warner, K.D., 321, 323
Warzecha, C.C., 256
Watanabe, N., 136
Webb, W.W., 63, 138–139
Weeks, K.M., 322
Wei, T., 107–108
Wei, W., 251–252
Weidenfeld, I., 129
Weigand, J.E., 332

Weil, T.T., 68, 148
Weinberg, Z., 319, 322
Weissleder, R., 90–91
Weissman, J.S., 124–125, 128
Wells, A.L., 52–53, 131
Wells, K.S., 90–91
Werb, Z., 89–90
Wessendorf, M.W., 61–62
Whipple, J.M., 249
Wickens, M., 238
Widengren, J., 139–140
Wieschaus, E.F., 66–67
Wilbertz, J.H., 124–153, 182
Wilkinson, M.F., 259, 292–293
Williams, C.C., 124–125, 243
Williams, K.R., 124–125
Williams, P.M., 52
Williams, R.M., 63
Wilson, C.G., 72
Wilson, P.D., 107–108
Wilson, S.C., 318, 327
Winkler, W., 318
Wippich, F., 124–153, 182
Witten, J.T., 271
Wittkopp, N., 292
Wohland, T., 165–166, 174–175
Wolberger, C., 160
Wolf, A.S., 239–243
Won, J.I., 209
Wong, A.C., 88–100
Wong, D., 243
Wong, E.H., 90–91
Wood, K.V., 130
Wooten, M.W., 57–58
Wouters, F.S., 204
Wright, A., 107–108, 126
Wright, D.W., 88–100
Wright, M., 297–298
Wu, B., 53–54, 126, 131, 161, 190
Wu, J.Y., 271
Wu, K.Y., 107–108, 194–195, 218, 317–318
Wu, M.-T., 8–10

X

Xayaphoummine, A., 196–197
Xia, T., 135–136
Xiang, A.P., 131

Xiao, J., 124–125
Xie, X., 91, 96, 99f
Xie, X.S., 124–125
Xing, Y., 256
Xoconostle-Cazares, B., 106
Xu, Q., 16
Xu, X., 149
Xu, Z., 131, 251–252
Xue, M., 88–89

Y

Yamada, K., 207
Yamada, T., 67, 70–74, 81
Yamashita, A., 292–293, 308
Yanagida, T., 77
Yanes, O., 316
Yang, J., 322
Yang, L., 106
Yang, M., 88–89
Yang, W., 136–137, 139
Yang, Y., 90
Yanushevich, Y.G., 311
Yao, T.P., 128–129
Yaping, E., 53–54
Ye, W., 88–89
Yeo, G., 256
Yewdell, J.W., 124–125
Yilmaz, F., 14
Yokomaku, T., 207
Yoon, T.Y., 136
Yoon, Y.J., 53–54, 136–137, 190
Yoshida, H., 3, 106
Yoshihisa, T., 216–218
Yoshimura, H., 66–84
You, M., 316–332
You, Y., 90
Youk, H., 3–4
Young, D.L., 238–239, 243, 250
Yu, B., 271
Yu, J., 124–125
Yu, Y., 106
Yunger, S., 139

Z

Zamore, P.D., 70–72
Zaraisky, A., 124–125
Zavadil, J., 292
Zawilski, S.M., 53–54

Zearfoss, N.R., 106
Zeitelhofer, M., 110
Zenklusen, D., 3–4, 52–54, 126, 161, 194–195
Zhang, L., 131, 256, 259–260, 266
Zhang, M.Q., 256
Zhang, W., 106
Zhang, X., 88–89, 238–239, 243, 250
Zhang, Y., 90
Zhang, Z., 130, 190, 256
Zhao, S., 88–89

Zheng, S., 270–288
Zhiyentayev, T., 3, 5
Zhu, H., 53–54
Zhu, J., 256
Zhu, L., 90
Zhuang, X., 2–48
Zimprich, C., 130
Zimyanin, V.L., 148
Zipfel, W.R., 63
Zuker, M., 128
Zünd, D., 292–293

SUBJECT INDEX

Note: Page numbers followed by "f" indicate figures, and "t" indicate tables.

A
Acceptor photobleaching method
  Förster resonance energy transfer (FRET), 205
  region of interest (ROI), 205
Activatable cell-penetrating peptides (ACPPs), 90–91
Acute brain slices, 61–62, 63f
Agarose gel electrophoresis, 113
Agroinfiltration, 115–116
Allosteric Spinach sensors
  metabolites, intracellular dynamics of, 318
  structure of, 317f, 318
Alternative exon, 257, 258f, 259, 261
Alternative splicing, fluorescent protein-based quantification of
  exon skipping, 256
  high-throughput sequencing, 256
  in mammalian cell lines, 256
  materials
    equipment, 261
    software, 261
  method description
    overall scheme, 257–259
    p53-inducible gene 3 (PIG3), reporter for, 259–260
  procedure
    cell culture and transfection, 262
    flow cytometry, 265–266
    fluorescence microscopy, 263–264
    genetic constructs preparation, 261–262
  protein isoforms, 256
  protein–protein interactions, 256
Aptamer-ligand complex, 194–195
Array-derived complex oligonucleotide pools, 22
Autocorrelation, 163
Automated fluidics system, 33
Averaging correlation methods
  autocorrelation functions, 168f, 170
  convergence of, 184–185, 184f
  decision chart, 171, 171f
  (see also Correlation function)
  fluorescence-to-RNA conversion factor, 170–171

B
Back-splicing, 216–218
Barcode assignment, for codebook, 19–20
Baseline correction and renormalization
  correlation functions
    properties of, 172
    standard error on, 172–174
  fluctuation analysis technique, 174
  freedom, degree of, 174
  global mean estimation methods, 172–174
  RNA transcription, 172–174
  transcriptional time traces, 172
  transcription site (TS), 174
Bayesian information criterion (BIC), 181
β-actin, 52–53, 54f
β-actin messenger RNA (mRNA)
  actin cytoskeleton networks, 66–67
  proteins, 66–67
  pumilio homology domain (PUM-HD), design of, 74–75
  single-molecule dynamics of, 82, 83f
  visualization, 73–74
BIC. See Bayesian information criterion (BIC)
Bimolecular fluorescence complementation (BiFC), 107–108
Bit-flipping probability, 43–44
Bootstrap technique, 167, 173f, 174–176
Boric acid buffer (BAB), 55
Brain slice imaging
  materials, 61
  protocols
    acute brain slices preparation, 61–62
    imaging mRNA, 62–63
Broccoli, 218, 229–231
Bulge-helix-bulge (BHB), 222

C

Caffeine, 308
Calling rate, 7
Carlo simulation, 168f
Cassette exon, 257, 258f
Cell-based high-throughput screens
  materials and equipment, 281
  protocol, 281–282
Cell culture and transfection, NMD quantification of
  Lewis lung carcinoma, 299
  liposome-based transfection, 297–298
  nucleofection, 298–299
Cell fixation and permeabilization, 29–30
Cell heterogeneity, 266
Cellular imaging, tricRNAs
  HEK293T cells, 233–234, 234f
  materials and equipment, 233
  workflow, 233
Cellular metabolites, 316
  fluorescence imaging approaches, 316
  quantification of, 316
  riboswitches, 318
Chemiluminescence assays, 287
Chimeric human promoter/fly tRNA constructs, 231–232
Circular RNA (circRNA) expression, tRNA splicing
  cellular roles, 216
  engineering and imaging, 216–218
  expression of, 216
  high throughput sequencing, 216
  tricRNA vectors, design and generation
    cellular imaging of, 233–234
    external promoters, addition of, 222–225
    in-gel imaging of, 229–232
    in vivo expression of, 226–229
    parental tRNA gene, isolation and mutagenesis, 218–222
    vector, notes on, 219–221
Cis-regulatory sequences, RNA-ID reporters
  cloning sequences, 245–248
  flow cytometry, 249
  library, analysis of, 250
  steps, 244, 244f
  yeast, 248–249

Cloning repetitive sequences
  E. coli, 209
  genetic instability, 209
  initial repetitive aptamer sequences, 209–210, 210f
  intracellular multiaptamer genetic tag (IMAGEtags), folding of, 209, 210f
  longer and original, 210–211, 210f
  synthetic oligonucleotides, 209–210, 210f
Cloning sequences, RNA-ID reporter
  annealing, 247
  inserts preparation, 245–247
  variation, 248
  vector preparation, 245, 246f
Codebook design
  advantage of, 18
  binary barcodes, 13
  extended Hamming code, 18
  oligonucleotide probe design
    barcode assignment, 19–20
    modified Hamming distance 2 (MHD2) encoding scheme, 19
    modified Hamming distance 4 (MHD4) encoding scheme, 19
CO_2 incubator, 60–61
Colorimetric readout approach, 5
Compact vector, 219–221
Computing correlation functions, flowchart for, 172–174, 173f
Concatameric cDNA, 228–229
Confidence interval/uncertainty, 173f, 174–176
Confidence ratio and background counts, 44–46, 45f
Confocal microscopy
  hairpin DNA functionalized gold nanoparticles (hAuNPs), 95
  intracellular uptake and fluorescence emission of, 99, 100f
  vs. matrix metalloproteinase (MMP), 99, 101f
Constant Hamming Weight codes, 8
Contransfections, 309
Control RNA
  Förster resonance energy transfer (FRET), 202–203
  tobramycin, 197, 200f
Control vector, 257, 262

Correlation function, 162f
  analytical expressions
    advantages of, 179
    mechanistic models, 178f, 179
  averaging methods, 168f, 170–171, 171f
  baseline correction and renormalization, 172–174, 173f
  in biophysics field, 163
  bootstrapping, 173f, 174–176
  and computing, 165–176
  Dirac functions, 164
  error bars, 173f, 174–176
  fluorescence traces, mean subtraction of, 167–170, 168f
  geometrical feature
    principle of, 177, 178f
    RNAs, fluorescence time profiles of, 177
  hybrid Monte Carlo approach
    analytical method, 173f, 179–180
    disadvantage of, 179–180
    single-RNA transcription, 180
  interpretation of
    data fitting and model discrimination, 180–181
    modeling, 176–180, 178f
  Poisson process, 176–177
  properties of, 172
  single, 165–167
  standard error on, 172–174
  time-delay points, correct weighting of, 172
  transcriptional kinetics, 164
  uncertainty, 173f, 174–176
Cortex, 58, 63f
Cross-correlation, 163
Cut and paste restriction enzyme cloning, 218

D

Data acquisition
  and analysis, 203–205
  fluorescence signals, wells and plates, 283–284
  high-content fluorescence imager, 282
  imaging resolution and speed, 282–283
  materials and equipment, 284
  protocol, 284
  signal-to-noise ratio, 282
  typhoon imager, 282

Data fitting and model discrimination.
  See also Correlation function
  auto-and cross-correlation functions, 181
  Bayesian information criterion (BIC), 181
Decoding barcodes, 11f, 42–43
Detrending, fluctuation analysis, 185
DFHBI. See 3,5-Difluoro-4-hydroxybenzylidene imidazolinone (DFHBI)
3,5-Difluoro-4-hydroxybenzylidene imidazolinone (DFHBI), 317–318
Dirac functions, 164
Discrete particles, 140–141
DMEM. See Dulbecco's modified Eagle's medium (DMEM)
DNA oligonucleotide synthesis, 92–93t, 93–94
Double-stranded DNA (dsDNA), 80, 323–324
Drosophila, translating RNA imaging by coat protein knockoff (TRICK)
  coat proteins, 149
  controls, 151–152
  embryonic axis determinants, 148
  fluorescent signals, quantification of, 150f, 152
  genomic rescuing construct, 148–149
  *hsp*83 promoter, 149
  imaging and analysis, 149–151
  immuno-fluorescence techniques, 151–152
  MS2-tagged reporter mRNAs, 148
  oocytes, 149, 150f
  oskar promoter, 148–149
dsDNA. See Double-stranded DNA (dsDNA)
Dual-fluorescence minigene reporters, construction of
  alternative exon, 276
  considerations, 278
  GFP/RFP ratio, 278
  intronic sequence, selection of, 276
  materials and equipment, 277
  protocol, 277–278
Dual-fluorescence splicing reporter, 272, 273f
Dulbecco's modified Eagle's medium (DMEM), 75

E

ECM. *See* Extracellular matrix (ECM)
Electron multiplying charge-coupled device (EMCCD), 56, 60, 60f
Encoding probes
  array-based synthesis, 22
  assembly and screening of, 9f, 20
  hybridization of, 30–31
  readout sequences, 20
  template molecule, 21, 21t
Endogenous RNA molecules *vs.* modified RNAs, 67, 184
Endogenous target exon, 287
Error bars, 173f, 174–176.
  See also Correlation function
Error-correcting encoding scheme, 44–46
Error-robust and-correcting codes
  binary barcodes, fraction of, 6f, 7
  calling rate, 7
  encoding schemes, 7–8
  Hamming distance (HD), 7–8
  misidentification rate, 6f, 7–8
  single-molecule fluorescence in situ hybridization (smFISH), 6–7
Error tracking, 187
Escherichia coli
  cloning repetitive sequences, 209
  metabolite concentration
    data analysis of, 330–331
    imaging dynamics of, 328–330
    materials and equipment, 328–329
    procedures, 329–330, 330f
    thiamine pyrophosphate (TPP), 330–331
  sensor-expressing
    materials and equipment, 327–328
    procedures, 328
Excitation optical system. *See also* Single-molecule RNA, visualization of
  design and principle, 77–78, 78f
  inclined illumination, 77–78, 78f
  red fluorescent proteins (RFPs), 77–78
Exon–exon junction site, 259
Exon skipping, 256, 258f, 259
Expression cassettes, 294–295, 309–310

Extended Hamming code, 18
External promoters
  cloning
    materials and equipment, 223, 224f
    workflow, 225
  pol III promoters, 223
  site-directed mutagenesis, 223f
  tricRNAs, expression of, 222–223
Extracellular matrix (ECM), 89–90

F

FACS. *See* Fluorescence-activated cell sorting (FACS)
False discovery rate (FDR), 286
Fast Fourier transform (FFT) algorithm, 166–167
Finite-duration time traces, 169
5S ribosomal RNA, 69–70
Flow cytometry, 95, 99f
  cell analysis of
    critical, 302–303
    cytomics FC500, 303
    detection parameters, 302–303
    FACSAria III, 302
    fluorescence-activated cell sorters (FACS), 301–302
    HEK293 cells, 302f
    Katushka, 301–302, 302f
    TagGFP2, 301–302, 302f
  cell samples, 265
  control strains, 249
  results, 265–266
    digital data, 303
    Export–Text, 303
    green/red ratio, 303
    pNMD+-expressing cells, 303
    quadrants tool, 303
    well-expressing cells, 303, 305
  sample preparation for, 301
  strains, growth of, 249
Flow system, assembly and operation of, 33–36
Fluctuation analysis, transcriptional kinetics dissection
  correlation functions
    averaging methods, 170–171
    baseline correction and renormalization, 172–174, 173f

and computing, 165–176
fluorescence traces, mean subtraction
  of, 167–170, 168f
interpretation of, 176–181, 178f
single, 165–167
time-delay points, correct weighting
  of, 172
uncertainty, error bars, and
  bootstrapping, 173f, 174–176
issues and pitfalls
  data, biased selection of, 188
  fluctuations, technical sources of,
    185–188
  MS2 and PP7 cassettes, 182–184
  single trace interpretation, 184–185,
    184f
  validation by complementary
    measurements, 188–189
technical sources of, 185–188
transcriptional time traces
  correlation functions, 164
  definitions and terminology, 163–164
Fluctuations, 185–188
Fluorescence
imaging, 75, 88–90
microscopy
  fluctuation analysis, 185
  intracellular multiaptamer genetic tag
    (IMAGEtags), 194–195
  probes, 88–90
  sensor characteristics
    imaging approaches, 316
    kinetic measurements, 327
    materials and equipment, 325
    procedures, 320f, 325–327
  spot identification of, 40–41
  trace mean subtraction of, 167–170, 168f
Fluorescence-activated cell sorting (FACS),
  275, 279, 301–302
Fluorescence correction and normalization,
  285
Fluorescence in situ hybridization (FISH),
  52, 145
Fluorescence microscopy
  adherent cells, 300–301
  auxiliary readout, 300
  confocal fluorescence microscope,
    300–301

critical, 301
green-to-red ratio, 301
HEK293 cells, 263, 263f, 300f
Katushka, 300–301
numerical aperture (NA), 301
TagGFP2, 300–301
Fluorescence recovery after photobleaching
  (FRAP), 80
Fluorescence-to-RNA conversion factor,
  163–164
Fluorescent live-cell imaging
  cell preparation for, 76
  molecular beacon method, 68–69
  RNA localization, 68
  signal-to-noise ratio, 68–69
  stem–loop structure, 68–69
Fluorescently labeled oligonucleotide
  probes, 2
Fluorescent protein (FP), 125–126, 316
Fluorescent protein reconstitution
  technique. See also Split fluorescent
    protein reconstitution approach
  dissection of, 72
  limitations of, 72–73
  live-cell RNA imaging, 76
  protein–protein interactions, 72
  pumilio homology domain (PUM-HD),
    design of, 72
  reconstruction of, 72
  target RNA, 72
Fluorescent RNA aptamer, 218, 229
Fluorogenic riboswitches
  imaging metabolite concentration
    dynamics, 316–318, 317f, 319f
  metabolite sensor types, 317–318, 317f
Spinach riboswitch sensors
  in vitro identifying, 323–327
  metabolites, live-cell imaging of,
    327–331, 330f
  structure-guided design of, 319–323,
    320f
Fluorophore-binding pocket, 318
Fluorophore-conjugated ligands, 195–196
Förster resonance energy transfer (FRET)
  acceptor photobleaching method, 205
  cell preparation, 202–203
  control RNA, 202–203
  cross talk signals, 204

Förster resonance energy transfer (FRET)
(Continued)
efficiency of, 205
messenger RNA (mRNA) detection, 194–195
optical imaging, 90
signal acquisition, 196–198, 198–199f
Forward light-scattering (FSC), 264f, 265
Frameshifts, 256–257, 258f
FRAP. See Fluorescence recovery after photobleaching (FRAP)
FRET. See Förster resonance energy transfer (FRET)
Full-length fluorescent protein, 72–73, 81f

G

Gain-of-function, 275, 280
Gateway cloning, 219–220. See also Primer design
Gel-documentation systems, 231
Gene expression visualization.
See also Intracellular multiaptamer genetic tag (IMAGEtags)
data acquisition and analysis, 198–200f, 203–205
flow cytometry analysis, 91, 97–98, 99f
Förster resonance energy transfer (FRET) acceptor photobleaching method, 205
cell preparation for, 202–203
materials and instrumentation, 201–202, 201f
operator blinding, 199–200
yeast cell, transformation of, 202
Genetic constructs
alternative splicing, regulation of, 261–262
cassette alternative exon, 261
fluorescence reporter, 261
fluorescent indicators, replacement of beta-globin gene, 310
far-red fluorescent protein Katushka, 310
green fluorescent protein TagGFP2, 310
inducing fragment, 311
PCR primers, 262
pCtrl-GOI, 262
pSpl-GOI, 262

vector backbone changes
chimeric NMD target, 309–310
critical, 309
independent expression cassettes, 309
lentiviruses, 310
transposon vector, 310
two-promoter fragment, 310
GFP. See Green fluorescent protein (GFP)
GLN4, 239f, 240
Global mean estimation methods, 172–175
Gold–thiol bond formation, 88–89
Green fluorescent protein (GFP), 238, 239f, 257, 258f, 259, 271–274, 273f
Green-to-red ratio, 257

H

Hairpin DNA (hDNA)
DNAse degradation, 88–89
oligonucleotide synthesis, 92–93t, 93–94
sequences, 92, 92t
Hairpin DNA functionalized gold nanoparticles (hAuNPs)
applications of, 89–90
matrix metalloproteinase (MMP) and subtypes, 89–92
mechanism of, 88–89, 89f
messenger RNA (mRNA) imaging protocol
cell culture studies, 95
confocal microscopy, 95
DNA hairpin (hDNA), 92–93t, 93–94
flow cytometric analysis, 95
materials and instrumentation, 92–93, 92–93t
oligonucleotide synthesis, 92–93t, 93–94
synthesis and characterization of, 94–95
Hamming distance (HD), 7–8, 10–12
Hanks' balanced salt solution (HBSS), 75
hAuNPs. See Hairpin DNA functionalized gold nanoparticles (hAuNPs)
hDNA. See Hairpin DNA (hDNA)
Head-to-tail fashion, 216–218
HEK293T cells, 226–227, 231–234, 234f
HeLa cells
image acquisition, 146–147
image analysis, 147–148
live-cell imaging, 145–146

Subject Index

Hemocytometer, 59
High-throughput screening, 266–267
High-fidelity PCR amplification, 111
Hippocampal dissection, 57, 57f
Hippocampus, 58, 63f
Homologous riboswitches, 322
Human fibroblast cells (IMR90s), 10–12, 11f
Human transcriptome. *See* Transcriptomics
Hybridization, 3. *See also* Multiplexed error-robust fluorescence in situ hybridization (MERFISH)
Hybrid Monte Carlo approach.
  See also Correlation function
  analytical method, 173f, 179–180
  disadvantage of, 179–180
  single-RNA transcription, 180

I
Identifying regulators of alternative splicing (IRAS)
  cell-based high-throughput screens
    materials and equipment, 281
    protocol, 281–282
  cis-regulatory elements, 270–271, 287
  data acquisition
    fluorescence signals, wells and plates, 283–284
    materials and equipment, 284
    protocol, 284
    resolution and speed, imaging, 282–283
  data analysis, 285–287
  dual-fluorescence minigene reporters, construction of
    considerations, 278
    materials and equipment, 277
    protocol, 277–278
  dual-fluorescence splicing reporter, 272
  dual-output minigene reporters, 272–274
  dual-output reporter, 271–272
  fluorescence-activated cell sorting (FACS), 275
  fluorescence correction and normalization, 285
  gain-of-function genetic screen, 275, 276f
  green fluorescent protein (GFP), 272–274, 273f
  library and array construction
    materials and equipment, 280
    protocol, 281
  limitations, 287–288
  MAPT exon, 272
  method design, 272–275
  minigene reporters, 271
  myriad genes, 270–271
  open-reading frame (ORF) translation, 274–275
  pflareA minigene, 272–275
  pre-mRNA splicing patterns, 270–271
  red fluorescent protein (RFP), 272–274, 273f
  screening, 274–275
  single-output reporters, 271
  splicingdependent readouts, 272–274, 274f
  splicing regulators, 270–274, 273f
  stable cell clones, generation of
    considerations, 279–280
    materials and equipment, 279
    protocol, 279
  trans-acting protein factors, 270–271
  validation, 287
Image offset correction, 41–42
IMAGEtags. *See* Intracellular multiaptamer genetic tag (IMAGEtags)
Imaging and time-lapse image acquisition
  materials, 117
  N. benthamiana, 116–117
  phototoxic effects, 116–117
  protocol, 117–119
Imaging protocol
  of multiplexed error-robust fluorescence in situ hybridization (MERFISH), 37
  readout probe hybridization, 37
Immunoprecipitation, 75–76
In gel imaging of tricRNAs
  Broccoli, 229
  DFHBI-1T, 231, 232f
  fluorescent RNA aptamers, 229
  materials and equipment, 230
  workflow, 230
Inhibitors, nonsense-mediated mRNA decay (NMD)
  caffeine, 308
  RNAi, 308–309
  wortmannin, 308

In situ hybridization techniques, 68.
    See also Hybridization
Institutional Animal Care and Use
    Committee (IACUC), 57–58
Internal ribosome entry site (IRES), 131
Intracellular multiaptamer genetic tag
    (IMAGEtags)
  cloning repetitive sequences, 209–211,
    210f
  fluorescence, 194–195
  Förster resonance energy transfer (FRET)
    signal acquisition, 196–198,
    198–199f
  ligands, synthesis of, 200–201f, 207–209
  messenger RNA (mRNA) transcription,
    194–195
  potential component of, 196–197
  real time-qualitative PCR (RT-qPCR)
    measurement, 205–207
  steady-state RNA levels, 197
  tandem aptamers, 195–196
  technology of, 195f
  visualizing gene expression, 199–205,
    201f
  yeast cells, 197, 199f
Intron-modified tRNA, 225
In vitro identifying. See also Spinach
    riboswitch sensors
  fluorescence sensor characteristics
    materials and equipment, 325
    procedures, 320f, 325–327
  RNA-based sensor variants
    materials and equipment, 324
    procedures, 324–325
In vitro template amplification, 24–26, 24t
In vitro transcript (IVT), 231
In vitro transcription reaction, 26
In vivo expression, tricRNAs
  product analysis
    northern blotting, 228
    RT-PCR, 228–229
    sequencing, 229
    RNA isolation, 228
  transfection
    cell line, 226–227
    reagent, 227
In vivo RNA visualization in plants, MS2
    tagging

cis-acting sequences, 109–110
fluorescent proteins (FPs), 107–108
imaging and time-lapse image acquisition
    materials, 117
    protocol, 117–119
invasive delivery methods, 107–108
natural expression pattern, 110
N. benthamiana, transient expression in
    materials, 115
    protocol, 115–116
origin of assembly (OAS), 108–109
plasmid constructs
    materials, 111–112
    protocol, 112–114
plasmodesmata (PD), 106
RNA molecules, 106
single-molecule detection, 109–110
symplastic pathway, 106
tumor-inducing (Ti) plasmid, 110
two-component approach, 107–108
yeast, 108–109
Iterative method, 162f, 165

J

Jigsaw puzzle-shaped cells, 116–117.
    See also Nicotiana benthamiana

K

Katushka-encoding region, 259–260
KineFold, 196–197
Kinetic competition, 160–161

L

Lewis lung carcinoma (LLC), 299
Library and array construction
    identifying regulators of alternative
        splicing (IRAS), 280
    loss-of-function screens, 280
    materials and equipment, 280
    paralogous splicing regulators, 280
    protocol, 281
    short hairpinRNA (shRNA), 280
Library, RNA-ID
    fluorescence-activated cell sorting
        (FACS), 250
    generating, 250
    purifying transformants, 250

LIC. *See* Ligation-independent cloning (LIC)
Ligand-binding domain, 316
Ligands, tobramycin
  ligand–fluorophore conjugate, 207
  sulfonated/nonsulfonated forms, 201*f*, 207
  synthesis of
    cyanine5 (Cy5) and cyanine3 (Cy3) tobramycin, 207–208
    cyanine5 (Cy5) conjugated ligands, 201*f*, 207
    cyanine5 (Cy5)-PDC, 208
    cyanine3 (Cy3)-PDC-Gly, 208–209
Ligation-independent cloning (LIC), 239–241, 241*f*
Linker sequences, Spinach forms
  metabolite concentration, 323–324
  structure-guided design of
    crystal structure of, 323
    thiamine pyrophosphate (TPP), 320*f*, 323
Lipofuscin, 61–62
Liposome-based transfection, 297–298
Live-cell imaging approach, 67, 88–89, 233–234
Localization-based super-resolution microscopy, 40
Long time-lapse experiments
  cell motion tracking, 140
  fluorophores, photo-oxidation of, 139–140
  motorized Piezo stages, 140
  photobleaching, 139–140
  reactive oxygen species (ROS), 139–140
  stage stability, 140
  translation regulation, 139–140
Loss-of-function, 278–280, 287

M

Mammalian cells
  coat proteins expression
    fluorescence-activated cell sorting (FACS), 131
    inducible promoters, 131
    optimal imaging, 130
    sequential excitation, 130
    single-molecule two-color imaging, mRNAs, 130
    single-particle tracking (SPT), 130
    spectrally distinct fluorophores, 129–130
    tetramethylrhodamine-based dyes, 130
  controls, 132
  primary cells, considerations and challenges
    coat protein (CP)–fluorescent protein (FP) constructs, 131
    internal ribosome entry site (IRES), 131
    primary neurons, 131
    stable transgene integration, 131
  translating RNA imaging by coat protein knockoff (TRICK)
    cell-to-cell variation, 129
    constitutive expression, 128–129
    inducible expression, 128–129
    reporter messenger RNA (mRNA), 128–129
    site-specific integration, 129
    transient transfection, 129
Mammalian gene collection (MGC), 280–283
Matrix metalloproteinase (MMP)
  optical imaging
    activatable cell-penetrating peptides (ACPPs), 90–91
    in breast cancer cell lines, 89–90
    extracellular matrix (ECM), proteolysis of, 89–90
    fluorophore–quencher pairs, 90–91
    near-infrared (NIR), 90–91
    potential biomarkers, 90
    role of, 90
    subtypes, strategy for
      messenger RNA (mRNA) transcripts, 91
    phenotypic fingerprints, 91–92
MCP. *See* MS2 bacteriophage coat protein (MCP)
MCP×MBS mice, 52, 54*f*
MERFISH. *See* Multiplexed error-robust fluorescence in situ hybridization (MERFISH)

Messenger RNA (mRNA)
  cell culture studies, 95
  confocal microscopy, 95
  correlation function, 160
  DNA hairpin
    and oligonucleotide synthesis,
      92–93t, 93–94
    sequences, 92, 92t
  flow cytometric analysis, 95
  Förster resonance energy transfer
    (FRET), 194–195
  materials and instrumentation
    hairpin DNA (hDNA) sequences,
      92, 92t
    target sequences, 92, 93t
  Spinach riboswitch approach, 318
  target regions, design of, 14
Metabolite-binding aptamer, 318
Metabolite live-cell imaging.
  See Escherichia coli
MGC. See Mammalian gene collection
  (MGC)
Micro-Manager, 60
Microscopy
  imaging modality
    collimated/parallelized beam, 132–133
    confocal illumination scheme, 133–134
    diffraction-limited spot, 133–134
    highly inclined laminated optical sheet
      (HILO), 134–135
    laser scanning confocal microscopy
      (LSCM), 133–134
    live-cell imaging, 134–135
    rapid photobleaching, 133–135
    spinning-disk microscopy, 133–134
    wide-field microscopy, 132–133
  light source
    fluorophores, 135
    gas lasers, 135–136
    illumination source, 135–136
    light-emitting diodes (LEDs), 136
    xenon-based arc lamps, 136
  signal detection
    fluorophore emission spectra,
      136–137
    messenger RNA (mRNA) particles,
      136–137
    multiline dichroic mirror, 136–137
    multiplying charge coupled device
      (EMCCD), 137
    postimaging registration, 137
    scientific complementary metal oxide
      semiconductor (sCMOS), 137
    single-molecule RNA, 77
      excitation optics, 77
      target RNA, 77
    total internalization reflection
      fluorescence (TIRF) microscopy, 77
    temperature and CO_2 control
      humidity level, 137–138
      pH, 137–138
      plexiglas incubation box, 137–138
Minigene reporters, 271–275, 287
Misidentification rate, 6f, 7–8
Mitochondrial messenger RNA, localization
  fluorescence recovery after
    photobleaching (FRAP), 80, 81f
  mitochondrial genome, 80
MMP. See Matrix metalloproteinase (MMP)
Modified Hamming distance 2 (MHD2)
  code, 12–13
Modified Hamming distance 4 (MHD4)
  code, 10–12, 11f
Molecular beacon method, 88–89, 96
  oligonucleotide and synthetic
    fluorophores, 68–69
  target RNA, 68–69
Movement protein, 108–109
mRNA. See Messenger RNA (mRNA)
MS2 bacteriophage coat protein (MCP), 52
  correlation function, 161, 162f
  location of
    elongation kinetics, measure of,
      183–184
    endogenous vs. modified RNAs, 184
    measurement sensitivity, 182
    noncoding regions, 182
    open reading frames (ORFs), 182
    and PP7 Cassettes, 182
    signal intensity, 182
    single-RNA sensitivity, 182–183
    temporal resolution, 182
    transcriptional time traces, 161, 162f
MS2 binding site (MBS), 52
MS2-GFP system, 52
MS2 phage coat protein (MCP), 108–109

Subject Index

MS2 tagging, 108–109, 118f
Multiple-tau algorithm, 162f, 165–166
Multiplexed error-robust fluorescence in situ hybridization (MERFISH)
  automated fluidics system, 33
  cells, fixation and permeabilization of, 29–30
  combinatorial barcoding and sequential readout, 4–5, 6f
  conceptual approach, 5, 6f
  data analysis, 39–47, 45f
  decoding barcodes, 39–40, 42–43
  encoding probes, hybridization of, 30–31
  error-robust and-correcting codes, 6–8, 6f
  flow system, assembly and operation of, 33–36
  fluorescent spots, identification of, 40–41
  hybridization and imaging, 5
  image offsets, correction of, 41–42
  imaging process for, 31–39, 34f
  materials, 32–33
  microscope requirements, 36–37
  oligonucleotide probe design
    codebook, 18–20
    encoding probes, assembly and screening of, 9f, 20
    priming regions, 20–22, 21t
    readout probes, 16–18, 17t
    target regions, 9f, 14–16
  performance of, 10–13, 11f
  probe construction
    encoding probes, purification of, 26–27
    in vitro template, amplification of, 24–26, 24t
    in vitro transcription, 26
    reagents, 23–24
    reverse transcription, encoding probes, 26–27
  required materials, 28–29
  RNase contamination, 27
  sample preparation and staining, 27–31
  schematic diagram of, 34, 34f
  single-molecule fluorescence in situ hybridization (smFISH), multiplexing of, 4–5
  two-stage hybridization, 8–10, 9f
Multiprotein exon junction complex (EJC), 292–293
Mutant pumilio homology domain (PUM-HD), 74–75

N

NADH dehydrogenase 6 (ND6) messenger RNA, 80, 81f
Near-infrared (NIR), 88–91
Neural dissection solution (NDS), 55
Neuron culture imaging
  materials, 55–56
  protocols
    dissection, 57–58
    imaging mRNA, 60
    maintaining neuron culture, 59
    poly-D-lysine (PDL) coating, 56–57
    seeding neuron cells, 58–59
Nicotiana benthamiana, 111, 114–116
NIR. See Near-infrared (NIR)
Nonhomogeneous illumination, 186
Nonsense-mediated mRNA decay (NMD), quantification of
  cell culture and transfection
    Lewis lung carcinoma, 299
    liposome-based transfection, 297–298
    nucleofection, 298–299
  flow cytometry
    cell analysis of, 301–303
    results, analysis of, 303–305
    sample preparation for, 301
  fluorescence microscopy, 300–301
  genetic constructs, modifications of
    fluorescent indicators, replacement of, 310
    inducing fragment, introduction of, 311
    vector backbone changes, 309–310
  inhibitors of
    caffeine, 308
    RNAi, 308–309
    wortmannin, 308
  materials
    equipment, 297
    reagents, 296
    software, 297
  multiprotein exon junction complex (EJC), 292–293

Nonsense-mediated mRNA decay (NMD), quantification of (Continued)
  physiological transcripts, 292
  quantitative ratiometric assessment of, 294–295
  RNA isolation and quantitative RT-PCR, 305–308
  single-cell level, 293
  splicing-dependent pathway, 292–293
  C-truncated proteins, 292
Northern blot hybridization, 311
Northern blotting, 228
N-terminal fluorescent protein fragment, 73–74
Nuclear localization sequence (NLS), 53
Nucleic Acid Database Project, 319
Nucleofection, 298–299
  human keratinocyte cell culture (HaCaT), 298–299
  human T-lymphocyte jurkat cell culture, 298
Numerical aperture (NA), 56–57

O

Oligonucleotide probes, 68–69
  basic components of, 9f, 13
  design of
    codebook, 13, 18–20
    encoding probes, assembly and screening of, 20
    priming regions, 20–22, 21t
    readout probes, 16–18, 17t
    target regions, 14–16
Oligonucleotide synthesis. See DNA oligonucleotide synthesis
Open-reading frame (ORF), 182, 271–272, 274–275
Optical imaging
  fluorophore-labeled substrates, 90–91
  Förster resonance energy transfer (FRET), 90
  matrix metalloproteinase (MMP), 90–91
  near-infrared (NIR), 90–91
  proteolytic cleavage of, 90–91
Optimal thresholds, (optional) iterative identification of, 47
ORF. See Open-reading frame (ORF)
Orphan red channel trajectories, 148

Oskar protein, 152
Overhang trimming. See Iterative method
Overhang wrapping, 166–167

P

Paralogous splicing regulators, 280
Parental tRNA gene
  choosing restriction sites, mutagenesis of, 222
  isolation of
    cloning method, 218
    primer design, 219
    vector, 219–221
Paromomycin, 242–243
PCP. See PP7 coat protein (PCP)
Pearson correlation coefficient, 10–13, 11f
Per-bit error rate, 43–44, 45f
Phosphate-buffered saline (PBS), 75
p53-inducible gene 3 (PIG3)
  exons, 259
  Katushka-encoding region, 259–260
  pCtrlPIGv2 vectors, 259–260, 259f
  pSplPIGv2 vectors, 259–260, 259f
  splicing reporter, 259–260
Plasmid constructs, SL-tagged RNA
  materials, 111–112
  protocol
    destination vector, generation of, 112–113
    entry vector generation, 113–114
    SL-tagged RNA expression vector, 114
    RNA visualization, 111
Plasmodesmata (PD), 106
Poisson process, 176–177
Poly-D-lysine (PDL) solution, 55
Poly-lysine–polyethylene glycol copolymer, 90–91
PP7 coat protein (PCP), 161, 162f
  elongation kinetics, measure of, 183–184
  endogenous vs. modified RNAs, 184
  measurement sensitivity, 182
  noncoding regions, 182
  open reading frames (ORFs), 182
  and PP7 Cassettes, 182
  signal intensity, 182
  single-RNA sensitivity, 182–183
  temporal resolution, 182
Precursor tRNAs, 221, 221t

Subject Index

363

Premature termination codon, 292
Pre-messenger RNA (mRNA), 160
Primer design, 219
Primers, 24–25, 24t
Priming region design, 9f, 13, 20–22.
  See also Encoding probes
Probe construction design
  multiplexed error-robust fluorescence in situ hybridization (MERFISH)
    encoding probes, purification of, 26–27
    in vitro template, amplification of, 24–26, 24t
    in vitro transcription, 26
    required reagents, 23–24
    reverse transcription, encoding probes, 26–27
  pumilio homology domain (PUM-HD)
    8-base sequence, 73–74
    fusion protein, 73–74
    16-base RNA sequence, 73–74
    target RNA, 73–74
Protein-based probes
  fluorescence microscope observation, 67–68
  living cells, 67–68
  RNA monitoring, 70
  target messenger RNA (mRNA), 69
Proximal nucleotides, 223
Pumilio homology domain (PUM-HD), 67
  crystal structure of, 70–72, 71f
  development of
    design principle, 73–75
    materials, 75
    methods, 75–76
  dissociation constant of, 70–72
  limitation of, 70–72
  principle of
    fluorescent protein reconstitution method, 72–73
    RNA-binding protein domain, 70–72, 71f
  RNA visualization
    mitochondrial messenger RNA, 80–81, 81f
    NADH dehydrogenase 6 (ND6) messenger RNA, 80–81, 81f
    single β-actin messenger RNA molecule, 81–82, 83f

Q

Quantitative ratiometric assessment, nonsense-mediated mRNA decay (NMD) activity
  β-globin, 295
  excitation and emission spectra, 295
  expression cassettes, 294–295
  far-red katushka, 295
  fluorescent proteins, 294–295, 294f
  genetic constructs, 294–295
  green TagGFP2, 295
  green-to-red ratios, 295
Quantitative reverse transcriptase polymerase chain reaction (qRT-PCR), 52

R

Rapid tRNA decay (RTD), 243
Readout probe design, 9f, 13
  binding efficiency, 16
  potential off-target binding sites, 16
  validated sequences, 16, 17t
Real time-qualitative polymerase chain reaction (RT-qPCR)
  intracellular multiaptamer genetic tag (IMAGEtags)
    measurement of, 205–207
    live cell imaging, 198
    tracking transcription methods, 194–195
Red fluorescent protein (RFP), 77–78, 238–239, 257, 258f, 259
Red–green cross-correlation, 163
Region of interest (ROI), 205
Resampling frequency, 166, 172
Reverse transcription polymerase chain reaction (RT-PCR) analysis, 75–76
Reverse transcription reaction
  of encoding probes, 26
  single-stranded DNA, 26
RFP. See Red fluorescent protein (RFP)
RNA-based sensor variants, 324–325
RNA binding proteins (RBPs), 70–72, 107–108
RNA dynamics, monitoring of
  in cultured cells, 68–70
  in living cells
    imaging, 66–67

RNA dynamics, monitoring of (Continued)
  labeling, 67–68
  microscopy setup, 76–79, 78f
  single-molecule RNA visualization, 76–79, 78f
  pumilio homology domain (PUM-HD) based
    development of, 73–76
    principle of, 70–73
    RNA visualization, 80–82, 81f, 83f
RNA-encoding barcode, 44
RNAi, 308–309
RNAi depletion, 280, 287–288
RNA-ID, identifying and characterizing regulatory sequences
  applications of
    ASC1, 243
    CGA codons, 243
    flow cytometry, 242–243, 242f
    frameshifting events, 242–243
    mutation effects, 241, 243
    paromomycin, 242–243
    rapid tRNA decay (RTD), 243
    trans-acting factors, 241, 243
    translation, 241
  cis-regulatory sequences, analysis of
    cloning sequences, 245–248
    flow cytometry, 249
    library, analysis of, 250
    yeast, 248–249
  extensions of, 251–252
  features of, 239–241
    GLN4, 240
    green fluorescent protein (GFP), 239–240
    HA epitope, antibody detection of, 239–240
    ligation-independent cloning (LIC), 239–241, 241f
    LTN1, deletion of, 240
    red fluorescent protein (RFP), 239–240
    Renilla luciferase, 240
  verification of, 250–251
RNA imaging, 36, 234–235
RNA isolation
  and quantitative RT-PCR
    Alternative Splicing, 266

  critical, 305
  fluorescence nonsense-mediated mRNA decay (NMD) reporter, 305–308
  Katushka, 307
  phenol–chloroform extraction, 305
  plasmid DNA, 305
  TagGFP2, 307
  tricRNA expression, 228
RNA-labeling technique, 161–162
RNA localization, 106–108
RNA polymerase II (Pol II), 160
RNA probes, 322
RNA reporters, 194–195
RNA Spinach riboswitch sensors.
  See Spinach riboswitch sensors
RNA synthesis and processing, 160–161
RNA trafficking, 197
RNA transport, 106
RNA visualization, monitoring localization of
  mitochondrial messenger RNA, 80–81, 81f
  NADH dehydrogenase 6 (ND6) messenger RNA, 80–81, 81f
  single β-actin messenger RNA molecule, 81–82, 83f
ROI. See Region of interest (ROI)
RT-qPCR. See Real time-qualitative polymerase chain reaction (RT-qPCR)

S

Saccharomyces cerevisiae, 202
Sampling distribution, 173f, 175
Sampling noise, 184–185, 184f
Sequential hybridization and imaging readout approach, 5
Shine–Dalgarno (SD) sequence, 320f, 321–322
Short hairpin RNA (shRNA), 308–309
Side light-scattering (SSC), 264f, 265
Signal bleed, 136–137
Signal colocalization, 137
Signal-to-noise ratio, 116
Single β-actin messenger RNA molecule
  cytosolic RNA, 81–82
  and monitoring localization, 81–82, 83f

Subject Index

mutant pumilio homology domain (PUM-HD), 81
single Gaussian function, 81–82
Single cells
  spatial distribution of, 2
  transcriptome, 3
Single correlation functions
  Fourier transforms, 166–167
  iterative method, 162f, 165
  multiple-tau algorithm, 162f, 165–166
Single-molecule detection and tracking, 109–110
  analysis
    centroid estimation, 141–142
    fluorescent channel, 141–142
    Gaussian intensity distribution, 141–142
    maxima-based detection, 141–142
    motion-based models, 142–143
    precise particle detection and tracking methodology, 141–142
    robust particle tracking methods, 142–143
    trajectories, 142–143
    two-color-labeled mRNAs, 143
    unprocessed data, 141–142
  considerations for
    adequate calibration techniques, 138–139
    exposure times, 139
    frame rates optimizing, 139
    high localization precision, 138–139
    light, diffraction of, 138–139
    messenger RNA (mRNAs), diffusion coefficients of, 139
    phototoxicity, 138
    point spread function (PSF), 138–139
    rapid photobleaching, 138
    signal-to-noise ratio (SNR), 139
Single-molecule 4D imaging, 3, 161–162
Single-molecule fluorescence in situ hybridization (smFISH)
  fluorescence, 5
  hybridization, 16 rounds of, 10–12, 11f
  multiplexing of, 4–5
  RNA species, 3
Single-molecule RNA, visualization of
  excitation optical system
    design and principle of, 77–78, 78f
    inclined illumination, 77–78, 78f
    red fluorescent proteins (RFPs), 77–78
  fluorescent protein reconstitution method, 76
  methods, 79
  and microscopy system, 77
  excitation optics, 77
  target RNA, 77
  total internalization reflection fluorescence (TIRF) microscopy, 77
  pumilio homology domain (PUM-HD), 76
Single mRNA imaging
  β-actin mRNA, 52–53
  brain slice imaging
    materials, 61
    protocols, 61–63
  fluorescence in situ hybridization, 52
  local protein synthesis, 53–54
  MCP×MBS mice, 52, 54f
  MS2 bacteriophage coat protein (MCP), 52
  MS2 binding site (MBS), 52
  MS2-GFP labeling technique, 52
  neuron culture imaging
    materials, 55–56
    protocols, 56–60
  nuclear localization sequence (NLS), 53
  quantitative reverse transcriptase polymerase chain reaction, 52
  three-dimensional (3D) cultures, 54
  two-dimensional (2D) cultures, 54
Single-particle tracking (SPT), 130, 138, 145, 147
Single-RNA synthesis
  dynamics, 160–161
  postinitiation kinetics of, 188
  and splicing, 160, 172–174
Single-stranded DNA oligonucleotides, 8–10
Single-stranded linkers, 323
Single trace interpretation. *See* Fluctuation analysis, transcriptional kinetics dissection
16-base RNA sequence, 73–74
Small-molecule ligands, 195–196

smFISH. *See* Single-molecule fluorescence in situ hybridization (smFISH)
Solid-phase oligonucleotide synthesis, 22
Spinach and broccoli aptamers, 194–195
Spinach riboswitch sensors
  3,5-difluoro-4-hydroxybenzylidene imidazolinone (DFHBI)-binding pocket of, 317f, 318
  in vitro identifying fluorescence sensor characteristics, 320f, 325–327
  RNA-based sensor variants, 324–325
  metabolites, live-cell imaging
  data analysis of, 330–331, 330f
  imaging dynamics, in live *E. coli*, 328–330, 330f
  sensor-expressing *E. coli*, 327–328
  thiamine pyrophosphate (TPP), 327–330, 330f
  RNA-based structure, development of, 318, 319f
  structure-guided design
    of linker sequences, 320f, 322–323
    and transducer sequences, 320f, 321–322
Splicing, 257, 258f
  ratio, 285–286
  regulators, 270–271, 274–275, 280, 286–287
  reporter, 259–260, 272
Splicing-dependent NMD reporter, 311
Split fluorescent protein reconstitution approach, 69–70
  live-cell RNA imaging probes, 69–70
  RNA-binding protein domain, 70
Spot tracking, 140–141, 141f
Spreadsheet processing software, 266
SPT. *See* Single-particle tracking (SPT)
Stable cell clone generation
  cell line, 278–279
  considerations, 279–280
  loss-of-function screens, 278–279
  materials and equipment, 279
  protocol, 279
  signal-to-noise ratio, 278–279
Stem-loop cassettes, 145
Stereoscope, 58
Switching sequences, Spinach forms

Shine–Dalgarno (SD) sequence, 320f, 321–322
  structure-guided design of
    algorithmic prediction, 322
    G-quadruplex structure, 321
Synonymous codons, 238, 241–242
Synthetic fluorophores, 68–69
Synthetic oligonucleotides, 209–210, 210f

T

Tansfectable cell clone, 280. *See also* Stable cell clone generation
Target region design
  goal of, 14
  single-molecule fluorescence in situ hybridization (smFISH) measurement, 9f, 14
  thermodynamic stability, 14
Target RNA sequence
  fluorescent protein reconstitution technique, 72
  molecular beacon method, 68–69
  pumilio homology domain (PUM-HD) mutant, design of, 73–75
Template amplification. *See* In vitro template amplification
Thiamine pyrophosphate (TPP)
  E. coli cells, 330–331
  linker sequences, 320f, 323
  metabolites, live-cell imaging, 327–330, 330f
  Spinach riboswitch sensor, 320f, 323
  transformed bacteria images, 330, 330f
30-nt targeting region, 9f, 13
Three-dimensional (3D) cultures, 54
Time-delay points, correct weighting of, 172
Time traces, finiteness of, 167, 168f
TIRF microscopy. *See* Total internalization reflection fluorescence (TIRF) microscopy
Tobacco mosaic virus (TMV), 107–109, 118f
Tobramycin
  control RNA, 197, 200f
  cyanine3 (Cy3) conjugated ligands, 201, 201f
  cyanine5 (Cy5) conjugated ligands, 201, 201f

Total internalization reflection fluorescence (TIRF) microscopy, 77
TPP. *See* Thiamine pyrophosphate (TPP)
Trace-to-trace variations, 170–171
Tracked two-colored mRNA particles, colocalization of
  algorithms, 144
  automated fashion, 143
  calibration, 143
  diffraction-limited spot, 143
  spatial transformation, 143
  spatiotemporal overlap, 144
  trajectories, 144
Trans-acting factors, 241, 243
Transcription, 53–54
Transcriptional time traces, 172
Transcription reaction. *See* In vitro transcription reaction
Transcription site (TS), 161, 162*f*
  baseline correction and renormalization, 174
  data collection/analysis, 174
  single-molecule imaging, 161–162
Transcriptomics, 3, 7
Transducer sequences, Spinach forms
  Shine–Dalgarno (SD) sequence, 320*f*, 321–322
  structure-guided design of
    algorithmic prediction, 322
    G-quadruplex structure, 321
Transformation, yeast, 248
Transient expression, *N. benthamiana*
  agrobacterium cultures, 114
  materials, 115
  protocol, 115–116
  RNA subcellular localization, 114–115
Translating RNA imaging by coat protein knockoff (TRICK)
  analysis
    controls, 145
    single-molecule detection and tracking, 141–143
    tracked two-colored mRNA particles, 143–144
  data collection
    long time-lapse experiments, 139–140
    single-molecule detection and tracking, 138–139

  in *Drosophila*
    controls, 151–152
    imaging and analysis, 149–151
    fluorescent proteins (FPs), 124–125
  in HeLa cells
    image acquisition, 146–147
    image analysis, 147–148
    live-cell imaging, preparation of cells, 145–146
  in mammalian cells
    coat proteins, expression of, 129–131
    controls, 132
    primary cells, considerations and challenges, 131
    reporter transcripts, expression of, 128–129
    messenger RNA (mRNA) translation, 124
  microscopy
    imaging modality, 132–135
    light source, 135–136
    signal detection, 136–137
    temperature and CO_2 control, 137–138
  reporter messenger RNA (mRNA), design of, 125–128
    fluorescent microscopy, 125–126
    fluorescent protein (FP), 125–126
    MS2 bacteriophage coat protein (MCP), 125–126, 127*f*
    NLS-MCP-RFP, 126–127
    NLS-PCP-GFP, 126–127
    nuclear localization sequence (NLS), inclusion of, 125–126, 127*f*
    PP7 bacteriophage coat protein (PCP), 126
    RNAbiosensor, 126–127
    RNA stem-loop, 126
    translatable PP7 stem-loops, 126–128, 127*f*
    translation, act of, 126
    western blot analysis, 128
  single translation, 125
  spatial information, 124–125
Translation, 241
tricRNA vetcors, design and generation
  cellular imaging of, 233–234

tricRNA vetcors, design and generation (*Continued*)
　external promoters, addition of cloning, 223–225
　　pol III promoters, 223
　　genes, introns, 216–218, 217*f*, 220*f*, 221
　　in-gel imaging of, 229–232
　　in vivo expression of
　　　product analysis, 228–229
　　　RNA isolation, 228
　　　transfection, 226–227
　　parental tRNA gene, isolation of
　　　cloning method, 218
　　　primer design, 219
　　　vector, 219–221
　　parental tRNA gene, mutagenesis of
　　　choosing restriction sites, 222
　　splicing, 216–218, 217*f*, 228
　　vector, notes on, 219–221
TRIzol reagent, 228
tRNA splicing endonuclease (TSEN), 216–218
TS. *See* Transcription site (TS)
Tumor-inducing (Ti) plasmid, 110
2D histogram, 41–42
Two-dimensional (2D) cultures, 54
Two-photon excitation, 62–63
Two-photon microscopy, 63, 63*f*
Two-stage hybridization
　DNA oligonucleotides, 8–10
　encoding probes, 8–10, 9*f*
　readout barcodes, 8–10
　targeting region, 8–10, 9*f*
Typhoon imager, 282, 283*t*
Typhoon laser scanner, 231

U

U6* lane, 231, 232*f*

V

Van der Waals force, 70–72, 71*f*
Vector
　materials and equipment, 219
　workflow, 220
viral ribonucleoprotein complex (vRNP), 108–109
Volumetric imaging, 187

W

Watson–Crick interaction, 68–69
White noise statistics, 186
Wiener–Khinchin theorem, 166
Wortmannin, 308

Y

Yeast, 248–249
　cell transformation
　　gene expression visualization, 202
　　S. cerevisiae, 202
　chromosome, 249

Z

Zinc-dependent endopeptidases, 89–90

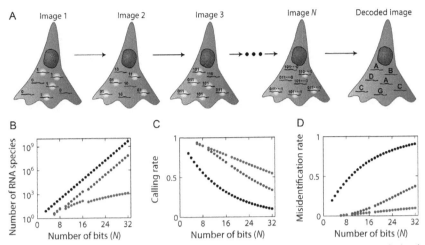

J.R. Moffitt and X. Zhuang, Fig. 1 Multiplexed, error-robust fluorescence in situ hybridization (MERFISH). (A) A schematic depiction of the process by which binary barcodes associated with each labeled RNA in a sample is readout. The presence or absence of fluorescence in each round of hybridization and imaging determines whether the barcode associated with each RNA in the sample has a "1" or "0" in the corresponding bit in the measured binary barcode. These barcodes are then used to identify each RNA, eg, species, A, B, C, etc. (B) The number of RNA species that can be encoded with a binary barcode vs the number of bits N in those barcodes. A binary encoding scheme that utilizes all possible binary barcodes of length N is depicted in black. A binary encoding scheme that utilizes a subset of all possible binary barcodes that are all separated by at least a Hamming Distance of 4—an encoding scheme known as the Extended Hamming Code—is depicted in *blue*. A binary encoding scheme that utilizes a modified Hamming Distance 4 encoding scheme, which consists of all barcodes from the Extended Hamming Code that have a Hamming Weight of 4, ie, only 4 "1" bits, is depicted in *magenta*. (C) The fraction of the binary barcodes that can be properly decoded into the correct RNA species—the calling rate—in the presence of modest readout errors as a function of the number of bits in the barcode for the same encoding schemes depicted in (B). (D) The fraction of binary barcodes that are misidentified as the wrong barcode and, thus, are decoded as the wrong RNA—the misidentification rate—as a function of the number of bits for the same encoding schemes depicted in (B). Panels (C) and (D) were calculated assuming an average "1" → "0" error rate of 10% and a "0" → "1" error rate of 4%, which correspond to the typical error rates observed in MERFISH measurements. *Reproduced with permission from Chen, K.H., Boettiger, A.N., Moffitt, J.R., Wang, S., & Zhuang, X. (2015). Spatially resolved, highly multiplexed RNA profiling in single cells. Science, 348, aaa6090.*

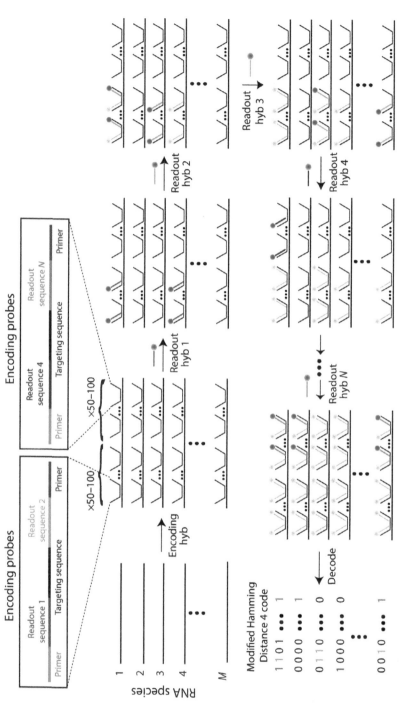

J.R. Moffitt and X. Zhuang, Fig. 2 See legend on opposite page.

J.R. Moffitt and X. Zhuang, Fig. 2 Schematic depiction of the hybridization process used for MERFISH. Cellular RNAs are hybridized with a set of oligonucleotide probes, which we term *encoding probes*. These encoding probes contain a targeting sequence which directs their binding to the specific RNA. They also contain two readout sequences. For an experiment utilizing N-bit binary barcodes, N different readout sequences will be used with each bit assigned a different unique readout sequence. The specific readout sequences contained by an encoding probe to a given RNA are determined by the binary barcode assigned to that RNA: only the readout sequences assigned to bits for which this barcode contains a "1" are used. Each encoding probe also contains PCR priming regions used in its construction. To increase the signal from each copy of the RNA, multiple encoding probes, each with a different target region, are bound to the same RNA. The length of this *tile* of probes is typically between 50 and 100 probes. To identify the readout sequences contained on the encoding probes bound to each RNA, N rounds of hybridization and imaging are performed. Each round uses a unique, fluorescently labeled probe whose sequence is complementary to the readout sequence for that round. The binding of these fluorescent probes determines the bits which contain a "1," allowing the measurement of the specified binary code. *Modified with permission from Chen, K.H., Boettiger, A.N., Moffitt, J.R., Wang, S., & Zhuang, X. (2015). Spatially resolved, highly multiplexed RNA profiling in single cells. Science, 348, aaa6090.*

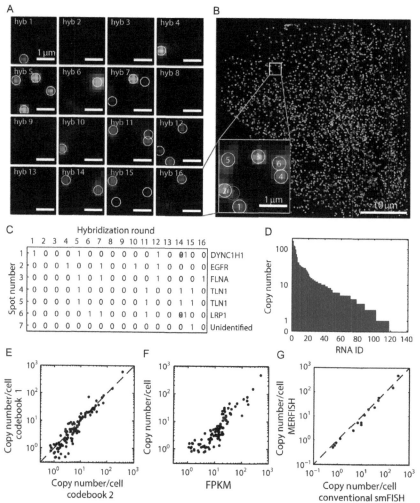

J.R. Moffitt and X. Zhuang, Fig. 3 Example MERFISH data for a 16-bit MHD4 code. (A) smFISH images from each of 16 rounds of hybridization of a small field of view of a Human fibroblast (IMR90) stained with encoding probes utilizing an 16-bit MHD4 code that encodes 140 RNAs. The label depicts the readout hybridization round corresponding to each image. Circles correspond to the locations of identified fluorescent spots. (B) A single 40-μm square field of view with all measured barcodes marked. The *color* of each marker represents the measured barcode. (*Inset*) An overlay of the small section of this field of view depicted in (A) with each set of overlapping spots labeled. White circles correspond to sets of spots that represent a barcode that can be decoded into an RNA while *red* represents a set of spots for which the measured barcode does not represent an RNA. (C) The measured binary barcodes for each set of spots in the small field of view depicted in (A) with the identity of the RNA represented by that barcode. Error correction was required for two barcodes in hybridization

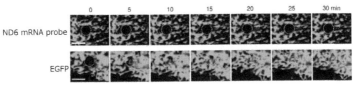

H. Yoshimura and T. Ozawa, Fig. 3 FRAP experiment of the ND6 mRNA probe and EGFP in mitochondria. The fluorescence of the ND6 mRNA probe in the bleached region did not fully recover, whereas that of EGFP almost recovered within 30 min. This result indicates restriction of ND6 mRNA in mitochondria. The *circles of white dotted lines* indicate the photobleaching region with the diameter of 4 μm. Scale bar: 5 μm.

J.R. Moffitt and X. Zhuang, Fig. 3—Cont'd round 14 and is represented by *red crosses*. (D) The number of RNAs of each species identified in the single field of view depicted in (D). Approximately 2000 RNAs were measured in this single field of view, and in a single measurement, ~100 such fields of view containing 250,000 RNAs can be measured. (E) The average RNA copy number per cell measured with one implementation of the 16-bit MHD4 code vs the average copy number per cell for every RNA measured with another 16-bit MHD4 code in which each RNA was assigned a different barcode. The *dashed line* represents equality. (F) The average RNA copy number per cell vs the abundance as measured with bulk RNA-seq (FPKM). (G) The average copy number per cell measured via MERFISH vs that measured using conventional smFISH for 15 different RNAs. The *dashed line* represents equality. *Panels reproduced with permission from Chen, K.H., Boettiger, A.N., Moffitt, J.R., Wang, S., & Zhuang, X. (2015). Spatially resolved, highly multiplexed RNA profiling in single cells. Science, 348, aaa6090.*

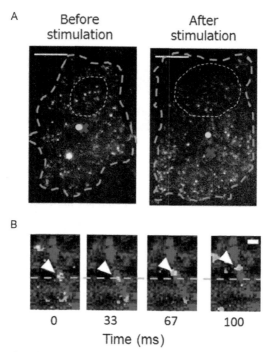

H. Yoshimura and T. Ozawa, Fig. 4 Single-molecule monitoring of β-actin mRNA in living cells. (A) Localization alteration of β-actin mRNA upon serum stimulation. The *blue* and *white broken lines* represent the edges of the cell and the nucleus, respectively. The *yellow spots* indicate the centroid of the cell, and the *red crosses* show the median points of all fluorescent spots of β-actin mRNA probes. Scale bar: 8 μm. (B) Directed motion of a single β-actin mRNA molecule along a microtubule. The *green* and *red signals* represent β-actin mRNA probes and microtubules, respectively. The β-actin mRNA indicated by the *white arrow head* moved directionally along a microtubule. The *blue broken line* indicates the initial position of the spot. Scale bar: 800 nm.

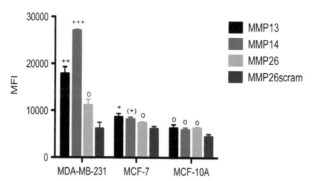

S.R. Jackson et al., Fig. 3 Flow cytometry analysis of the gene expression of select MMPs in cells using hAuNPs. The *symbols above the columns* refer to relative expression of MMP mRNA by RT-PCR from previously published studies (Hegedus et al., 2008; Kohrmann, Kammerer, Kapp, Dietl, & Anacker, 2009). 0—no expression, (+)—very weak expression, +—weak expression, ++—moderate expression, +++—high expression.

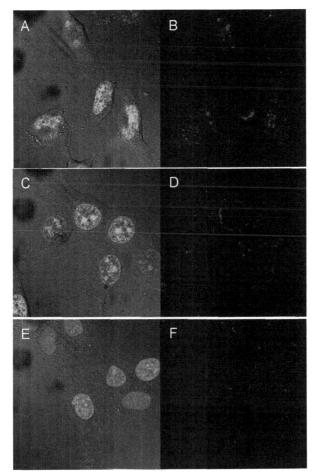

S.R. Jackson et al., Fig. 4 Intracellular uptake and fluorescence emission of hAuNPs across cell lines. MMP14-specific hAuNP emission (*blue*) decreases from MDA-MB-231 cells (A and B) to MCF-7 (C and D) to MCF-10A (E and F). *Red* fluorescence is GAPDH-specific hAuNP. *Green* is Syto13 nuclear counterstain. 100× magnification.

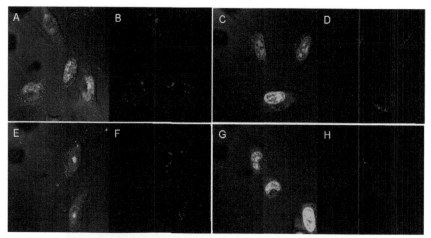

S.R. Jackson et al., Fig. 5 Comparison of MMP hAuNP activation in MDA-MB-231 cells. Blue fluorescence is specific for MMP14 (A and B), MMP13 (C and D), MMP26 (E and F), and the negative control MMP26scram (G and H). Red fluorescence is GAPDH-specific hAuNP. Green is Syto13 nuclear counterstain. 100× magnification.

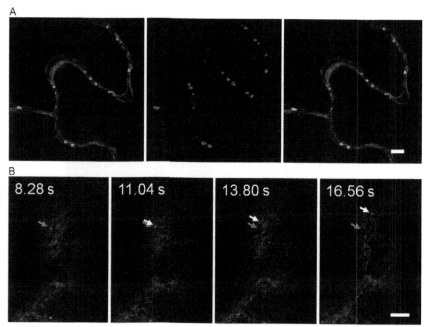

E.J. Peña and M. Heinlein, Fig. 1 Visualization of TMV MP:mRFP mRNA by MS2 tagging in N. benthamiana epidermal cells. (A) MP:mRFP mRNA (labeled with NLS:MCP:GFP, green, left) and MP:mRFP (red, middle) coincide to small dots (presumably PD) at the cell wall (yellow, right). Size bar, 5 μm. (B) Example of video frames showing the colocalization of MP:mRFP and MP:mRFP mRNA (labeled by NLS:MCP:GFP) in mobile particles present in the cortical cytoplasm. Red and green channels are merged. The particles appear in yellowish color. MP:mRFP is shown also in association with microtubules, which are seen as red-colored filaments. Gray arrows indicate the mRNA particle at the location within the first frame and white arrows indicate the particle location in each subsequent time frame. Size bar, 5 μm.

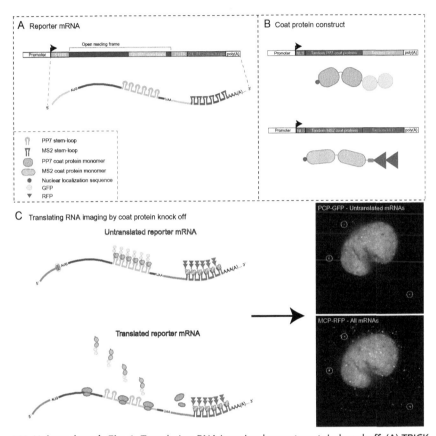

J.M. Halstead et al., Fig. 1 Translating RNA imaging by coat protein knockoff. (A) TRICK reporter transcript contains translatable PP7 stem-loops in the open reading frame and MS2 stem-loops in the 3′-UTR. (B) PP7 and MS2 bacteriophage coat proteins are fused to spectrally distinct fluorescent proteins (eg, NLS-PCP-GFP and NLS-MCP-RFP). The addition of nuclear localization sequences (NLS) results in accumulation of unbound fluorescent proteins in the nucleus. (C) Untranslated mRNAs are bound by both fluorescent fusion proteins while translated mRNAs are labeled with only NLS-MCP-RFP. Dual-color RNA imaging can distinguish single molecules of untranslated mRNAs that are fluorescent in both *green* and *red channels* (*yellow circles*) from those that have been translated and are detected in the *red channel* alone.

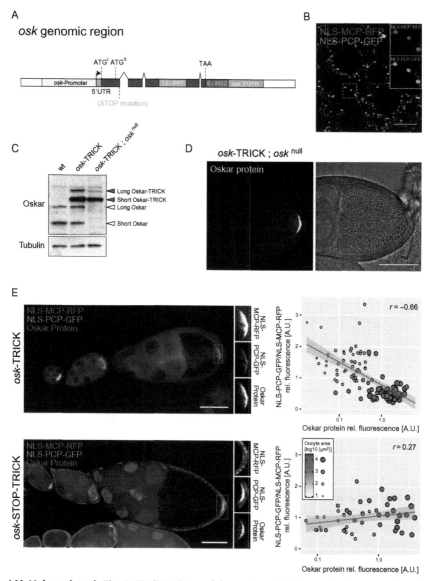

J.M. Halstead et al., Fig. 3 TRICK in *Drosophila* oocytes. (A) Schematic of a genomic *osk*-TRICK reporter construct. The alternative translational start sites producing the long and short Oskar isoforms (ATG^L and ATG^S), the insertion site of 12 × PP7 in the coding region, 6×MS2-binding sites right after the stop codon (TAA), and the position of the stop codon mutation used for control experiments are indicated. (B) Imaging of individual mRNPs in the ooplasm of an egg-chamber expressing *osk*-TRICK mRNA, NLS-MCP-RFP and NLS-PCP-GFP using FP-booster, scale bar 5 μm. (C) Insertion of 12 × PP7-binding

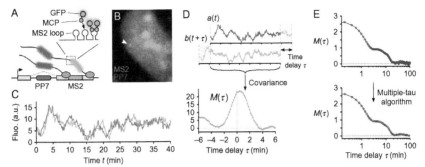

A. Coulon and D.R. Larson, Fig. 1 Transcriptional time traces and correlation function. (A) The MS2 and PP7 RNA-labeling technique consists in inserting, in one or two gene(s) of interest, two DNA cassettes (MS2 and PP7; here at two different locations in the same gene). They produce stem-loop structures in the nascent RNAs, which are bound by an MS2 or PP7 coat protein (MCP and PCP) fused to a fluorescent protein (eg, GFP and mCherry, respectively). (B) The transcription site (*arrow*) appears as a bright diffraction-limited spot in the nucleus in both fluorescence channels. (C) Recording its intensity fluctuations then yields a signal that is proportional to the number of nascent RNAs on the gene over time. (D) Using this signal as an example, the computation of a correlation function (here the covariance function) consists in shifting one signal relatively to the other and calculating the covariance between the values of the overlapping portions of the two signals as a function of the time-delay shift (Eqs. 1 and 2). (E) To analyze fluctuations at multiple timescales in a signal, computing the correlation function with the multiple-tau algorithm yields a somewhat uniform spacing of the time-delay points on a logarithmic scale (simulated data as in Fig. 2A). *Panels (B) to (D): Data from Coulon, A., Ferguson, M. L., de Turris, V., Palangat, M., Chow, C. C., & Larson, D. R. (2014). Kinetic competition during the transcription cycle results in stochastic RNA processing. eLife, 3. http://doi.org/10.7554/eLife.03939.*

J.M. Halstead et al., Fig. 3—Cont'd sites does not disturb translation of *oskar* mRNA. Western blot analysis of ovarian samples from wild-type flies, flies expressing *osk*-TRICK and *osk*-TRICK in an *osk*^{null} background. (D) Immunostaining against Oskar protein in an egg-chamber expressing *osk*-TRICK in an *osk*^{null} background shows exclusive synthesis of Oskar protein at the posterior pole of the oocyte. (E) Quantification of fluorescent signals from NLS-PCP-GFP/NLS-MCP-RFP, Oskar protein immunostaining and oocyte area (color- and size-coded) of individual oocytes. The correlation of the TRICK reporter readout with Oskar protein and oocyte area observed in *osk*-TRICK expressing egg-chambers (*upper* plot) is abolished by the introduction of the STOP mutation prior the PP7-binding sites (*lower* plot). Pearson correlation coefficient (*r*), scale bars 50 μm.

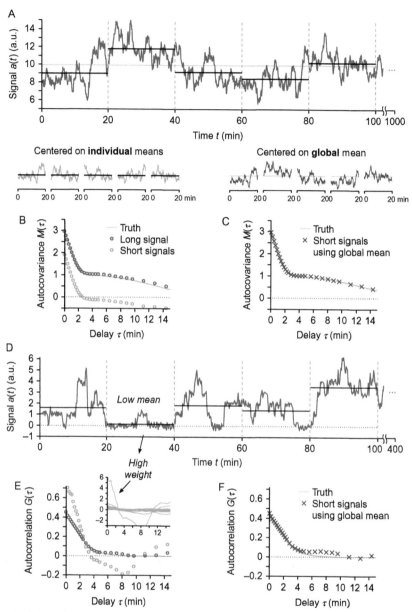

A. Coulon and D.R. Larson, Fig. 2 Biases due to the finiteness of time traces. (A) Shown is a portion of a signal used to illustrate the effect of inaccurate mean estimation. This 1000-min-long signal is partitioned into a set of 20-min-long signals. The true mean of the signal (ie, calculated on the long trace) is shown in *gray* and the inaccurate means of individual short traces are shown in *black*. (B) The autocovariance $M(\tau)$ of the entire signal shown in (A) is close to the expected curve (*red circles* vs *gray curve*). When averaging the autocovariances computed on each one of the 20-min-long traces, the resulting autocovariance deviates from the expected curve by a constant offset (*green circles*). (C) Performing the same calculation using a global estimation of the mean of the

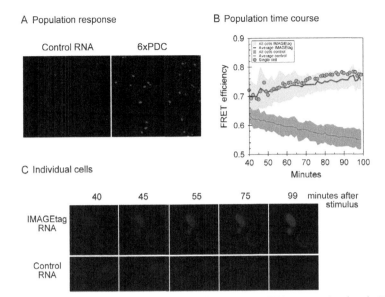

J. Ray et al., Fig. 3 IMAGEtags track temporal changes in RNA expression levels. Yeast cells were grown in 2% raffinose containing SD-Uracil and induced with 2% galactose. (A) FRET images are shown of two populations of yeast expressing either control RNA or 6×PDC IMAGEtags under the control of the GAL1 promoter. (B) Quantification of the FRET from 8 cells expressing control RNA and 13 cells expressing IMAGEtag RNA. The control cell FRET decreased with time due to cumulative exposure to laser light. To obtain the FRET efficiency values for IMAGEtag-expressing cells, the average control FRET emissions per cell at each time point were subtracted from the FRET emissions of individual IMAGEtag-expressing cells. The data are represented as the range of values (areas of *color*) with the mean value as a *line* within the range of values. The *circles* show the time course of increase in FRET for the single cell-expressing IMAGEtags shown in (C). (C) Images of a budding yeast cell that expresses IMAGEtags (*upper panel*) or that expresses control RNA (*lower panel*) are shown as a function of time after addition of galactose to the culture. Shown are merged DIC and FRET images.

A. Coulon and D.R. Larson, Fig. 2—Cont'd signals (ie, once over all the short signals) resolves the issue. (D) Another long signal is shown and partitioned into small sections to illustrate another artifact that may arise when averaging correlation functions $G(\tau)$. (E) The average (*green circles*) of the autocorrelation functions obtained on the 20-min-long sections of the signal shown in (D) deviates from the expected curve. This is due to an inaccurate weighting of the individual curves that occurs when averaging autocorrelation functions $G(\tau)$. As illustrated by the inset, the section that has a very low mean in (D) dominates the average. (F) As in (C), estimating the mean globally over all the signals solves the weighting problem. Both examples shown are simulated signals (A: Gaussian noise shaped in the Fourier domain, D: Monte Carlo simulation of transcription with Poisson initiation, distributed transcript dwell time and additive Gaussian noise). The "truth" curves in *gray* in (B), (C), (E), and (F) are the theoretical curves for both simulated situations.

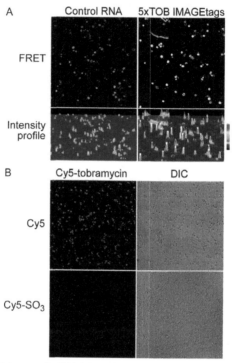

J. Ray et al., Fig. 4 Tobramycin IMAGEtags and the impact of ligand conjugate charge on cell uptake. (A) Yeast cells expressing either the control RNA or 5×TOB IMAGEtags were grown in 1% raffinose containing SD-Uracil and incubated for 30 min with Cy3-tobramycin and Cy5-tobramycin and 2% galactose. The cells were washed and FRET images were acquired. Quantification of the images is shown by the intensity profiles below each image. (B) Yeast cells expressing TOB IMAGEtags were incubated with either Cy5-tobramycin or Cy5-SO_3-tobramycin for 90 min, washed and then imaged in the Cy5 channel. The DIC images show the locations of the cells in the field. The DIC for the Cy5-SO_3 has been overlaid by the image to its left from the Cy5 channel.

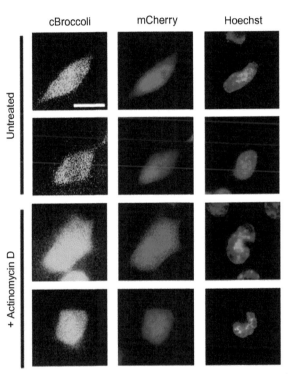

C.A. Schmidt *et al.*, **Fig. 7** HEK293T cells expressing Broccoli-containing tricRNAs can be imaged, with mCherry as a positive control for transfection. Scale bar is 20 μm.

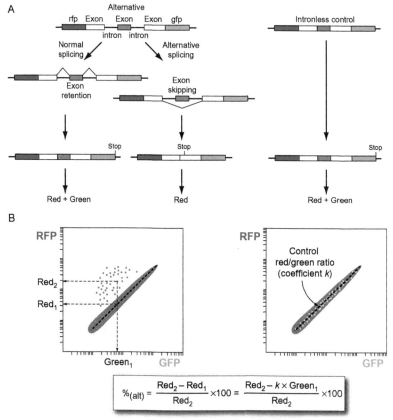

N.G. Gurskaya et al., Fig. 1 Outline of the method of analysis of alternative splicing of a cassette exon using two fluorescent proteins. (A) Scheme of genetic constructs. *Left*: A fragment of a gene of interest (consisting of the alternative exon and two adjacent introns and exons) is cloned between rfp- and gfp-coding regions in such a way that normal splicing (exon retention) results in a full-length protein (*red* and *green* fluorescence), whereas alternative splicing (exon skipping) leads to the frameshift and appearance of a stop codon before GFP (only *red fluorescence*). *Right*: A control intronless construct provides reference *red-to-green ratio* characteristic for the 100% normally spliced transcript. (B) Schematics of flow cytometry analysis of the splicing reporter in *green* and *red channels*. Cells expressing the control plasmid (*right plot*) form a diagonal distribution (*gray*) and make it possible to determine a reference *red-to-green ratio* (*k*) characteristic for a particular model and detection parameters. Cell expressing the target transcript undergoing alternative splicing (*left plot*) can be found to the left from the control diagonal. Percentage of the alternative transcript can be estimated for each cell from its fluorescence intensities in *green* and *red channels* (Green$_1$ and Red$_1$) using the equation shown below the plots.

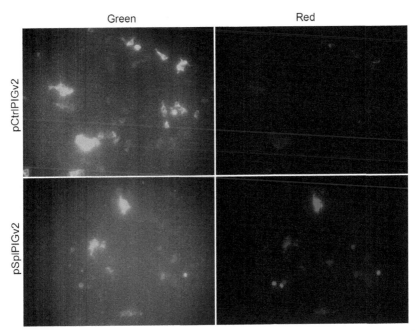

N.G. Gurskaya et al., **Fig. 3** Wide-field fluorescence microscopy of HEK293 cells expressing the alternative splicing reporter. *Top images*: pCtrlPIGv2-expressing cells. *Bottom images*: pSplPIGv2-expressing cells. Cells were imaged with BZ9000 fluorescence microscope (Keyence) using 20× objective and standard filter sets for *green* (excitation at 450–490, emission at 510–560 nm; *left panels*) and *red* (excitation at 540–580, emission at 600–660 nm; *right panels*) *channels*. The same settings were applied for the two cell samples.

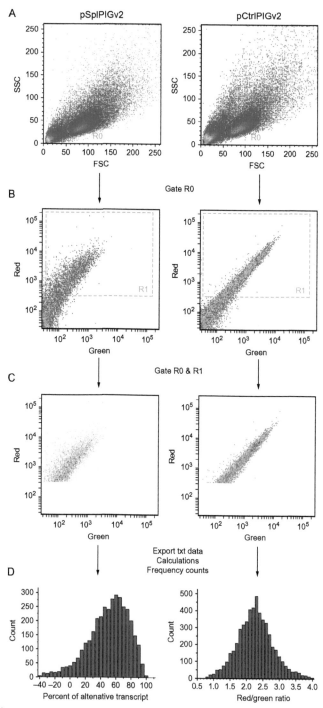

N.G. Gurskaya et al., Fig. 4 See legend on opposite page.

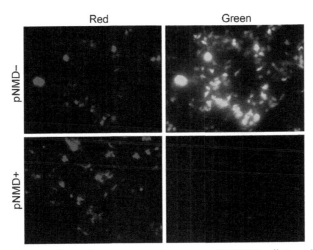

N.G. Gurskaya et al., Fig. 2 Fluorescence microscopy of HEK293 cells transiently transfected with pNMD− or pNMD+ vectors. Pictures were taken in *red* (*left*) and *green* (*right*) *channels* using the same settings for the two cell samples.

N.G. Gurskaya et al., Fig. 4 Flow cytometry analysis of HEK293 cells expressing pSplPIGv2 (*left*) or pCtrlPIGv2 (*right*). Analysis was done by FACSAria III (Becton Dickinson) using 488 nm laser and 530/30 nm detector for TagGFP2, and 561 nm laser and 670/14 nm detector for Katushka. (A) *Dot plots* in forward scattering (FSC) and side scattering (SSC) channels. Region R0 was used to select live cells. (B) *Dot plots* in *green* and *red channels*. Region R1 was used to select well-expressing cells. (C) *Dot plots* of cells from R1 region. Digital data from these plots were exported in the text format to perform statistical analysis shown in (D). Cells showing negative percentage of alternative transcript arise from the area below the *middle line* of the control pCtrlPIGv2 diagonal.

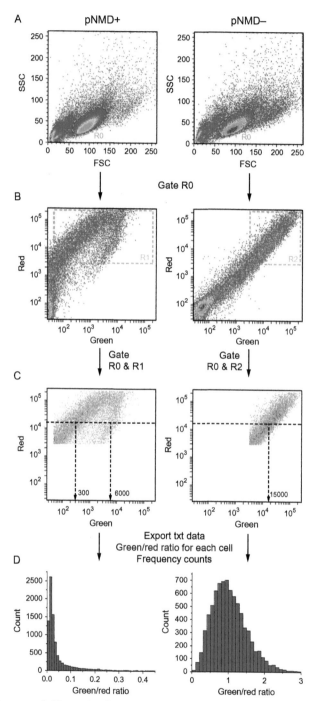

N.G. Gurskaya et al., **Fig. 4** See legend on opposite page.

N.G. Gurskaya et al., Fig. 4 Flow cytometry analysis of HEK293 cells expressing pNMD+ (*left*) or pNMD− (*right*) using FACSAria III. (A) *Dot plots* in forward scattering (FSC) and side scattering (SSC) channels. Region R0 was used to select live cells. (B) *Dot plots* in *green* and *red channels*. Regions R1 and R2 were used to select well-expressing cells. (C) *Dot plots* of cells from R1 and R2 regions. *Dashed lines* illustrate simplified ("by eye") assessment of NMD activity by comparison of *green signals* in pNMD+ and pNMD− samples at the equal level of the *red signal*. Note strong cell heterogeneity resulting in a major subpopulation with a high NMD activity (15,000/300 = 50-fold) and a minor subpopulation with a low NMD activity (15,000/6000 = 2.5-fold). Full data from these plots were exported in the text format to perform statistical analysis shown in (D).

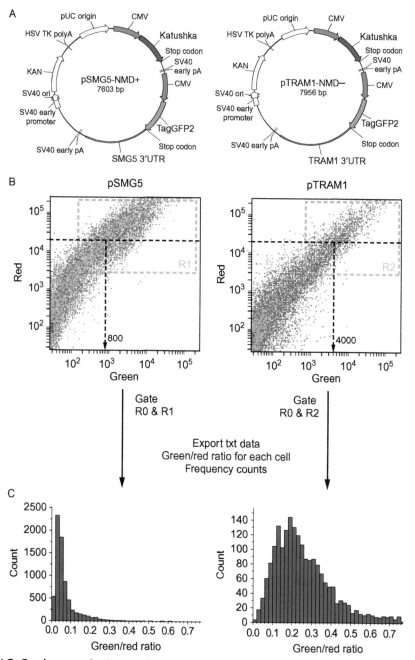

N.G. Gurskaya et al., **Fig. 5** Splicing-independent NMD reporter. (A) Scheme of plasmids pSMG5-NMD+ and pTRAM1-NMD−. (B) Flow cytometry analysis of HEK293 cells expressing pSMG5-NMD+ (*left*) or pTRAM1-NMD− (*right*) using FACSAria III. Dashed lines show simple way to estimate NMD activity by comparison of green signals in pNMD+ and pNMD− samples at the equal level of the red signal. (C) Statistical analysis of data for cells from regions R1 and R2. See Fig. 4 legend for details.

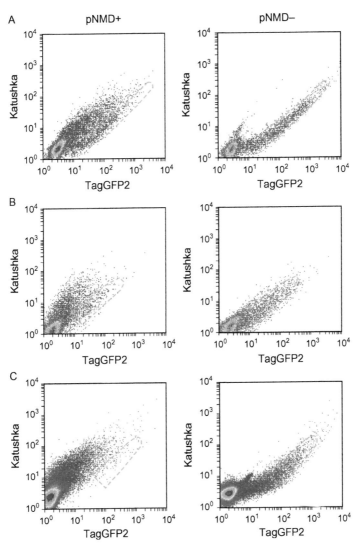

N.G. Gurskaya et al., Fig. 6 Flow cytometry analysis of Jurkat (A), HaCaT (B), and LLC (C) cells expressing pNMD+ (*left*) or pNMD− (*right*) using Cytomics FC500. To simplify comparison, diagonal areas of the control pNMD− cells are marked with *cyan dashed lines*.

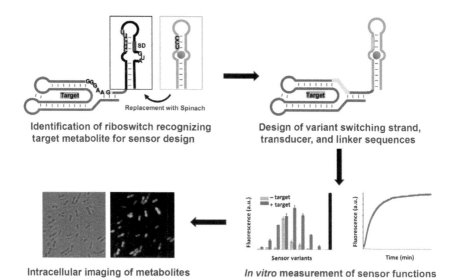

J.L. Litke et al., Fig. 2 Scheme for development of RNA-based structural switching sensor. A riboswitch for the target metabolite is identified and the expression platform is replaced with the Spinach sequence. The switching sequence of the riboswitch is identified that can interact with Spinach's transducer sequence or a slightly modified transducer sequence. Variants of putative sensor with changes in the switching strand, transducer, and linker regions are tested in vitro for properties such as signal response, affinity, and kinetics of binding. Cloning of best sensor variants into vector for bacterial T7-induced expression allows imaging of metabolite concentration changes in the presence of DFHBI.

A

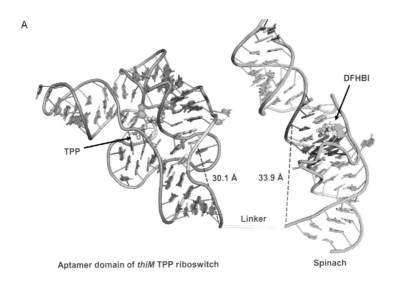

B

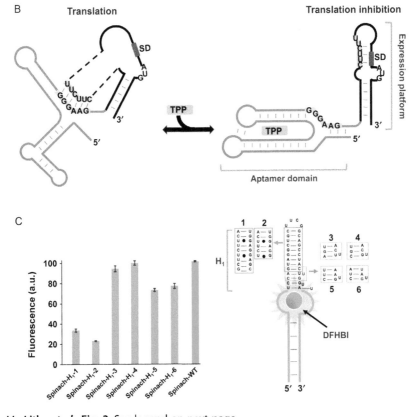

C

J.L. Litke et al., Fig. 3 See legend on next page.

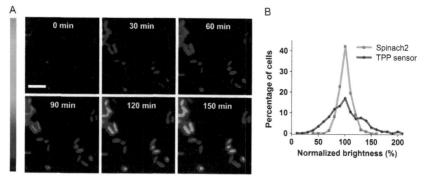

J.L. Litke et al., Fig. 4 (A) Representative images of bacteria transformed with TPP Spinach riboswitch. Time course of fluorescence over 3 h demonstrates intracellular sensor function. (B) Comparison of distributions of fluorescence signal from bacteria expressing TPP Spinach riboswitch and the corresponding Spinach sequence by itself. Demonstrates that variations in cell-to-cell fluorescence are due to differences in TPP concentrations rather than differences in expression from T7 promoter. *Fluorescence images adapted from You, M., Litke, J. L., & Jaffrey, S. R. (2015). Imaging metabolite dynamics in living cells using a Spinach-based riboswitch. Proceedings of the National Academy of Sciences of the United States of America, 112(21), E2756–E2765. http://doi.org/10.1073/pnas.1504354112. Copyright (2015) National Academy of Sciences, USA.*

J.L. Litke et al., Fig. 3 (A) Model of the *thiM*-based Spinach riboswitch sensor for TPP. Riboswitch in *blue* with its switching sequence in *yellow*, while Spinach in *green* with transducer sequence in *gray*. Spinach U•A•U base triple is shown in *cyan* (overlap with most proximal residue of transducer sequence) and G-quadruplex shown in *purple*. TPP and DFHBI are shown in *ball-and-stick* representation. The distance from the switching sequence and transducer sequence to the base of *thiM* and Spinach, respectively, is quite similar, which would not be apparent in a two-dimensional representation. (B) Mechanism of translational control by the *thiM* riboswitch in response to TPP. The transducer sequence shows complementarity to the 5′-AGGAGG-3′ Shine–Dalgarno sequence and switching sequence of the riboswitch's aptamer domain. (C) H_1 stem of Spinach tolerates certain modifications with only modest fluorescence changes. Such mutants expand the modularity of the approach and the range of transducer sequences that can be used for Spinach-based sensor design. *Structural model, sensor mechanism, and Spinach stem diagrams originally printed in You, M., Litke, J. L., & Jaffrey, S. R. (2015). Imaging metabolite dynamics in living cells using a Spinach-based riboswitch. Proceedings of the National Academy of Sciences of the United States of America, 112(21), E2756–E2765. http://doi.org/10.1073/pnas.1504354112. Copyright (2015) National Academy of Sciences, USA.*

Edwards Brothers Malloy
Ann Arbor MI. USA
June 1, 2016